MACMILLAN / McGRAW - HILL

MATHEMATICS IN ACTION

Alan R. Hoffer   Steven L. Leinwand   Gary L. Musser

Martin Johnson   Richard D. Lodholz   Tina Thoburn

MACMILLAN/McGRAW-HILL SCHOOL PUBLISHING COMPANY

New York / Chicago / Columbus

CONSULTANTS

Zelda Gold, Mathematics Advisor, Los Angeles Unified School District, Panorama City, California • Audrey Friar Jackson, Math Specialist K-6, Parkway School District, Chesterfield, Missouri • Susan Lair, Department Chairperson, Wedgwood Middle School, Fort Worth Independent School District, Texas • Gail Lowe, Principal, Conejo Valley United School District, Thousand Oaks, California

ACKNOWLEDGMENTS

The publisher gratefully acknowledges permission to reprint the following copyrighted material:

Chart of "Tallest Trees Found in the United States." Excerpted from AMERICAN FORESTS MAGAZINE, a publication of the American Forestry Association, 1516 P Street N.W., Washington, D.C. 20005. Used by permission.

"Olympic Swimming—100-Meter Freestyle Records," from THE WORLD ALMANAC, 1989. Copyright © Newspaper Enterprise Association. New York, N.Y. 10166. Used by permission.

COVER DESIGN B B & K Design Inc.   **COVER PHOTOGRAPHY** Scott Morgan

ILLUSTRATION   Ray Alma; 447 • Phil Anderson; 6, 7 • George Baquero; 77, 121, 163, 245, 291, 318, 319, 331, 373, 419, 434, 461, 505 • Rick Berlin; 230, 304, 305, 308 • Rose Mary Berlin; 286, 287 • Craig Berman; 136, 137 • Lloyd Birmingham; 162, 228, 236, 237 • Rick Brown; 2, 3 • Circa 86, Inc.; 173, 215, 216, 217 • William Colrus; 93 • Jack E. Davis; 26, 27 • Bruce Day; 144, 150, 151, 406, 407 • Jim Deigan; 272, 273 • Steve DeStefano; 222 • Cathy Diefendorf; 310, 311, 390 • Bert Dodson; 366, 367, 418 • Eldon Doty; 60, 61, 87, 110, 111, 337, 408, 409, 438, 439, 471, 484 • Leslie Dunlap; 444, 496 • Steve Duquette; 234, 235 • Julie Durell; 490 • Kim Wilson Eversz; 102, 103 • Barbara Friedman; 156, 157, 278, 279 • Ignacio Gomez; 178, 179, 260, 261, 268, 269 • Gershom Griffith; 362, 388 • Patricia Gural-Hinton; 324, 325 • Marika Hahn; 364 bottom • Meryl Henderson; 116 • Al Hering; 206 • Dennis Hockerman; 196, 197, 276, 277 • Marilyn Janovitz; 176, 177 • Terry Kovalcik; 148, 149, 262, 263 • Ron Le Hew; 194, 195 • Jeff Lewin; 322 • Benton Mahan; 383 • Wallop Manyum; 22, 23 • Claude Martinot; 356, 357 • Bill Mayer; 488, 489 • Kimble Mead; 232, 233 • Patrick Merrell; 72, 73, 90, 91, 108, 109, 146, 147 • Mike Quon Design Office, Inc.; 142, 143, 220, 221, 320, 321 • MKR Design; Handmade props • Leo Monahan; 1 • Sal Murdocca; 96, 97, 352, 353 • Ann Neumann; 326 • Patrick O'Chapin; 12, 13, 58, 59 • Hima Pamoedjo; 104, 177 right, 186, 202, 203, 258, 259, 289, 354, 355, 364 top, 502 • Alex Pietersen; 207 • Jan Pyk; 392, 393 • Marcy Dunn Ramsey; 515 • Chris Reed; 64, 65 • Suzanne Roz; 16, 17 • Phil Scheuer; 8, 9, 412, 413, 478, 479 • Bob Shein; 358, 359, 476 • David Shelton; 192, 314, 315 • Blanche Sims; 188, 189 • Terry Sirrell; 190, 191 • Steve Smallwood; 365 • Joel Snyder; 131, 217, 255 • Peter Spacek; 448, 449 • Leslie Stall; 404 • Arvis Stewart; 134 • Wayne Anthony Still; 270 • Susan Swan; 10 • George Ulrich; 337 • Vantage Art, Inc.; 328 • Joe Veno; 24, 25 • Josie Yee; 159, 415, 453, 455 • Rusty Zabransky; 36, 74, 118, 160, 204, 242, 288, 289, 348, 370, 416, 458, 502 • Ron Zalme; 4, 5, 138, 346, 442, 446, 494 • Jerry Zimmerman; 34, 35, 100, 101, 398, 399 • Robert Zimmerman; 394, 395

PHOTOGRAPHY   Bruce Coleman, Inc./A.J. Deane, 185; Frank Oberle, 373B • Esto Photographics/Ezra Stoller, 303 • Folio, Inc./John Keith, 324R, Al Messerschmidt, 486 • The Image Bank/Gary Cralle, 275; Mel DiGiacomo, 492R; Gerard Mathieu, 324B; Roger Miller, 492; Co Rentmeester, 345; Weinberg-Clark, 325; A.T. Willett, 493 • The Image Works/Mark Antman, 474 • International Stock Photography Ltd./Richard Pharaoh, 39T • Michal Heron, 54, 284, 285, 386, 387, 397, 400, 433, 454, 456 • Lawrence Migdale, 68R • Monkmeyer Press/Nancy Lehmann, 317C • Warren Ogden, 31, 160 • Stephen Ogilvy, 18, 19, 30, 53, 66, 94, 113, 267, 396, 415, 436, 457, 482, 483, 502, 503 • Omni-Photo Communications/Ken Karp, 39, 54, 76, 112, 159, 154, 155, 205, 182, 322, 410, 411, 414, 416, 417, 441; John Lei, 49, 52, 95, 106, 158, 226, 227, 238, 239, 266, 328, 344, 401, 432, 440, 452, 455, 459, 500, 501 • Photo Edit/Robert Brenner, 481; Tony Freeman, 183; Alan Odie, 280 • Photo Researchers/M.H. English, 350; Explorer, 317R; Tom McHugh, 203; Lawrence Migdale, 316C; Hans Namuth, 482 • Rainbow/Larry Brownstein, 63 • Photori/Mauritius, 402B; Leonard Rue, 403 • Rainbow/Dan McCoy, 312BL • Research Plus, 288 • John Running, 330 • Joseph Sachs, 118, 458 • Shostal/Superstock/David Forbert, 313; August Upitis, 312TC • Elliott Varner Smith, 282 • Starlight/NASA JSC, 224; Roger Ressmeyer, 225 • Stock Boston/Donald Dietz, 267 • The Stock Market/Peter Beck, 307; Roy Morsch, 373T; Dave Wilheim, 373M • Viesti Associates, Inc./Dan Barba, 20; Larry Kolvoord, 21; Skye Mason, 33 inset, 368 both; Ginny Ganong Nichols, 32 all, 62L, 71, 114 both, 140, 141, 198, 199 all, 240, 241 Courtesy of The Fernbank Science Center, Atlanta Georgia, 369, 402L, 450L, 451T, 480, 498; Ken Ross, 70 both, 115, 152, 153, 266 both, 316L, 450R, 451B, 499; Kevin Vandivier, 33T, Joe Viesti, 62R, 317L • Bill Waltzer, 75 • West Light/Bill Ross, 324M • Woodfin Camp/John Blaustein, 312TL; Robert Frerck, 283.

Macmillan/McGraw-Hill School Division
866 Third Avenue
New York, New York 10022

Printed in the United States of America
ISBN 0-02-108504-8
9 8 7 6 5 4 3 2 1

CONTENTS

DECISION MAKING
Choosing a Job to
Earn Money,
pages 11 and 36

DECISION MAKING
Planning a Vacation,
pages 51 and 74

DECISION MAKING
Planning a Lunch,
pages 89 and 118

CHAPTER
4

Measuring Time, Capacity, and Mass 133
MATH CONNECTIONS GRAPHING / TEMPERATURE / PROBLEM SOLVING

CHAPTER
5

Multiplication and Division Facts . . . 175
MATH CONNECTIONS AREA / GRAPHING / PROBLEM SOLVING

DECISION MAKING
Planning a Trip to the Museum,
pages 133 and 160

DECISION MAKING
Adopting a Pet,
pages 175 and 204

DECISION MAKING
Planning a Trip to Mars,
pages 219 and 242

DECISION MAKING
Giving a Fruit Basket,
pages 257 and 288

CHAPTER
9

Understanding Fractions and Mixed Numbers 343

MATH CONNECTIONS LENGTH / PROBLEM SOLVING

DECISION MAKING
Planning a Hiking Trip,
pages 385 and 416

DECISION MAKING
Choosing the Right Gift,
pages 431 and 458

CHAPTER
12

Multiplying and Dividing by 2-Digit Numbers 473

MATH CONNECTION PROBLEM SOLVING

DECISION MAKING
Planning a Car Wash,
pages 473 and 502

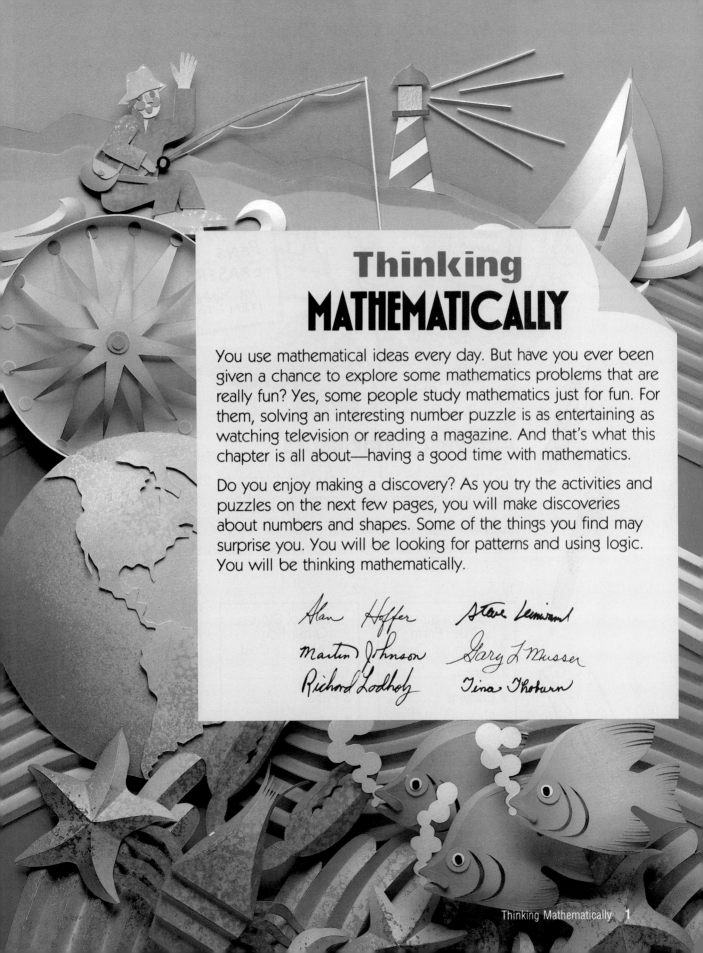

Thinking
MATHEMATICALLY

You use mathematical ideas every day. But have you ever been given a chance to explore some mathematics problems that are really fun? Yes, some people study mathematics just for fun. For them, solving an interesting number puzzle is as entertaining as watching television or reading a magazine. And that's what this chapter is all about—having a good time with mathematics.

Do you enjoy making a discovery? As you try the activities and puzzles on the next few pages, you will make discoveries about numbers and shapes. Some of the things you find may surprise you. You will be looking for patterns and using logic. You will be thinking mathematically.

Alan Hoffer   *Steve Leinwand*

Martin Johnson   *Gary L Musser*

Richard Lodholz   *Tina Thoburn*

Applying Mathematics

You use mathematics to solve problems all the time. Think about buying and selling items in a school store.

Suppose you are running the school store. Copy and complete the following order forms from other students.

1.

```
    School Store
    Order Form
 #   Item     Cost
 2 Pencils  _____
 3 Erasers  _____
 Total Cost _____
```

2.

```
    School Store
    Order Form
 #   Item     Cost
 1 Pencil   _____
 3 Pens     _____
 2 Erasers  _____
 Total Cost _____
```

3.

```
    School Store
    Order Form
 #   Item     Cost
 3 Pens     _____
 3 Pencils  _____
 Total Cost _____
```

4. Make up your own order and find its cost.

5. A friend of yours spent exactly 30¢ in the store. What possible purchases could your friend have made?

6. Can you spend exactly 50¢? Why or why not?

Suppose you are the clerk in the school store. Solve the following problems.

7. Mike pays for 1 pencil, 3 pens, and 1 eraser with 2 quarters.
How much does Mike spend?
How much change does Mike get?
What coins would you give him to make the change?

8. Alissa buys one of each item with a dollar.
How much does she spend?
How much change does she get?
What coins would you give her to make the change?

9. Lonnie needs 3 pencils and 2 pens. He has a quarter. What problem does Lonnie face?
What can Lonnie do to solve his problem?

Now suppose you are a customer in the store.

10. What is the greatest amount you could spend at one time in the school store? (*Hint:* Remember the sign.)

11. List all the different orders you can make of 3 or more items that have a total cost of 25¢ or less.

12. Suppose you had exactly 26¢. Could you spend the 26¢ exactly in the school store? How? Is there more than one way to spend exactly 26¢ in the store? What other ways could you spend exactly 26¢?

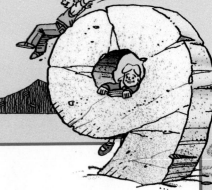

PLAYING ···WITH··· NUMBERS

Using Number Concepts

First, make a set of 9 playing cards, one card for each of the numbers 1 to 9. Use blank cards or cut sheets of paper into 9 equal-sized pieces.

| 1 | 2 | 3 | 4 | 5 | 6 | 7 | 8 | 9 |

Next, on another sheet of paper, draw a board like the one at the right. The spaces should be about the same size as your cards.

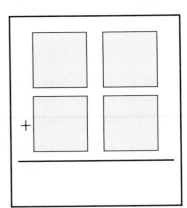

Play the games.

1. Pick any 4 of your cards. Place them on your board so that you get the greatest sum. How did you arrange the numbers you picked?

2. Now pick the 4 cards that would give you the greatest sum. How would you arrange these cards? What is the greatest possible sum?

3. Pick the 4 cards that would give you the least possible sum. How would you arrange these cards? What is the least possible sum?

4. Can you use 4 cards to get a sum of exactly 75 in this game?

   How many different ways can you get 75 using 4 cards?
   How many different ways can you get 75 using 3 cards?

5. Try making up some of your own questions using the addition arrangement. Try out your questions with a partner.

Now make a subtraction board like the one to the right.

6. Pick any 3 cards. Arrange them on the board to get the greatest possible difference.

7. Use these same 3 cards to get the least possible difference.

8. Which 3 cards would you pick to get a difference closest to 10? What is the difference? Is there another way?

9. Discuss your reasoning on Problems 6, 7, and 8 with a partner.

10. Now pick the 3 cards that will give you the greatest possible difference. What is this difference? How do you know that it is the greatest possible difference?

11. Pick the 3 cards that will give you the least possible difference. What is this difference? How do you know that it is the least possible difference?

12. Make up your own questions using the subtraction arrangement. Try them out with a partner.

TRIANGULAR RECTANGLES

Visual Reasoning

Cut out two identical squares from two sheets of different-colored paper. Then fold each square in half to make a triangle and cut along the fold. You should have four triangles—two of each color.

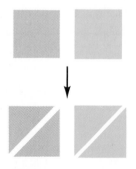

Now experiment with your triangles. Put them together to make a rectangle. Then move the triangles around to make another rectangle with a different design. Each time you create a new design, draw a picture of it. For example:

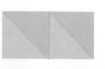

Be careful; [image] and [image] are are not different designs.

You can make the design on the first rectangle simply by turning the second rectangle.

1. Work with a partner. Compare your designs. Did you both find the same designs? Can you find other designs working together?

2. How many different rectangular designs can you make?

3. What other shapes and designs can each of you make using your four triangles? Do you have to decide on some rules for how to arrange the triangles?

4. Work with your partner and use all eight of your triangles to create shapes and designs. See what you can discover. Do you think you will make more shapes and designs than you did with four triangles? Why?

Measuring

Here is a funny-looking ruler. It is 8 units long, but it has only three marks.

8 units

| | 1 | | | 5 | 6 | | |

Suppose you may use the ruler only once for each length you measure. How can you measure any length from 1 to 8 units?

For example, to measure 1 unit you can use the part from the end of the ruler to the 1.

1. Which part can you use to measure 5 units? 6 units? 8 units?

To measure 4 units you can use the part from 1 to 5.

2. Which part can you use to measure 2 units? 3 units?

Suppose you have another ruler, 10 units long.

10 units

| ? |

You want to be able to measure all lengths from 1 to 10 units. What is the least number of marks you need to make on the ruler?

Copy the ruler and the number grid. Each time you put a mark on the ruler, use the grid to cross off the lengths you can measure.

| 1 | 2 | 3 | 4 | 5 | 6 | 7 | 8 | 9 | 10 |

3. If you make a mark at 1 unit, which lengths can you measure?

4. Suppose you make another mark at 4 units. Now which other lengths can you measure?

5. Which lengths are left for you to measure?

6. Where will you put the next mark? Which lengths can you measure?

7. Do you need another mark? If so, where will you put it? How many marks did you use in all?

8. Can you find a different set of marks that will also work? What are the marks?

MONDAY, MONDAY

WE BOTH HAVE BASEBALL PRACTICE ON THURSDAY. IT'S OUR FAVORITE DAY!

Collecting and Interpreting Data

GETTING STARTED

Some days are more popular than others. For example, Saturday is much more popular than Monday.

1. Why do you think this is true?

2. What is your favorite day of the week? Why?

3. Guess which day or days will be more popular with the students in your class. Which day or days will be least popular?

COLLECTING AND DISPLAYING DATA

A better way to answer these questions is to collect some actual data.

4. Take a class survey. Ask each student his or her favorite day. Then ask about the least favorite day. Use tally marks to record the answers. Copy the survey form below or design your own.

5. Total your results.

	FAVORITE DAY		LEAST FAVORITE DAY	
	Tally	Total	Tally	Total
Monday				
Tuesday				
Wednesday				
Thursday				
Friday				
Saturday				
Sunday				

I LOVE SUNDAYS BEST 'CAUSE OF SUNDAY DINNER!

DISPLAYING DATA

6. Use the data in your table. Copy and complete the two bar graphs below.

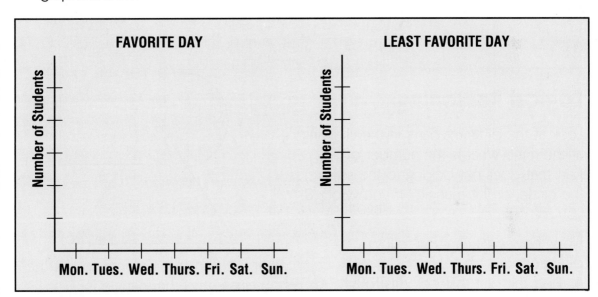

FAVORITE DAY

Number of Students

Mon. Tues. Wed. Thurs. Fri. Sat. Sun.

LEAST FAVORITE DAY

Number of Students

Mon. Tues. Wed. Thurs. Fri. Sat. Sun.

ANALYZING DATA

7. What do the bar graphs show you? What are the results of your survey?

8. Which day was the second favorite? Was it a close second? What does this tell you?

9. Were your own favorite and least favorite days the same as the survey results? Were your guesses about the other students' choices correct?

10. Can you think of some ways to use what you learned from this survey?

Daffy Directions

Logical Reasoning

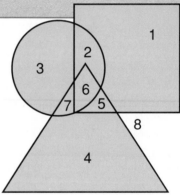

Look at the different parts in the drawing at the right. What is the number of the part that does not belong to the square, the circle, or the triangle? If you wrote 8, you are correct.

Now, how well can you follow directions?

A. Find the part that belongs to all three figures. If the part is numbered 6, write an O. If not, write an M.

B. Find the part that belongs to the square and the circle, but not the triangle. If it is numbered 5, write a B. If not, write a C.

C. Find the part that belongs to the circle and the triangle, but not the square. If it is numbered 7, write an I. If not, then write a T.

D. If the part that belongs to just the square and the triangle is numbered 2, write FF. If not, then write LL.

E. If the sum of all the numbers is greater than 33, write AG. If not, then write ED.

F. Unscramble all the letters you wrote. Complete this sentence:

I am __ __ __ __ __ __ __!

G. Make up your own directions. Try them out on a friend.

10

Understanding Numbers and Money

MATH CONNECTIONS: STATISTICS • PROBLEM SOLVING

Lemonade

Small cup 50¢

Medium cup 75¢

Large cup 95¢

1. What is happening in this picture?
2. What information do you see?
3. How can you use this information?
4. Tell a story about the picture.

Building Tens and Hundreds

Mr. Massaro lays tile. He keeps his tiles in:

 Squares of 100   Strips of 10   ▫ Single tiles

He has these red tiles left from his last job. How many red tiles does he have?

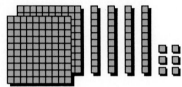

WORKING TOGETHER

1. Use your **place-value** models to show how many red tiles Mr. Massaro has.

2. Record on a place-value chart how many of each model you used.

Hundreds	Tens	Ones
■	■	■

3. Complete: ■ hundreds ■ tens ■ ones

4. How many red tiles does Mr. Massaro have? Write the number.

5. Read the number aloud. Then write the number in words.

Mr. Massaro has 12 strips of white tiles.

6. Use your place-value models to show this. Write how many of each model you used.

■ hundreds ■ tens ■ ones

7. Regroup your models to show hundreds. Tell how you did this.

■ hundreds ■ tens ■ ones

8. Write how many of each model you have now.

9. How many white tiles does Mr. Massaro have? Write the number.

10. Read the number aloud. Then write the number in words.

11. How does place-value help you read and write numbers?

12. How is regrouping tens as hundreds the same as regrouping ones as tens?

PRACTICE

Write the number.

13. **14.**

15. 7 tens 4 ones   **16.** three hundred twenty   **17.** 16 tens 5 ones

Write the word name.

18. 216   **19.** 705   **20.** 9 hundreds 6 tens

Regroup.

21. 3 hundreds 3 tens 14 ones = 3 hundreds ■ tens ■ ones

22. 18 tens 9 ones = ■ hundred ■ tens 9 ones

23. 2 hundreds 9 tens 17 ones = 3 hundreds ■ tens ■ ones

Solve.

24. Gene has 5 squares and 6 strips of yellow tiles. He also has 4 single yellow tiles. How many yellow tiles does he have?

25. Kerry has 4 squares and 2 strips of blue tiles. How many blue tiles does she have?

26. Rearrange the digits 5, 7, and 8 to make as many different three-digit numbers as you can.

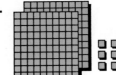

Critical Thinking

27. Joshua regrouped 1 hundred 2 tens 4 ones as 12 tens 4 ones. Tell how he did this.

UNDERSTANDING A CONCEPT

Thousands

A. Jacques owns the Crafty Card store. He sells postcards in boxes of 100, in sets of 10, and as singles. During one week Jacques sells 12 boxes, 4 sets, and 6 single cards. How many postcards does he sell?

You can record what Jacques sells on a place-value chart.

Hundreds	Tens	Ones
12	4	6

Think: 10 hundreds = 1 thousand

Regroup the hundreds to show thousands.

Thousands	Hundreds	Tens	Ones
1	2	4	6

You read this as one thousand, two hundred forty-six. Jacques sells 1,246 postcards.

B. The value of a digit depends on its place in a number.

The value of the 1 in the ten-thousands place is 10,000.

Thousands Period			Ones Period		
H	T	O	H	T	O
	1	0	5	7	3

1. What is the value of the 5? How do you know?

2. In which place is the 0? What does this mean?

You can use these values to write the number in **expanded form:** 10,000 + 500 + 70 + 3.

You read 10,573 as ten thousand, five hundred seventy-three.

3. Where do you place a comma when you write a number? Why?

TRY OUT Complete the table.

Place Value						Number	Word Name
Thousands Period			Ones Period				
H	T	O	H	T	O		
4. ■	■	■	■	■	■	398,028	5. ■
	6	3	7	4	5	6. ■	7. ■
8.		■	■	■	■	8,571	9. ■

Write the number.

10. 5,000 + 900 + 20 + 9   **11.** 6,000 + 70 + 1

12. one thousand, four hundred two

PRACTICE

Name the place and the value of the digit 5 in the number.

13. 6,415   **14.** 5,146   **15.** 1,564   **16.** 451   **17.** 54,614

Write the number in expanded form.

18. 210   **19.** 9,999   **20.** 65   **21.** 4,028   **22.** 100,702

Write the word name.

23. 61,032   **24.** 5,460   **25.** 204,204   **26.** 913,722

Write the number.

27. eight hundred twenty-five thousand, four hundred twenty-one

28. 4,000 + 200 + 60 + 5   **29.** nine hundred thousand, one

30. 5 ten thousands 9 thousands 2 hundreds 8 ones

Solve.

31. Last week Jacques sold 8,641 postcards. Write this number in expanded form.

32. During one month he sold one hundred sixty-three thousand, nine hundred forty-five cards. Write this number.

UNDERSTANDING A CONCEPT

Millions

999999
Meals served

A. The counter in the sign at the main store of Fred's Fast Food chain shows the number of meals served. What number will show when one more meal is served?

The counter is like a place-value chart.

After one more meal is served, 10 hundred thousand meals will have been served.

Millions Period			Thousands Period			Ones Period		
H	T	O	H	T	O	H	T	O
			9	9	9	9	9	9

Think: 10 hundred thousand = 1 million

When one more meal is served, the counter will show 1,000,000.

1. Write this number in words.

2. What is the greatest possible number you can write using this place-value chart? How do you know?

B. You can use what you learned about place value to read and write greater numbers.

Read: seven hundred nine million, forty-eight thousand, three hundred twenty

Millions Period			Thousands Period			Ones Period		
H	T	O	H	T	O	H	T	O
7	0	9	0	4	8	3	2	0

Write: 709,048,320

3. In which place is the 7? What is its value?

4. What number is 10,000 greater? 1,000,000 greater?

TRY OUT

Write the letter of the correct answer.
Match the number to the clue.

5. 10,032,850

6. 31,208,050

7. 13,208,050

a. The 3 is in the one millions place.

b. The value of the 3 is 30,000.

c. There are 3 thousands.

d. There are 3 ten millions.

PRACTICE

Write the digit in the ten-thousands place
and in the ten-millions place.

8. 534,509,390   **9.** 67,272,388   **10.** 48,078,010   **11.** 206,923,200

12. 45,528,200   **13.** 137,175,438   **14.** 29,713,010   **15.** 305,708,403

Write the number.

16. forty million, three hundred two thousand, one hundred ten

17. sixteen million, thirty thousand, nine hundred fourteen

18. one hundred ninety million, six hundred forty-five

19. two hundred five million, thirty thousand, eleven

20. 3 ten millions, 4 millions, 2 ten thousands, 9 tens, 8 ones

21. 100 greater than 3,451

22. 1,000 less than 3,674,261

23. 10,000,000 less than 276,892,371

24. 1,000,000 greater than 14,567,890

25. 100,000,000 greater than 434,275,211

26. 1,000,000 less than 1,789,204

27. What number is 100,000 greater than 99,999,999? In which places did the digits change?

Solve.

28. Tio's Taco Tower has sold 34,639,751 tacos. In what place is the 4 in the number?

29. There are 32,645 Paula's Pita Palaces across the country. What is the value of the 3 in the number?

CALCULATOR

1. Tell how 5,861,000 was changed to 5,961,000 on the display.

2. How would you change the display to 5,951,000?

3. How would you change the display to 6,951,000?

4. How would you change the display back to 5,861,000?

How Big Is a Hundred Million?

How tall would a stack of 100,000,000 sheets of paper be? Would it be as tall as your school? as tall as a mountain?

You can use what you know about numbers to estimate the height of the stack.

WORKING TOGETHER

Make a stack of 100 sheets of paper.

1. What object in your classroom is about this tall?

2. How many stacks of 100 do you need to make a stack of 1,000?

3. How would you estimate the height of a stack of 1,000 sheets of paper? What object in your classroom is about this tall?

4. How could you estimate the height of a stack of 10,000 sheets of paper? of 100,000 sheets? of 1,000,000 sheets?

5. How do your estimates compare with your height? with the height of your classroom? with the height of your school?

6. Estimate the height of 100,000,000 sheets of paper. Tell how you made the estimate. Can you think of something this tall?

SHARING IDEAS

7. How did you use place value to help you estimate?

8. Compare your method for estimating the heights of the stacks with those of others. How were the methods the same? How were they different?

ON YOUR OWN

Choose at least one activity. You may wish to try more than one.

9. About how many pages are in all the books in your school or neighborhood library? How did you find this number?

10. Estimate how long it would take you to count to 100,000,000. How did you make this estimate?

11. About how many steps do you take to get to school? 100? 1,000? 1,000,000? Tell how you estimated this number.

12. Look in newspapers and magazines for large numbers. Write about how you think these large numbers were counted.

P ROBLEM SOLVING

Using the Five-Step Process

Anna is preparing for the new school year. She has $10.00 and wants to buy a five-subject notebook for $3.99, book covers for $2.99, and a package of pens and pencils for $1.99. Will Anna have enough money to pay for these?

Study how the five-step process is used to solve this problem.

UNDERSTAND	
What do I know?	I know that Anna has $10.00 and that the items she wants to buy cost $3.99, $2.99, and $1.99.
What do I need to find out?	I need to find out whether Anna has enough money to buy the items.

PLAN	
What can I do?	I can estimate the total cost and compare it with the money she has.

TRY	
Let me try my plan.	*Think:* $3.99 is about $4. $2.99 is about $3. $1.99 is about $2.
	$4 + $3 + $2 = $9
	The total cost is less than $10.00. So Anna has enough money.

CHECK	
Did I answer the question?	Yes. Anna has enough money.
Does my answer make sense?	Yes. When I estimated I used prices a little greater than the real cost of the items.

EXTEND	
What have I learned?	I learned that I can solve some problems by estimating.

Remember:
What do I know?
What do I need to find out?
What can I do?
Did I answer the question?
What have I learned?

PRACTICE

Apply what you have learned about the five-step process to solve the problem.

1. Anna wants to buy knee pads for $4.99, soccer shoes for $8.98, and a sweatband for $1.99. Is $15.00 enough money to pay for them?

2. Ramón had 8 pencils. He bought 6 more. How many pencils does he have now?

3. Mary has 9 pages to read in her science book and 7 pages to read in her social studies book. How many pages does she have to read in all?

4. There are 18 students in art class. Only 6 students have finished their projects. How many students have not finished their projects?

5. Ronnie saw 9 friends at school in the morning and 8 friends in the afternoon. How many friends did he see that day?

6. During a game of tag, Jill tagged 9 players, and Ed tagged 6 players. How many more players did Jill tag than Ed?

UNDERSTANDING A CONCEPT

Counting Money

A. Kenata has a half dollar, a quarter, 2 nickels, and 4 pennies. Can she buy the notebook? First Kenata needs to know how much she has. She can count to find the amount.

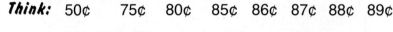

Think: 50¢   75¢   80¢   85¢   86¢   87¢   88¢   89¢

Kenata has 89¢. She can buy the notebook.

1. How does Kenata know she can buy the notebook?

2. What coins can she use to pay the exact price?

3. What is another way you could pay the exact price?

B. Jan has one each of these bills.

| ten-dollar bill | five-dollar bill | one-dollar bill |
| $10.00 | $5.00 | $1.00 |

4. How much money does Jan have?

5. *What if* Jan also had 2 quarters, 3 dimes, and 2 pennies? How much money would she have in all?

TRY OUT   Write the amount.

6.

7. 1 nickel, 4 pennies

8. 1 quarter, 2 dimes, 3 nickels

9. 3 five-dollar bills, 2 quarters, 3 dimes

Sidebar (coin reference)

half dollar
50¢
$.50

quarter
25¢
$.25

dime
10¢
$.10

nickel
5¢
$.05

penny
1¢
$.01

PRACTICE

Write the amount. Write amounts less than 1 dollar in two ways.

10.

11.

12. 1 half dollar,
2 nickels,
3 pennies

13. 2 one-dollar bills,
1 quarter, 2 dimes,
6 pennies

14. 1 ten-dollar bill,
1 five-dollar bill,
1 nickel

Name the bills and coins needed to make the amount.

15. $1.10, with 11 coins

16. $1.42, with 1 bill and 9 coins

Solve.

17. Alice buys 3 postcards for 75¢.
She gives the clerk 4 coins.
She does not get any change.
What coins does she give the
clerk?

18. Lucy buys a glass of orange
juice for 85¢. Are 3 dimes and
a half dollar enough to pay
for it?

19. Jaime sees a sweater for
$18.00. He says, "I have 5 bills.
That is just enough to buy it."
What bills does Jaime have?

20. *Write a problem* about money
using the following information:
Adam has a five-dollar bill,
2 quarters, and 1 penny.

Critical Thinking

21. Erik is buying a book that costs $7.29.
He wants to give the clerk the exact
amount using the fewest coins and bills.
Which coins and bills should Erik
give the clerk? How did you
find your answer?

Making Change

Millie works in a restaurant. A customer pays her $3.00 for a meal that costs $2.49. Millie counts out loud as she gives change.

Millie counts: $2.49 $2.50 $2.75 $3.00

Millie gives:

1. Why does Millie begin counting at $2.49 and stop counting at $3.00?

2. How much change does Millie give to the customer?

3. How could Millie have given change using only 2 coins?

Suppose the customer handed Millie a $5 bill.

4. Tell how Millie would count out the change.

5. How much change would Millie give?

TRY OUT   Name the coins and bills given as change.

6. Cost: $3.48
   Amount given: $5.00

7. Cost: $.37
   Amount given: $1.00

8. Cost: $7.34
   Amount given: $10.00

PRACTICE

Name the coins and bills given as change. Tell how much change would be given.

9. Cost: $.65
Amount given:
$1.00

10. Cost: $.83
Amount given:
$1.00

11. Cost: $.39
Amount given:
$1.00

12. Cost: $1.19
Amount given:
$2.00

13. Cost: $3.59
Amount given:
$5.00

14. Cost: $2.64
Amount given:
$10.00

15. Cost: $9.55
Amount given:
$20.00

16. Cost: $4.29
Amount given:
$5.00

17. Cost: $5.81
Amount given:
$20.00

Solve.

18. Ed buys a pear for 35¢. He gives the clerk a dollar. There are no quarters in the cash register. How can the clerk give him his change?

19. Arlette buys a meal that costs $4.16. She gives Millie $5.00. What change should Arlette receive?

20. *Write a problem* about a purchase made by Leo and the change he received. Leo has $10.00.

Critical Thinking Find the amount given. Tell how you found it.

21. Cost: $3.98
Change: 2 pennies, 1 one-dollar bill
Amount given: ■

22. Cost: $1.78
Change: 2 pennies, 2 dimes
Amount given: ■

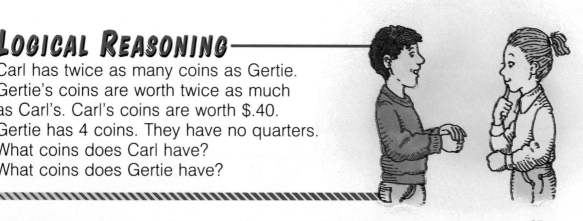

LOGICAL REASONING

Carl has twice as many coins as Gertie. Gertie's coins are worth twice as much as Carl's. Carl's coins are worth $.40. Gertie has 4 coins. They have no quarters. What coins does Carl have? What coins does Gertie have?

Spying by

375

Logical Reasoning

Number Spy is a game for two players. One player is the Codemaker. The other player is the Number Spy.

The Codemaker chooses a 3-digit number that does not repeat any digits. The Number Spy uses logical reasoning to find the secret number.

Each turn, the Number Spy guesses a 3-digit number. The Codemaker compares the guess with the secret number and gives a clue. The clue tells:

- how many of the digits are in the secret number, and
- how many of the digits are in the correct position.

Kate and Marty play Number Spy. Marty is the Codemaker. Kate is the Number Spy.

The chart on the next page describes their game. The third column shows how Kate uses reasoning to find the secret number.

NUMBERS

Kate's Guess	Marty's Clue	Kate's Reasoning
375	1 digit, no places	There is a 3 or a 7 or a 5 in the number.
786	no digits, no places	There is no 7, 8, or 6. There must be a 3 or a 5.
123	1 digit, no places	There could be a 1, 2, or 3.
125	no digits, no places	The 3 must be correct. There is no 1, 2, or 5. The only other possible digits are 0, 4, and 9.
349	3 digits, 1 place	I was right about the digits. Now I'll get the places.
394	3 digits, no places	The 3 can't be first.
493	3 digits, 1 place	Closer! I'll try switching the last two numbers.
439	3 digits, 3 places	Got it!

one digit, no places.

1. Before you begin playing, practice giving correct clues. Suppose the secret number is 682. What clue would you give for each of these guesses?

   a. 931   **b.** 614   **c.** 827

   d. 862   **e.** 801

2. Try playing Number Spy with a partner. Switch roles after each round. When you are the Spy, write the digits 0–9 on a piece of paper. Cross out digits when you know they are not in the secret number.

UNDERSTANDING A CONCEPT

Comparing and Ordering Numbers

A. Phyllis and Alex are marathon runners. Last year Phyllis ran 1,932 miles and Alex ran 1,682 miles. Who ran the greater distance last year?

Compare the numbers to solve the problem.

Step 1	Step 2
Line up the ones. Begin at the left and compare to find the first place where the digits are different.	**Compare the digits.**
1, **9** 3 2 1, **6** 8 2	**9 > 6 or 6 < 9** *Think:* 900 > 600 600 < 900

So 1,932 > 1,682 or 1,682 < 1,932.
Phyllis ran the greater distance.

1. Why don't you need to compare the digits in the tens places?

2. Compare 3,459 and 876. Which is greater? How do you know?

3. Compare $803.46 and $807.01. Which is less? How do you know?

B. You can compare $874; $8,742; $87; and $847 to order them from least to greatest.

Step 1	Step 2	Step 3
Line up the ones.	**Compare the other numbers.**	**Order the numbers from least to greatest.**
$ 874 → $8,742 $ 87 ← $ 847	$8**7**4 $8**4**7 *Think:* 7 > 4. So $874 > $847.	$87; $847; $874; $8,742

The number with the fewest digits is the least.

The number with the most digits is the greatest.

4. Tell how you would order the numbers from greatest to least.

TRY OUT

Compare. Use >, <, or =.

5. $21.36 ● $24.92   **6.** 7,451 ● 989   **7.** 12,131 ● 13,121

Write in order from greatest to least.

8. 881; 1,857; 5,281; 2,181   **9.** 432; 1,002; 39; 986

PRACTICE

Compare. Use >, <, or =.

10. 982 ● 321   **11.** 758 ● 1,298   **12.** $19.62 ● $18.62

13. $11.09 ● $10.90   **14.** 8,621 ● 2,394   **15.** 4,979 ● 4,991

16. 10,365 ● 8,365   **17.** 11,091 ● 11,901   **18.** 97,000 ● 79,999

Write in order from least to greatest.

19. 89; 1,361; 782; 4,321   **20.** 1,051; 269; 3,468; 892

21. 6,520; 520; 25; 652   **22.** $28.50, $8.50, $25.80, $19.90

Write in order from greatest to least.

23. 58, 97, 103, 49   **24.** 1,321; 3,211; 1,231; 1,123

25. $1.29; $12.90; $10.29; $1.92   **26.** 5,116; 516; 5,616; 561

Mixed Applications

27. In golf the lowest score wins. Rank these players' scores from first to fourth: Randy, 98; Debby, 87; Lucy, 93; Tad, 79.

28. In bowling the highest score wins. Which player won? May, 135; Luigi, 198; Agnes, 203; Phil, 145.

29. Jim buys a set of three golf balls for $3.75. He gives the clerk $5.00. What change does he receive?

Mixed Review

Write the number.

30. five hundred twenty-eight   **31.** 900 + 90 + 8   **32.** four thousand, two

UNDERSTANDING A CONCEPT

Rounding Numbers

A. Last year 62 students went to computer class. This year 69 students went. About how many students went each year?

You can round numbers to estimate about how many.

Round 62 to the nearest ten.   **Round 69 to the nearest ten.**

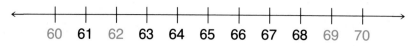

60 61 62 63 64 65 66 67 68 69 70

Think: 62 is between 60 and 70.   ***Think:*** 69 is between 60 and 70.
62 is closer to 60 than to 70.   69 is closer to 70 than to 60.
Round down.   Round up.

62 rounded to the nearest ten is 60.   69 rounded to the nearest ten is 70.
About 60 students went last year.   About 70 students went this year.

1. What is 64 rounded to the nearest ten?

When a number is halfway between two numbers, round up to the greater number.

2. What is 65 rounded to the nearest ten?

B. You can round numbers without using a number line.

Round 247 to the nearest hundred.

Step 1	Step 2	Step 3
Find the place to which you are rounding.	Look at the digit to the right of that place.	If it is 5 or greater, round up. If it is less than 5, round down. **Think:** 4 < 5 Round down.
247	2**4**7	**247** → 200

So 247 rounded to the nearest hundred is 200.

3. What is 247 rounded to the nearest ten? How did you round?

4. What is 1,578 rounded to the nearest thousand? How did you round?

5. Round $32.78 to the nearest ten dollars. How did you round?

Write the number.

6. Round 678 to the nearest hundred.

7. Round 1,323 to the nearest thousand.

8. Round $51.78 to the nearest dollar.

9. Round $55.00 to the nearest ten dollars.

PRACTICE

Round to the nearest ten or ten cents.

10. $.37 **11.** 42 **12.** $1.68 **13.** 25 **14.** 99 **15.** 328

Round to the nearest hundred or dollar.

16. $1.53 **17.** 241 **18.** $65.92 **19.** 367 **20.** 1,233 **21.** 78

Round to the nearest thousand or ten dollars.

22. 1,200 **23.** $38.42 **24.** 6,938 **25.** $92.99 **26.** 5,630 **27.** 3,501

Critical Thinking

28. Which of these numbers would round to 700 if rounded to the nearest hundred? Why?

634   734   658   758   690   750   599

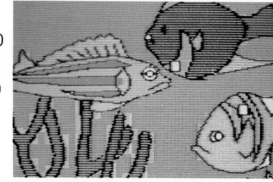

Mixed Applications

Solve. You will need to use the Databank on page 517 for Problems 31 and 32.

29. Mary spends $14.65 on computer disks. Count up to find her change from $20.00.

30. Jim buys 1,348 computer disks. How many disks to the nearest hundred does he buy?

31. During 1990 which computer company had the greatest profits? the least profits?

32. Order the companies from least to greatest according to their profits in 1991.

Mixed Review

Write in expanded form.

33. five hundred two **34.** 6,487 **35.** 3,092 **36.** 10,111

PROBLEM SOLVING

Strategy: Using Number Sense

Ari and Keesha want to know how many students can be seated in different places in their school. They use *number sense* to estimate the answer to the following problem.

1. If every seat in the classroom is filled, about how many students are in the classroom?

5?   25?   150?

> Let's see . . . How many students are there in my class?

> I know that when the lunchroom is full, 5 or 6 classes are eating lunch at the same time.

2. If someone is sitting at every table in the lunchroom, about how many students are in the lunchroom?

5?   25?   150?

> I'll ask Mrs. Jean in the school office how many students are enrolled in the school.

3. *What if* every seat in the school gym is filled during the championship basketball game? About how many students are watching the game?

25?   250?   25,000?

4. How did Ari and Keesha use number sense to solve the problems?

PRACTICE

Write the letter of the greatest possible answer.

5. About how many people can fit in a car?
a. 5 **b.** 25 **c.** 50

6. About how many office workers can fit in an elevator?
a. 5 **b.** 20 **c.** 200

7. About how many shoppers can ride on an escalator at the same time?
a. 5 **b.** 20 **c.** 90

8. About how many children can ride on a merry-go-round?
a. 5 **b.** 50 **c.** 500

9. About how many passengers can fly on a large jet?
a. 5 **b.** 25 **c.** 400

10. About how many riders can sit on a bus?
a. 2 **b.** 50 **c.** 200

11. About how much money do you need to pay for one person's lunch?
a. $5 **b.** $50 **c.** $500

12. About how much does it cost to buy 2 notebooks?
a. $.30 **b.** $3.00 **c.** $30.00

13. About how much money should you save if you want to buy a bicycle?
a. $6.00 **b.** $60.00 **c.** $600

14. About how much does a complete videogame system cost?
a. $.80 **b.** $8.00 **c.** $80.00

Using Tables

Erin is ordering records, cassettes, and compact discs for her store. She needs to know which type of recording is bought most often. This table shows last month's record sales for the three most popular groups.

Group	Record Sales
Broken Rocks	113
Up Beat	109
Music Masters	88

WORKING TOGETHER

The Broken Rocks also sold 345 cassettes and 213 compact discs. Up Beat sold 536 cassettes and 510 compact discs. Music Makers sold 423 cassettes and 315 compact discs.

Group	Number Sold		

1. Complete this table to show how many records, cassettes, and compact discs of each group were sold.

2. How did you label the column headings and list the groups?

3. How did this help you complete the table?

4. Which type of recording is bought most often?

SHARING IDEAS

5. Compare your table with those of others. How are the tables the same? How are they different?

6. Are there other ways to organize the data? How?

7. Which group is the most popular? How do you know?

8. Why is a table a useful way to record data? What makes the data easy to read?

PRACTICE

Use the table to answer
Problems 9–13.

Code	Price		
	Record Albums	Cassettes	Compact Discs
A	$3.99	$3.99	$8.99
B	$6.99	$6.99	$10.99
C	$8.99	$8.99	$14.99
D	$11.99	■	■
E	$14.99	$14.99	■

9. How are the prices organized?

10. The two most expensive compact discs sell for $16.99 and $20.99. Copy and complete the table to show these prices.

11. Look at the prices of records and cassettes. What might be the price of a code D cassette? How do you know?

12. Pablo has $10. What is the most expensive compact disc he can buy?

13. Sara wants to buy a code C record. Is $10 enough? How do you know?

14. The Inwood branch of the public library has 215 pop, 79 jazz, and 174 classical record albums. There are 187 pop, 102 jazz, and 241 classical record albums at the Elmsford branch. The Shore branch has 256 pop, 163 jazz, and 204 classical record albums. Make a table to show this data. Compare your table with those of others.

15. Ask other students to name their favorite type of music. Then ask them to name their favorite recording artist. Make a table to show their responses. Compare your table with those of others.

Mixed Review ▨▨▨▨▨▨▨▨▨▨▨▨▨▨▨▨▨▨▨▨▨▨▨▨▨▨▨▨▨▨▨▨▨▨▨▨▨

Compare. Use >, <, or =.

16. 867 ● 483   **17.** 78 ● 102   **18.** 45¢ ● $.45   **19.** 367 ● 365

Round to the nearest 10 or $.10.

20. 55   **21.** 81   **22.** 78   **23.** $.44

DECISION MAKING

Problem Solving: Choosing a Job to Earn Money

SITUATION

Laura needs $60 to buy a bicycle. She asks three friends how they have earned money.

PROBLEM

Which job is best for Laura?

DATA

Ben – had lemonade stand

Time spent: Worked when he wanted, usually on warm day. Up to 6 hours on weekend days and 2 hours after school.
Equipment: table, pitcher, cups, ice, lemonade, sign
Earnings: $1–$5 each hour
Note: Liked being his own boss.

Joey – did yard work (mowing, raking, weeding)

Time spent: 4 hours for each yard. Weekend days only. Nice days only.
Equipment: rake, lawn mower, trash bags – provided by lawn owner
Earnings: $10 per yard
Note: Jobs are not steady. Must find new jobs each week.

Rita – walked dogs

Time spent: half hour twice a day (before and after school) 5 days a week
Earnings: $2 for each dog each day
Note: Work is steady, but lots of responsibility

USING THE DATA

1. Walking one dog, Laura will make $2 each day. How much will she make in a week? How many weeks will it take Laura to make $60?

2. How many yards must Laura do to make $60? At two yards a week, how many weeks will it take her to earn $60?

3. If Laura makes $5 an hour selling lemonade, how much can she make in 1 weekday? in 5 weekdays?

4. If Laura makes only $1 an hour selling lemonade, how much can she make in 1 weekday? in 5 weekdays?

5. At $5 an hour, how long would it take Laura to make $60 selling lemonade? At $1 an hour, how long would it take her to make $60?

MAKING DECISIONS

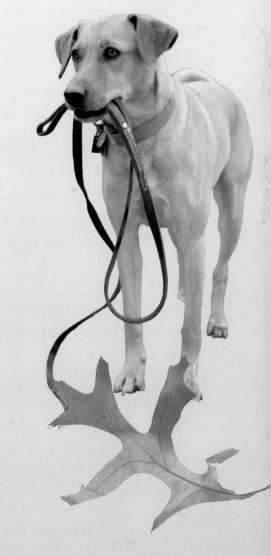

6. What things should Laura consider about each job before deciding which one is right for her?

7. At which job can Laura earn the most money in the least time?

8. Can Laura be sure she will make $5 an hour selling lemonade? Why or why not?

9. At which job can Laura be sure of the amount she will earn? Why?

10. *What if* Laura walks two or three dogs a day? How would this change the time it takes her to earn $60?

11. *What if* Laura's parents lend her money to start her lemonade stand? How would this change the time it takes her to earn $60?

12. Which job would you choose? Why?

Math and Social Studies

Our calendar is a **solar calendar**. It is based on the time it takes the earth to travel around the sun: 365 days, 5 hours, 48 minutes, and 46 seconds. We round the time to the nearest day. So our calendar year is 365 days long. The extra time is made up by adding a day every fourth year. This is called a leap year.

Some people, like the Aztec Indians, also based their calendar on the sun. Others, like North American Indians, measured time by the number of full moons before or after an event. A moon, or **lunar**, calendar month is based on the time it takes the moon to go from new to full and back to new. This takes almost 30 days. The Islamic calendar is still based on the moon. Its year is 354 days long.

How many more days are in the solar calendar than are in the lunar calendar?

Think: You know that 354 is 1 ten and 1 one less than 365. You could also count on from 354.

There are 11 more days in the solar calendar than there are in the lunar calendar.

ACTIVITY

1. Make up your own calendar. You can base it on the lunar year or the solar year. How many months will you have? How many days and weeks will be in each month? How many days will be in each week? What will you name the months and the days?

TECHNOLOGY

Computer Applications: Problem Solving

Why do people use computers? When is a problem better solved by using a computer?

Using a computer is often the easiest way to solve a problem when you need to:

- add, subtract, multiply, and divide a large column of numbers.
- graph data.
- draw geometric figures.
- re-create an experiment.

All of these activities can be done with paper and pencil or with a calculator. A computer is more useful when:

- there is a lot of data.
- the data need to be updated often.
- you need the answer quickly.

Throughout this book you will be given situations that you solve first without using a computer. You will then be asked how a computer could have been used to solve the same problem.

THINKING ABOUT COMPUTERS

Tell which method is best and why. Write *paper and pencil, calculator,* or *computer.*

1. You need to keep an inventory of ticket sales for the school play. The seating capacity at the auditorium is 75.

2. You need to keep an inventory of ticket sales for 10 movie theaters. The seating capacity at each theater is 300.

3. You need to make a graph to find the average rainfall in your city for last month.

4. You are the class treasurer. You need to plan budgets for the school year.

EXTRA PRACTICE

Building Tens and Hundreds, page 13...

Write the number.

1. 8 tens 7 ones   **2.** four hundred fifty   **3.** 19 tens 6 ones

4. two hundred five   **5.** 10 tens 9 ones   **6.** 12 tens 1 one

Regroup.

7. 5 hundreds 4 tens 17 ones = 5 hundreds ■ tens ■ ones

8. 15 tens 8 ones = ■ hundred ■ tens 8 ones

9. 3 hundreds 8 tens 28 ones = 4 hundreds ■ tens ■ ones

Thousands, page 15..

Name the place and value of the digit 7 in the number.

1. 4,987   **2.** 374   **3.** 2,713   **4.** 17,183

Write the word name.

5. 21,073   **6.** 6,000 + 100 + 50   **7.** 309,209

Write the number.

8. 3,000 + 400 + 50 + 3   **9.** sixty-four thousand, four

10. 8 hundred thousands 7 thousands 4 hundreds 2 ones

Millions, page 17 ....

Write the digit in the one thousands place and in the one millions place.

1. 637,149,875   **2.** 53,140,685   **3.** 80,656,030

4. 25,017,300   **5.** 240,372,311   **6.** 409,308,205

Write the number.

7. thirty million, two hundred five thousand, two hundred twelve   **8.** fourteen million, twenty thousand, eight hundred ten

9. 4 ten millions 2 millions 5 ten thousands 6 tens 2 ones

Problem Solving: Using the Five-Step Process, page 21 ....

Solve. Use the five-step process.

1. Gary wants to buy a pair of socks for $3.99, a package of wristbands for $1.99, and a cap for $5.99. Is $10 enough money to pay for these items?

2. There are 17 students in gym class. Eight students have finished their warm-up exercises. How many students have not finished their warm-up exercises?

3. Karla had 6 pennies. She found 8 more. How many pennies does she have now?

4. Mr. and Mrs. Anderson have 6 mollies and 5 guppies. How many fish do they have?

5. Frankie has 8 pages to read in his reading book and 5 pages to read in his social studies book. How many pages does he have to read?

6. Kevin has 15 stamps in his collection. Wanda has 9 stamps in her collection. How many more stamps does Kevin have than Wanda?

Counting Money, page 23 ....

Write the amount. Write amounts less than 1 dollar in two ways.

1. 2 quarters, 4 nickels, 7 pennies

2. 3 one-dollar bills, 3 quarters, 1 dime, 8 pennies

3. 2 ten-dollar bills, 1 five-dollar bill, 1 penny

4. 1 quarter, 4 dimes, 4 nickels, 4 pennies

5. 6 quarters, 5 dimes, 3 nickels, 9 pennies

6. 1 twenty-dollar bill, 2 five-dollar bills, 3 quarters

Name the bills and coins needed to make the amount.

7. $1.30 with 13 coins

8. $1.32 with 1 bill and 4 coins

9. $1.75 with 7 coins

10. $12.21 with 3 bills and 3 coins

11. $1.89 with 1 bill and 17 coins

12. $20.40 with 3 bills and 7 coins

EXTRA PRACTICE

Making Change, page 25

Name the coins and bills given as change. Tell how much change would be given.

1. Cost: $.35
Amount given:
$1.00

2. Cost: $.74
Amount given:
$1.00

3. Cost: $1.24
Amount given:
$2.00

4. Cost: $2.89
Amount given:
$5.00

5. Cost: $2.15
Amount given:
$10.00

6. Cost: $4.64
Amount given:
$20.00

Comparing and Ordering Numbers, page 29

Compare. Use >, <, or =.

1. 357 ● 253

2. 657 ● 2,175

3. $15.75 ● $12.57

4. $11.07 ● $10.10

5. 6,753 ● 3,789

6. 5,875 ● 5,987

Write in order from least to greatest.

7. 63; 1,255; 876; 2,309

8. 2,073; 409; 3,730; 794

9. $17.35; $7.35; $13.75; $11.73

10. $6.66, $16.21, $2.94, $9.30

Write in order from greatest to least.

11. $.35, $.76, $1.09, $.45

12. 529; 5,290; 5,092; 592

13. 1,416; 4,161; 1,136; 1,316

14. $7.38, $8.37, $7.57, $7.34

Rounding Numbers, page 31

Round to the nearest ten or ten cents.

1. $.43   **2.** 89   **3.** $1.47   **4.** 35   **5.** 98   **6.** 472

Round to the nearest hundred or one dollar.

7. $1.59   **8.** 312   **9.** $75.83   **10.** 489   **11.** 1,341   **12.** 89

Round to the nearest thousand or ten dollars.

13. 1,100   **14.** $47.39   **15.** 4,818   **16.** $81.89   **17.** 3,610   **18.** 4,502

Problem Solving: Using Number Sense, page 33.............................

Write the letter of the greatest possible answer.

1. About how many grapes are in a bunch?

   a. 5   **b.** 30   **c.** 500

2. About how many pencils can fit in a pencil case?

   a. 10   **b.** 60   **c.** 150

3. About how many pages are in a notebook?

   a. 5   **b.** 10   **c.** 50

4. About how many colors are in a rainbow?

   a. 7   **b.** 32   **c.** 100

5. About how many lines are on a loose-leaf page?

   a. 5   **b.** 30   **c.** 200

6. About how many people can ride in a car?

   a. 5   **b.** 25   **c.** 42

7. About how many peas are there in a pod?

   a. 6   **b.** 32   **c.** 68

8. About how many people can ride in a jumbo jet?

   a. 20   **b.** 125   **c.** 400

Using Tables, page 35 ....

Use the table to answer Problems 1–5.

1. How are the prices organized?

2. Look at the prices of the Top-40 and Rock albums. What might be the price of a Code D rock album? How do you know?

3. The two most expensive oldies albums sell for $14.99 and $24.99. Complete the table to show these prices.

	ALBUM PRICES		
Code	Top-40	Rock	Oldies
A	$7.99	$7.99	$8.99
B	$8.99	$8.99	$11.99
C	$10.99	$10.99	$12.99
D	$12.99	■	■
E	$15.99	$15.99	■

4. Christa has $10. What is the most expensive Rock album she can buy?

5. Rob wants to buy a Code C Top-40 album. Is $10 enough? How do you know?

PRACTICE PLUS

KEY SKILL: MILLIONS (Use after page 17.)

Level A ....

Write the digit in the ten thousands place.

1. 978,566   **2.** 64,830   **3.** 1,845,322   **4.** 29,763,076

Write the number.

5. six hundred thousand, twelve

6. four hundred thousand, five hundred seven

7. one million, two hundred ninety-one thousand, forty-two

8. ten million, sixty-seven thousand, three hundred forty

9. Ron's Rib Ranch has sold 54,879,310 ribs. In what place is the 4 in the number?

Level B ....

Write the digit in the ten millions place.

10. 310,430,780   **11.** 23,708,765   **12.** 134,956,743   **13.** 206,906,554

Write the number.

14. thirty million, two hundred three thousand, four hundred one

15. 9 hundred millions 6 hundred thousands 4 hundreds 3 tens

16. 6,000,000 + 30,000 + 800 + 40 + 9

17. 200,000,000 + 30,000,000 + 100,000 + 90

18. Sandy's Salad Bar sold 1,045,879 salads last year. What is the value of the 4 in the number?

Level C ....

Write the number.

19. one million greater than 13,678,987

20. ten thousand less than 846,999,560

21. 100,000,000 greater than 608,966,000

22. 10,000,000 less than 947,037,576

23. Harry's Hot Dog House sold 345,876,904 hot dogs. What is the value of the 3 in the number?

KEY SKILL: ROUNDING NUMBERS (Use after page 31.)

Level A ....

Round to the nearest ten or ten cents.

1. 22   **2.** 37   **3.** 65   **4.** $.51   **5.** $.14

Round to the nearest hundred or dollar.

6. $1.13   **7.** 250   **8.** $7.83   **9.** 926   **10.** 482

11. Last month 156 people went to the Computer Fair. How many people went, rounded to the nearest ten?

Level B ....

Round to the nearest ten or ten cents.

12. 88   **13.** 114   **14.** $8.46   **15.** 4,872   **16.** $23.58

Round to the nearest thousand or ten dollars.

17. 7,391   **18.** $85.78   **19.** 5,312   **20.** 2,764   **21.** $14.99

22. There are 2,065 software packages on display at the fair. How many packages are on display, rounded to the nearest hundred?

Level C ....

Round to the nearest hundred or dollar.

23. $81.58   **24.** 3,427   **25.** 7,611   **26.** 11,883   **27.** $410.39

Round to the nearest thousand or ten dollars.

28. 5,678   **29.** 91,023   **30.** $492.89   **31.** 629,346   **32.** 19,903

33. Which of these numbers would round to 1,600 if rounded to the nearest hundred?

1,541   1,651   1,574   1,599   1,700   1,629   1,612

34. Kerry spent $132.89 on computer disks. How much did he spend to the nearest ten dollars?

CHAPTER REVIEW

LANGUAGE AND MATHEMATICS

Complete the sentences. Use the words in the chart.

1. 14 tens 7 ones may be ■ as 1 hundred 4 tens 7 ones. *(page 12)*

2. The ■ of the 6 in the number 6,805 is 6,000. *(page 14)*

3. A ten-dollar bill, 2 quarters, 3 ■, 1 nickel, and 4 pennies is equal to $10.89. *(page 22)*

4. 839, ■ to the nearest hundred, is 800. *(page 30)*

5. *Write a definition* or give an example of the words you did not use from the chart.

CONCEPTS AND SKILLS

Write the word name. *(pages 12–15)*

6. 609   **7.** 760   **8.** 51,024   **9.** 308,803

Write the number. *(pages 12–17)*

10. 8 hundreds 4 tens 3 ones

11. three hundred twenty thousand, five hundred eleven

12. 1,000,000 less than 423,816,275 *(page 16)*

13. 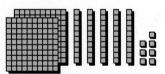   **14.**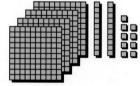

Write the number in expanded form. *(page 14)*

15. 713   **16.** 5,444   **17.** 26   **18.** 9,890

Write the amount. *(page 22)*

19. 1 half dollar, 1 quarter, 2 nickels, 2 pennies

20. 1 five-dollar bill, 1 one-dollar bill, 3 dimes, 1 penny

Name the coins and bills given as change. *(page 24)*

21. Cost: $.48
Amount given: $1.00

22. Cost: $6.29
Amount given: $20.00

Write in order from greatest to least. *(page 28)*

23. $34.00; $9.72; $82.70; $96.56

24. 3,247; 4,237; 3,742; 3,427

Round to the nearest 1,000 or $10.00. *(page 30)*

25. 2,540

26. $43.98

27. 1,921

Use the table to answer problems 28 and 29. *(page 34)*

28. Look at the prices of the different kinds of videos. What might be the price of a music video rented on Tuesday? How do you know?

29. Is $7 enough to rent two dramas?

COST OF RENTING VIDEO CASSETTES

Type of Video	Regular Price	Tuesday (Discount Day)
Music	$4.99	■
Drama	$4.99	$3.99
Cartoons	$3.99	$2.99

CRITICAL THINKING

30. Can you fit 100 pennies in a large shoe box? 1,000 pennies? Tell how you found your answer. *(page 32)*

31. Karen pays $8.37 for a kite. She uses the fewest coins and bills possible. Which coins and bills does she use? *(page 22)*

MIXED APPLICATIONS

32. Lisa has 34¢, Adam has 17¢, Josh has 12¢, and Mara has 22¢. Which is closest to their total amount of money? *(page 20)*
a. 80¢ **b.** 50¢ **c.** $5.00

33. If someone sits in each seat in the local theater, about how many people are watching the movie? *(page 32)*
a. 25 **b.** 250 **c.** 2,000

34. Pam buys orange juice for $.75. She gives the clerk $1.00. There are no quarters or dimes in the cash register. How can the clerk give Pam her change? *(page 22)*

CHAPTER TEST

Write the word name.

1. 5,009   **2.** 23,400

Write the number in expanded form.

3. three hundred ninety-one thousand, two hundred

4. one million, eighty-six thousand, nineteen

Write the amount.

5. 2 nickels, 3 pennies   **6.** 2 one-dollar bills, 2 quarters

Name the coins and bills given as change.

7. Cost: $4.62   Given: $5.00   **8.** Cost: $8.80   Given: $20.00

Compare. Use >, <, or =.

9. $15.02 ● $12.50   **10.** 565 ● 655   **11.** 82,812 ● 28,812

12. Write in order from least to greatest: 321; 3,021; 123; 3,210.

Round to the nearest ten.   Round to the nearest dollar.

13. 19   **14.** 104   **15.** $2.51   **16.** $38.29

Complete the table in Questions 17 and 18.

17. A new sports book costs $1.00 more than a new science book.

18. A used sports book costs the same as a new mystery book.

PRICE OF BOOKS

Type	Mystery	Science	Sports
New	$3.99	$3.99	**17.** ■
Used	$1.89	$2.59	**18.** ■

Solve.

19. If every seat in the theater is sold, about how many people are in the theater?

    a. 15   **b.** 250   **c.** 13,000

20. What if all the swings in the park were taken? How many children would be swinging?

    a. 1   **b.** 10   **c.** 100

ENRICHMENT FOR ALL

THE ABACUS

The abacus is an ancient counting board. Each column of beads stands for a place value.

A number is shown by pushing beads to the crossbar.
Each pushed-up bead below the crossbar is 1.
Each pushed-down bead above the crossbar is 5.
Add to find the digit in each place.

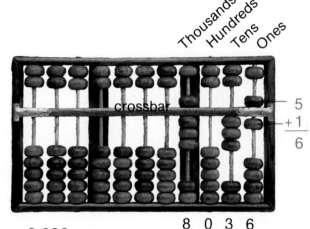

$$\begin{array}{r} 5 \\ +\ 1 \\ \hline 6 \end{array}$$

8 0 3 6

The abacus above shows the number 8,036.

1. How is 6 shown in the ones place?

2. How is 0 shown in the hundreds place?

3. **What if** you want to show the number 8,536? Which bead would you move?

4. Which beads would you move to show the number 8,076?

Write the number shown on the abacus.

5.
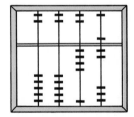

6.
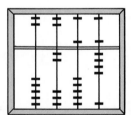

7.
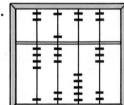

8.
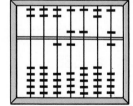

9.
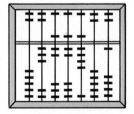

Draw a picture to show the number on an abacus.

10. 457   **11.** 3,604   **12.** 200,543   **13.** 2,076,098

CUMULATIVE REVIEW

Choose the letter of the correct answer.

1. Choose the word name for 203.

    a. two hundred thirty
    b. twenty three hundred
    c. two hundred three
    d. not given

2. Choose the number for six million one.

    a. 600,001   c. 60,000,001
    b. 6,000,001   d. not given

3. Which is 1,000 greater than 64,123?

    a. 65,123   c. 63,123
    b. 54,123   d. not given

4. How much is 1 quarter, 2 nickels?

    a. 27¢   c. 35¢
    b. 30¢   d. not given

5. Compare: 80,015 ● 80,510

    a. <   c. =
    b. >   d. not given

6. Compare: $4.52 ● $4.03

    a. <   c. =
    b. >   d. not given

7. Cost: $.68   Given: $.70
   Change?

    a. 2 pennies   c. 2 nickels
    b. 3 pennies   d. not given

8. With 4 one-dollar bills and 4 dimes, what can Amy spend?

    a. $3.99   c. $4.99
    b. $4.59   d. not given

9. Round 72 to the nearest ten.

    a. 60   c. 75
    b. 70   d. 80

10. Round 879 to the nearest hundred.

    a. 900   c. 870
    b. 880   d. 800

11. What is the expanded form for two hundred one?

    a. $200 + 10 + 1$
    b. $200 + 10$
    c. $200 + 1$
    d. not given

12. What is the value of 4 in 1,432?

    a. 4   c. 400
    b. 40   d. not given

13. Choose the greatest number.

    a. 538   c. 3,158
    b. 851   d. 3,518

14. Cost: $.59   Amount needed?

    a. $.25   c. $.55
    b. $.50   d. not given

Using Addition and Subtraction Facts

CHAPTER 2

MATH CONNECTIONS: LENGTH
• GRAPHING • PROBLEM SOLVING

Coyote Lake campgrounds
1 mile

Mountain Pass Pony Ride
1 mile

Wonder Falls Picnic Area
and Swimming Hole
3 miles

Frontier Town
5 miles

1. What is happening in this picture?

2. What information do you see?

3. How can you use this information?

4. Write a problem about the picture.

51

Meaning of Addition and Subtraction

This activity will help you learn about the properties of addition and subtraction.

WORKING TOGETHER

1. Pick any two numbers from 0 to 9. Write two word sentences using these numbers.

2. Ask an addition question.

   Write the number sentence that answers the question.

3. Ask a subtraction question that uses the sum.

   Write the number sentence that answers the question.

4. Ask a subtraction question to compare the numbers.

   Write the number sentence.

5. Now pick another pair of numbers and do this activity again. Stop when you have completed the activity for four pairs of numbers.

Example

Think: 9 and 6
I have 9 red cars.
You have 6 blue cars.

How many cars are there in all?

$$9 + 6 = 15$$

$$\begin{array}{r} 9 \leftarrow \textbf{addend} \\ +\ 6 \leftarrow \textbf{addend} \\ \hline 15 \leftarrow \textbf{sum} \end{array}$$

There are 15 cars in all.

If there are 9 red cars, how many blue cars are there?

$$15 - 9 = 6$$

$$\begin{array}{r} 15 \\ -\ 9 \\ \hline 6 \leftarrow \textbf{difference} \end{array}$$

There are 6 blue cars.

How many more red cars are there than blue cars?

$$9 - 6 = 3$$

There are 3 more red cars than blue cars.

Think: 6 and 3
You have 6 model cars.
I have 3 model cars.

6. Look at the addition question given in the example. Can you write another number sentence to answer this question? If so, write it.

7. Can you ask another subtraction question that uses the numbers in the example? If so, write it and the number sentence to answer it.

SHARING IDEAS

8. How do your questions and number sentences compare with those of others?

9. How do your questions and number sentences show that addition and subtraction are opposites?

10. If you change the order of the addends, does the sum change? How do you know?

11. What is the sum when you add 0 to a number?

12. What is the difference when you subtract 0 from a number? when you subtract a number from itself?

PRACTICE

Add or subtract.

13.	**14.**	**15.**	**16.**	**17.**	**18.**	**19.**
0	5	2	10	4	5	1
+ 1	− 0	+ 8	− 8	− 4	+ 2	+ 7

20. $8 - 7$ **21.** $2 + 7$ **22.** $6 + 6$ **23.** $4 + 5$ **24.** $13 - 4$

25. $9 - \blacksquare = 9$ **26.** $\blacksquare + 5 = 5$ **27.** $3 - \blacksquare = 0$ **28.** $6 + \blacksquare = 3 + 6$

29. $\blacksquare + 0 = 2$ **30.** $4 - \blacksquare = 4$ **31.** $\blacksquare - 0 = 0$ **32.** $9 + 1 = \blacksquare + 9$

33. How much more than 3 is 6?   **34.** 9 and 5 more equals what number?

DEVELOPING A CONCEPT

Fact Families

Angie and Rhoda are using a spinner to learn about fact families. They take turns spinning. On the first round Angie spins a 4 and an 8. The girls use the numbers to write the fact family.

Angie records the related addition sentences on a chart.

Rhoda records the related subtraction sentences on another chart.

Addend	+	Addend	=	Sum or Total
4	+	8	=	12
8	+	4	=	12

Total	−	Number Taken Away	=	Number Left
12	−	8	=	4
12	−	4	=	8

WORKING TOGETHER

Copy the charts above.

1. The number sentences on both charts belong to the same fact family. Tell how they are related.

2. **What if** Angie spins a 7 and a 0? Write the related number sentences that make up the fact family on your charts.

3. **What if** Rhoda spins a 2 and a 2? Write the related number sentences on your charts.

Use a spinner and repeat this activity four more times. See how many related number sentences you can write. Record them on your charts.

4. Did you write the same number of related sentences for every fact family? Why or why not?

SHARING IDEAS

5. Compare your charts with those of other students. How are they the same? How are they different?

6. How did the addition and subtraction properties help you write fact families?

PRACTICE

Find the sum or difference. Write the related number sentences that make up the fact family.

7. 5 + 3	**8.** 11 − 8	**9.** 11 − 9	**10.** 8 + 7	**11.** 16 − 9	**12.** 7 + 7

13. $10 - 5$ **14.** $9 + 8$ **15.** $0 + 3$ **16.** $10 - 4$ **17.** $6 - 6$

Write the fact families for the set of numbers.

18. 6, 2, 8   **19.** 5, 1, 6   **20.** 0, 9, 9   **22.** 4, 4, 8   **22.** 3, 7, 10

Complete the fact. Find the missing number.

23. $8 + \blacksquare = 14$ **24.** $6 - \blacksquare = 4$ **25.** $\blacksquare + 5 = 8$ **26.** $9 + \blacksquare = 18$

27. $\blacksquare + 5 = 11$ **28.** $1 + \blacksquare = 1$ **29.** $\blacksquare - 2 = 7$ **30.** $17 - \blacksquare = 9$

Critical Thinking

Look at each set of numbers. Can you write a fact family? Why or why not?

31. 0, 7, 8   **32.** 1, 2, 3   **33.** 3, 3, 3   **34.** 9, 1, 10

Mixed Applications

35. Tom and Leah brought the same number of fruits to a picnic. Tom brought 6 plums and 9 mangos. Leah brought 6 mangos. How many plums did she bring?

36. Alma ordered 2 pounds of pecans by mail from Georgia. She used all of them to make a pecan pie. How many pounds of pecans did Alma have left?

37. Fizzville is 9 miles from Beat City. Alec rides the bus from Beat City for 5 miles. How much further is it to Fizzville?

38. Joe collected 1,213 bottle caps from all over the world. How many bottle caps did he collect to the nearest thousand?

Mixed Review

Round to the nearest ten or ten dollars.

39. 67   **40.** $29.85   **41.** 691   **42.** 7,555   **43.** $8.81

Mental Math: Addition Facts

Jason, Robert, and Joyce use mental math to help them learn addition facts.

A. To find 7 + 4, Jason uses **counting on.**

7 + 4 = ■

Think: Start at 7. Count on 4 more.
Count: 8, 9, 10, 11

So 7 + 4 = 11.

1. How can you use counting on to find 8 + 5? What is the sum?

2. Use counting on to find 3 + 9. How would changing the order of addends make it easier to count on?

B. Joyce and Robert use **doubles** to find 5 + 6.

5 + 6 = ■

Joyce thinks:
5 + 5 = 10.
I know that 6 is 1 more than 5.
5 + 5 + 1 = 11
So 5 + 6 = 11.

Robert thinks:
6 + 6 = 12.
I know that 5 is 1 less than 6.
6 + 6 − 1 = 11
So 5 + 6 = 11.

3. Use doubles to find 7 + 6.

C. Jason said that he could make a 9 a 10 to help him find 9 + 3.

$$\begin{array}{r} 9 \\ + 3 \end{array}$$
 Think: I add 1 to 9 to make 10.
        I take away 1 from 3 to make 2.
$$\begin{array}{r} 10 \\ + 2 \\ \hline 12 \end{array}$$

Since 10 + 2 = 12, Jason knows that 9 + 3 = 12.

4. Use this method to find 4 + 9.

5. 9 + 4   **6.** 5 + 4   **7.** 4 + 7   **8.** 2 + 5

PRACTICE

Add. Use mental math.

9. 3 + 3	**10.** 7 + 7	**11.** 0 + 2	**12.** 8 + 6	**13.** 5 + 8	**14.** 3 + 6

15. 4 + 3	**16.** 7 + 8	**17.** 7 + 1	**18.** 8 + 8	**19.** 8 + 0	**20.** 6 + 7

21. 2 + 3   **22.** 8 + 2   **23.** 2 + 4   **24.** 1 + 8   **25.** 5 + 7

26. 0 + 4   **27.** 6 + 3   **28.** 5 + 5   **29.** 5 + 9   **30.** 9 + 9

31. What is 4 plus 2?   **32.** What is the sum of 9 + 7?

33. 4 plus 8 is what number?   **34.** What is 7 and 5 more?

Critical Thinking

35. Tell how you could use what you know about adding doubles to help you find 6 + 8.

Mixed Applications

36. José's grandparents send him 7 stamps from Spain and 9 stamps from France. How many stamps does José get?

37. It is 5,663 miles from Paris to Los Angeles and 5,477 miles from Tokyo to Los Angeles. Which of these cities is closer to Los Angeles?

38. *Write a problem* for this fact: 3 + 4. Trade problems with other students and solve.

Mixed Review

Name the place and the value of the digit 4 in the number.

39. 409   **40.** 42,500   **41.** 491,999   **42.** 34,860   **43.** 574

UNDERSTANDING A CONCEPT

Mental Math: Subtraction Facts

Using mental math helped Joyce, Robert, and Jason learn addition facts. They decided to use mental math to help them learn subtraction facts.

A. To find 12 − 4, Robert uses **counting back.**

12 − 4 = ■

Think: Start at 12. Count back 4.
Count: 11, 10, 9, 8

So 12 − 4 = 8.

1. How can you use counting back to find 14 − 5? What is the difference?

B. Joyce subtracts from 10 and then adds to find the difference.

11 − 6 = ■ **Think:**
$$\begin{array}{r} 10 - 6 = 4 \\ +\ 1 \qquad +1 \\ \hline 11 - 6 = 5 \end{array}$$

2. Use this method to find 13 − 9.

C. Jason uses related facts to help him find the difference.

$$\begin{array}{r} 9 \\ -\ 4 \\ \hline \end{array}$$

Think: Since 5 + 4 = 9, then 9 − 4 = 5.

3. Use a related fact to find 12 − 6.

TRY OUT Subtract. Use mental math.

4. 7 − 3   **5.** 13 − 5   **6.** 11 − 7   **7.** 10 − 2

PRACTICE

Subtract. Use mental math.

8. 10 − 1	**9.** 13 − 6	**10.** 12 − 5	**11.** 14 − 8	**12.** 8 − 8	**13.** 11 − 5
14. 9 − 8	**15.** 18 − 9	**16.** 9 − 3	**17.** 7 − 2	**18.** 14 − 6	**19.** 16 − 9

20. 6 − 4   **21.** 8 − 1   **22.** 12 − 3   **23.** 14 − 9   **24.** 12 − 7

25. 12 − 9   **26.** 10 − 7   **27.** 16 − 8   **28.** 9 − 7   **29.** 13 − 8

30. 4 less than 11

31. the difference between 15 and 8

32. 5 take away 3

33. 13 is how many more than 7?

Mixed Applications

34. Jeff travels 8 miles from his house to the beach. He travels 3 miles from the beach to his friend Wendy's house. How many miles does Jeff travel?

35. The depth of the water at the docks is 7 feet at low tide. At high tide the depth is 15 feet. How many feet does the water rise?

36. Marsha found 7 shells by the seashore. Leon found 5 shells. How many more shells did Marsha find than Leon?

37. *Write a problem* for this fact: 7 − 4. Trade problems with other students and solve.

CHALLENGE

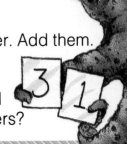

The ones digit of an even number is 0, 2, 4, 6, or 8. The ones digit of an odd number is 1, 3, 5, 7, or 9.

1. Pick two even numbers. Add them.

2. Pick two odd numbers. Add them.

3. Pick an even number and an odd number. Add them.

4. Repeat Steps 1–3.

5. Compare your sums. What does this tell you about adding even and odd numbers?

UNDERSTANDING A CONCEPT

Mental Math: Three or More Addends

A. Diego and Susan group addends in different ways to help them find sums.

Add: 4 + 3 + 2

Diego found the sum this way.

(4 + 3) + 2 = ■

7 + 2 = 9

Susan found the sum this way.

4 + (3 + 2) = ■

4 + 5 = 9

Parentheses show how they grouped the addends.

1. Does changing the way you group addends change the sum? Tell why.

2. How could you group the addends to find 5 + 2 + 4 in two ways?

B. You can group addends differently to help you add mentally.

Add: 3 + 6 + 7

Susan found the sum by making a 10.

$$\begin{array}{r} 3 \\ 6 \\ + 7 \end{array} \Big\} 10$$

Think: Look for sums of 10.
3 + 7 = 10; 10 + 6 = 16

So 3 + 6 + 7 = 16.

3. Find the sum of 2 + 4 + 8 + 4 by making a 10.

TRY OUT Write the letter of the correct answer.
Find the sum. Use mental math.

4. 9 + 1 + 3   **a.** 91   **b.** 10   **c.** 13   **d.** 9

5. 6 + 1 + 6   **a.** 13   **b.** 12   **c.** 11   **d.** 7

6. 5 + 4 + 5   **a.** 50   **b.** 5   **c.** 10   **d.** 14

7. 9 + 6 + 3   **a.** 9   **b.** 19   **c.** 18   **d.** 15

PRACTICE

Find the sum. Use mental math.

8.
```
   8
   7
 + 3
```

9.
```
   3
   6
 + 5
```

10.
```
   5
   4
 + 8
```

11.
```
   6
   2
 + 7
```

12.
```
   7
   9
 + 1
```

13.
```
   9
   2
 + 8
```

14.
```
   1
   4
 + 5
```

15.
```
   6
   7
 + 8
```

16.
```
   2
   3
   5
 + 4
```

17.
```
   8
   2
   1
 + 5
```

18.
```
   6
   3
   4
 + 2
```

19.
```
   1
   7
   9
 + 3
```

20. 5 + 3 + 6

21. 2 + 8 + 3

22. 4 + 7 + 6

23. 9 + 2 + 8

24. 1 + 1 + 4 + 5

25. 5 + 3 + 2 + 4

26. 5 + 3 + 3 + 5

Find the missing addend.

27. (4 + 5) + 6 = 5 + (■ + 6)

28. 8 + (2 + ■) = 2 + (6 + 8)

Mixed Applications

29. Joanne spent 5 hours on the beach on Monday, 5 hours on Tuesday, and 3 hours on Wednesday. How many hours did she spend on the beach?

30. Ken is flying on an 8-hour flight from Seattle to Honolulu. He has already been flying for 3 hours. How many more hours does he have to fly?

31. Jean has 3 leis. Nora has 5 leis. The twins have 4 leis each. How many leis do the sisters have in total?

32. An airline has 472 dinners ready for the next flight out. How many dinners are there to the nearest hundred?

CHALLENGE

Find the answer. Do the work in parentheses first.

1. (9 − 3) + 6   2. 3 + (7 − 5)   3. 8 − (4 + 3)

4. (4 + 5) + 3   5. 7 + (8 − 2)   6. 18 − (7 + 2)

7. Which of the exercises would have different answers if you moved the parentheses?

PROBLEM SOLVING

Finding Needed Information

Michael and his family are on vacation. They are going to Grand Canyon National Park in Arizona. They have been on the road for 2 days. Michael wants to find out how many days the trip will last.

1. What information does Michael know?

2. What does he need to find out?

Sometimes you do not have enough information to solve a problem. You need to look up additional information.

3. What additional information does Michael need?

Michael studied a map and figured out that it will take 4 more days to get to the Grand Canyon.

4. Does he have enough information now?

5. How many days will the trip take?

6. Michael's mother says she will take him to the movies every day it rains during their vacation. How many times will Michael go to the movies? What information does he need to solve the problem? Can he find it now?

7. Why is it important to list all the information needed to solve a problem?

AUGUST						
			1	2	3	4	5
6	7	8	9	10	11	12
13	14	15	16	17	18	19
20	21	22	23	24	25	26
27	29	30	31			

PRACTICE

Solve. Use mental math, or paper and pencil. If there is not enough information, write what is needed.

8. Michael is buying a poster at the Grand Canyon gift shop. The poster costs $8.98. How much change should he get?

9. Michael sees a small herd of wild burros in the park. He counts 17 burros. Then 8 run away. How many burros are left?

10. Michael hiked on 15 trails in the Grand Canyon. How many more trails than Michael did Michael's mother hike?

11. While hiking, Michael's sister saw 9 lizards, 5 raccoons, and 4 elk. How many animals did she see in all?

Strategies and Skills Review

Solve. You may need to use the Databank on page 517.

12. Michael and his sister hiked for 2 hours with only a 10-minute rest. Did they walk about 5 miles? 15 miles? 25 miles?

13. About how much more rain falls in Fairbanks, Alaska, than in Los Angeles, California, in a year?

14. Michael's father bought a package of frankfurters and 6 ears of corn for dinner. Did he spend about $2? $10? $50?

15. *Write a problem.* Leave out one piece of information needed to solve the problem. Ask others to tell what additional information is needed and how it can be found.

COUNTDOWN

Investigating Patterns

A. You can use the [=] key on a calculator to repeat calculations. You know that $5 + 5 + 5 + 5 = 20$.

What is the result when you enter this series on a calculator?

[5] [+] [5] [=] [=] [=]

Each time you press the [=] key, you add 5. The [=] key makes it easy to do repeated addition.

1. How could you use the [=] key to help you find the cost of six $6.85 records? Try it.

B. You can also use a calculator for repeated subtraction. Suppose you won a $300 prize. How many $43 video games could you buy?

You could subtract the cost of one game at a time:

[3] [0] [0] [−] [4] [3] [=] [−] [4] [3] [=] and so on.

Or you could use constant subtraction:

[3] [0] [0] [−] [4] [3] [=] [=] [=] [=] and so on.

2. How do you know when you have your answer?

Many problems can be solved using repeated addition or subtraction. Can you find more than one way to find each answer?

3. Suppose you have $35. How many $7.98 cassette tapes can you buy?

4. One class has raised $250 by selling cases of pens for $15.50 each. How many more cases of pens does the class need to sell to reach its goal of $325?

5. How many 5-minute pinball games can you play in 1 hour (1 hour = 60 minutes)? Would $2.75 be enough to play that many games if each game cost 25 cents?

C. Do you remember what to do when you make a mistake entering a number on a calculator? You should use the ⌊C⌋ key. Try this experiment:

What if you want to add 37 and 98? Press the following keys:

⌊3⌋ ⌊7⌋ ⌊+⌋ ⌊9⌋ ⌊9⌋

Now press ⌊C⌋ once. When you press this key once, it erases only the last number you entered. You will see 37 in the display. This tells you that you have erased the 99. Now enter ⌊+⌋ ⌊9⌋ ⌊8⌋ and press ⌊=⌋.

What if you want to add 54 and 89? Press the following keys:

⌊5⌋ ⌊4⌋ ⌊−⌋ ⌊8⌋ ⌊9⌋

If you press ⌊C⌋ once, you will erase only the 89, not the subtraction. Now press ⌊C⌋ twice. When you press this key twice, it erases everything you have done. Then start again.

6. What happens when you enter ⌊7⌋ ⌊8⌋ ⌊+⌋ and then press ⌊C⌋ once? What happens when you enter ⌊7⌋ ⌊8⌋ ⌊+⌋ ⌊3⌋ ⌊2⌋ and then press ⌊C⌋ once?

7. Here's a calculator puzzle. Use only these keys.

⌊0⌋, ⌊1⌋, ⌊+⌋, ⌊−⌋, ⌊=⌋

Can you get 135 in the display? How? Can you find the method that uses the fewest keystrokes?

8. Can you get the display to show 478 using the same keys? Can you find the method that uses the fewest keystrokes?

Measuring Length: Metric Units

WORKING TOGETHER

Use the width of the tip of your index finger to measure the length of this crayon.

1. How many fingertips long is the crayon?

2. Compare your measurement with those of other students. Are they the same? Why or why not?

The width of the tip of your index finger is about 1 centimeter. The **centimeter (cm)** is a metric unit of length.

You can use a centimeter ruler to measure short lengths.

> Be sure to line up the left end of the crayon with the left end of the scale on the ruler.

3. Use a centimeter ruler to measure the length of the crayon.

4. What is the length of the crayon to the nearest centimeter?

5. Use the tip of your index finger to measure the length of your desk. Compare your measurement with those of others. Are they the same? Why or why not?

6. Use a centimeter ruler to measure the length of your desk. Compare your measurements with those of others. Are they the same? Why or why not?

Use a centimeter ruler to measure the following. Compare your measurements with those of others.

7. the length of your math book

8. the height of your desk

SHARING IDEAS

9. Why do you think it is better to use a centimeter ruler than the width of your finger to measure length?

10. What is meant by "measuring to the nearest centimeter"?

PRACTICE

Copy and complete this chart. Estimate. Then use your centimeter ruler to measure.

	Object	Estimate	Measurement
11.	Width of your math book	■	■
12.	Width of your shoe	■	■
13.	Depth of a windowsill	■	■
14.	Height of your chair	■	■

15. Choose three more objects to measure. Record your work on your chart.

Use your centimeter ruler. Draw lines with the following lengths:

16. 6 cm   **17.** 11 cm   **18.** 1 cm   **19.** 10 cm

Critical Thinking

20. You use centimeters to measure the width, length, height, or depth of an object. How are these measurements the same? How are they different?

Mixed Applications

Solve. You will need to use the Databank on page 517.

21. The Sears Tower in Chicago and the World Trade Center in New York are both 110 stories tall. Which one is taller?

22. *Write a problem* using the information in the Databank. Solve your problem. Trade problems with other students and solve them.

UNDERSTANDING A CONCEPT
Estimating Length: Metric Units

The **decimeter (dm),** the **meter (m),** and the **kilometer (km)** are other metric units used to measure length and distance.

The width of a cassette tape is about 1 decimeter.

The width of a door is about 1 meter.

A kilometer is about how far you might walk in 20 minutes.

This chart shows how the units are related.

10 centimeters (cm)	= 1 decimeter (dm)
10 decimeters (dm)	= 1 meter (m)
100 centimeters (cm)	= 1 meter (m)
1,000 meters (m)	= 1 kilometer (km)

Tell whether you would use m, cm, dm, or km to measure:

1. the distance from Dallas, Texas, to Chicago, Illinois

2. the width of your desk

3. the height of a building

4. How did you decide which unit to use to measure each?

Write the letter of the correct answer. Would you use cm, dm, m, or km to measure?

5. the length of a key   **a.** cm   **b.** dm   **c.** m   **d.** km

6. the distance across a state   **a.** cm   **b.** dm   **c.** m   **d.** km

7. the width of a napkin   **a.** cm   **b.** dm   **c.** m   **d.** km

8. the length of a swimming pool   **a.** cm   **b.** dm   **c.** m   **d.** km

PRACTICE

Tell whether you would use cm, dm, m, or km to measure.

9. the height of a house   10. the length of a car

11. the length of a worm   12. the distance across a lake

Write the letter of the best estimate.

13. the distance a car travels in an hour   **a.** 60 dm   **b.** 60 m   **c.** 60 km

14. the length of your thumb   **a.** 2 cm   **b.** 2 dm   **c.** 2 m

15. the width of a chair   **a.** 6 cm   **b.** 6 dm   **c.** 6 cm

16. the height of a flagpole   **a.** 3 dm   **b.** 3 m   **c.** 3 km

Complete the chart.
Use the units given.

Object	Estimate	Actual Measure
Length of your arm (cm)	17. ■	18. ■
Distance around your head (dm)	19. ■	20. ■
Length of your classroom (m)	21. ■	22. ■

Mixed Applications

23. Judy hiked 4 km on Sunday, 3 km on Monday, 5 km on Tuesday, and 6 km on Wednesday. How many km did she hike in all?

24. An oak tree can grow more than ten times the height of an adult. Which metric unit would be the best to use to measure the height of an oak tree?

25. The city of Troy is 9 km north of my home. The city of Lester is 7 km south of my home. How far is it from Troy to Lester?

26. A highway sign shows that it is 9 km north to Troy and 18 km north to Ashby. How far is it from Troy to Ashby?

PROBLEM SOLVING

Strategy: Drawing a Diagram

Helen is on a camping trip with her family. She and her brother, Hal, are going to cut a log into 4 pieces to use for firewood. How many cuts do they need to make?

1. What information do you know?

2. What do you need to find out?

Sometimes it helps to draw a diagram to understand the problem better.

3. Draw a diagram of a log. Show how to cut it into 4 pieces. How many cuts are needed?

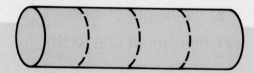

4. If you had not drawn a diagram of the log, would your answer have been different?

5. *What if* they wanted to cut the log into 8 pieces? How many cuts would they need to make? Draw a diagram and count the number of cuts.

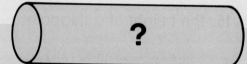

6. How does drawing a diagram help you solve problems like this?

PRACTICE

Draw a diagram to help you solve the problem.

7. Helen is using a red ribbon to mark the boundary of the campsite. She cuts the ribbon into 6 pieces. How many cuts does she need to make?

8. Four cars are parked bumper to bumper. How many bumpers are touching?

9. A farmer build a 100-foot fence to separate her farm from the campground. She puts posts into the ground every 10 feet. How many posts does she need in order to build her fence?

10. Hal is riding in the family car to the campsite. He notices that 1 telephone wire is always strung between 2 poles. How many wires are strung between 5 telephone poles?

Strategies and Skills Review

Solve. If there is not enough information, write what is needed.

11. Helen roasted 7 marshmallows, and Hal roasted 9. Their father roasted more than either of them. How many marshmallows did they roast in all?

12. One evening after a rain, Helen counted 17 fireflies. Hal counted 8 fireflies. How many more fireflies did Helen count than Hal?

13. Hal counts 12 campsites along the road. There is a picnic table between every neighboring campsite. How many picnic tables are there?

14. There is a light pole at each end of the road. There is also a light pole at every third campsite. How many light poles are there for the 12 campsites?

15. Helen bought 3 postcards and 3 stamps. Could these items cost $4? $14? $24?

16. **Write a problem** that can be solved with the help of a diagram. Ask others to solve it.

DEVELOPING A CONCEPT

Making Bar Graphs

Karen kept track of the fruit her family ate in one week. She used these steps to show the data on a **bar graph**.

Step 1 Choose a scale.

Step 2 Draw and label the sides.

Step 3 Draw the bars on the graph.

Step 4 Write a title above the graph.

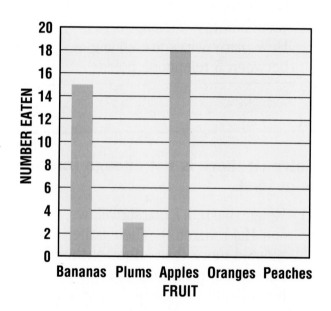

FRUIT MY FAMILY ATE	
Fruit	Number Eaten
Bananas	15
Plums	3
Apples	18
Oranges	16
Peaches	13

WORKING TOGETHER

1. Copy and complete Karen's bar graph.

2. Why did Karen count by 2s from 0 to 20 for the scale?

3. How tall did you make the bar for peaches? Why?

4. Which is the tallest bar? What does this tell you?

5. How did you title the graph?

SHARING IDEAS

6. Compare your bar graph with those of others. How are they the same? How are they different?

7. **What if** Karen's family had eaten 40 oranges? How could you change the scale to keep the graph the same size?

8. Which fruit was eaten the least? Is it easier to read this from the graph than from the table? Why?

9. Why is a bar graph a useful way to record and display data?

PRACTICE

10. Paula's Pasta Palace sells special fruit flavors of pasta. Paula wants to compare the sales of four of her most popular types of pasta. Copy and complete the bar graph below.

SALES OF PASTA	
Type	Boxes (to the Nearest Hundred)
Lemon	400
Berry	900
Orange	200
Cherry	700

11. Which type of pasta sold the least? Which type sold the most?

12. Paula also sold 107 boxes of peach pasta. Show this data on the graph. What do you have to do before you can graph the data?

13. Paula did not sell any boxes of prune pasta. How would you show this on the graph?

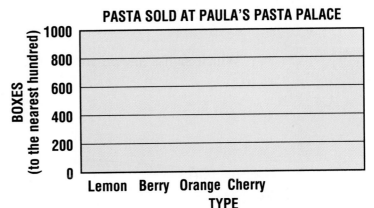

PASTA SOLD AT PAULA'S PASTA PALACE

14. This table shows the different pasta sauces sold in Paula's store last week. Make a bar graph to show this data. Compare your graph with those of others.

15. What does your graph tell you about sales of different sauces?

16. How can Paula use this data to help her order more pasta sauces for her store?

17. Collect data about the number of fruits eaten by other students during one week. Make a bar graph to show the results. Compare your graph with those of others.

SALES OF PASTA SAUCE	
Type	Number of Jars
Garlic	40
Tomato	65
Curry	10
Onion	30
Spinach	25

Mixed Review

Write the number.

18. 2 thousands 5 tens   **19.** 10,000 + 800 + 5   **20.** one hundred fourteen

DECISION MAKING

Problem Solving: Planning a Vacation

SITUATION

Judy and Frank DeSanto are on vacation at Coyote Lake Campgrounds with their parents and grandparents. They want to spend one day seeing some of the sights near the lake.

PROBLEM

Which sights should they see?

DATA

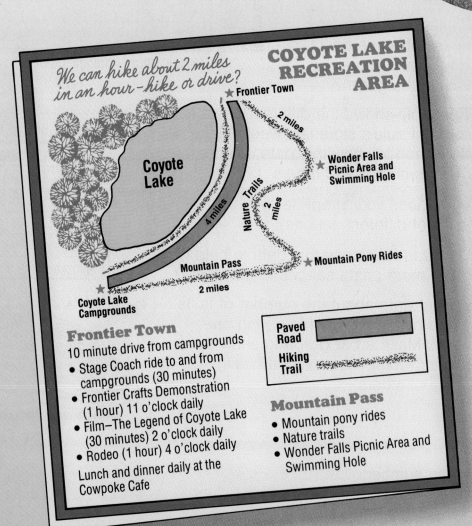

We can hike about 2 miles in an hour – hike or drive?

COYOTE LAKE RECREATION AREA

Frontier Town

Coyote Lake

Nature Trails

2 miles

4 miles

2 miles

Wonder Falls Picnic Area and Swimming Hole

Mountain Pony Rides

Mountain Pass

2 miles

Coyote Lake Campgrounds

Frontier Town
10 minute drive from campgrounds
- Stage Coach ride to and from campgrounds (30 minutes)
- Frontier Crafts Demonstration (1 hour) 11 o'clock daily
- Film–The Legend of Coyote Lake (30 minutes) 2 o'clock daily
- Rodeo (1 hour) 4 o'clock daily

Lunch and dinner daily at the Cowpoke Cafe

Paved Road

Hiking Trail

Mountain Pass
- Mountain pony rides
- Nature trails
- Wonder Falls Picnic Area and Swimming Hole

USING THE DATA

1. How many miles is it from the campgrounds to Frontier Town by Mountain Pass?

Tell how long it would take to hike to each site from the campgrounds. Include a 10-minute rest after each mile.

2. Pony Rides

3. Wonder Falls

4. Frontier Town by Mountain Pass

5. What are the different ways the DeSantos could get to Frontier Town?

MAKING DECISIONS

6. *Write a list* of the things the DeSantos should think about when planning their day trip.

7. *What if* the DeSantos want to see The Legend of Coyote Lake and take a pony ride? How should they travel to each place?

8. *What if* Frank wants to take a pony ride and Judy wants to see the rodeo? If they start at 9:00 A.M., do they have time for both? Why or why not?

9. *What if* the DeSantos want to take a pony ride and picnic at Wonder Falls? Do they still have time to see all the attractions at Frontier Town? Why or why not?

10. *Write a schedule* for a day at Coyote Lake. What would you see and do? How would you get to each attraction? Where would you eat?

Math and Health

In gym class you learn many different kinds of exercises. But what does exercising do for you?

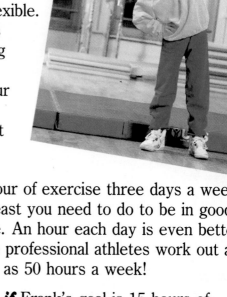

One type of exercise makes your muscles flexible. The warm-up exercises in gym class are this type. A second type of exercise builds strong muscles. This type involves pushing, pulling, and lifting. A third type of exercise helps your heart and lungs work well. Most sports that include running do this. All exercises burn fat to help you stay at a healthy weight.

An hour of exercise three days a week is the least you need to do to be in good shape. An hour each day is even better. Some professional athletes work out as many as 50 hours a week!

What if Frank's goal is 15 hours of exercise each week? He plays frisbee and baseball and rides his bike, and by Thursday has exercised 8 hours. How many more hours must he exercise to meet his weekly goal?

Think: Subtract. $15 - 8 = 7$

Frank must exercise 7 more hours.

ACTIVITY

1. Make a weekly exercise plan. Show your plan in a chart with these column heads: Day of the Week, Activity, and Number of Hours. At the bottom, write the Total Weekly Time. Try to include all three types of exercise.

Calculator: Challenge

Tom challenged Denise and Bruce to use their calculators to go from 15 to 100 in three steps, using only the number keys and the 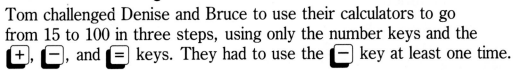 [+], [−], and [=] keys. They had to use the [−] key at least one time.

This is what Denise did:

She entered 15.

Calculator Display

| 15. |

Then she pressed the keys below:

Step 1 [+] 70 [=] | 85. |

Step 2 [−] 5 [=] | 80. |

Step 3 [+] 20 [=] | 100. |

This is what Bruce did:

He entered 15.

Calculator Display

| 15. |

Then he pressed the keys below:

Step 1 [+] 35 [=] | 50. |

Step 2 [+] 75 [=] | 125. |

Step 3 [−] 25 [=] | 100. |

Are there other ways to do it? Try it. Show two more ways to go from 15 to 100 in three steps.

USING THE CALCULATOR

Use number keys and only the [+], [−], and [=] keys.

1. Enter 12. Go to 30 in three steps. Use the [−] key at least once.

2. Enter 15. Go to 40 in four steps. Use the [−] key at least twice.

3. Enter 25. Go to 50 in four steps. Use the [−] key at least twice.

4. Enter 15. Go to 50 in five steps. Use the [−] key at least twice.

5. Enter 20. Go to 60 in five steps. Use the [−] key at least twice.

6. Enter 60. Go to 100 in six steps. Use the [−] key at least 3 times.

EXTRA PRACTICE

Meaning of Addition and Subtraction, page 53

Find the sum or difference.

1.	2.	3.	4.	5.	6.
0 + 8	8 − 0	3 + 2	10 − 6	7 − 7	1 + 6

7. $6 - \blacksquare = 6$ 8. $\blacksquare + 7 = 7$ 9. $4 - \blacksquare = 0$ 10. $3 + \blacksquare = 4 + 3$

11. $\blacksquare + 0 = 5$ 12. $1 - \blacksquare = 1$ 13. $\blacksquare - 9 = 0$ 14. $8 + 2 = \blacksquare + 8$

Fact Families, page 55

Find the sum or difference. Write the related
number sentences that make up the fact family.

1. $4 + 7 = \blacksquare$ 2. $12 - 6 = \blacksquare$ 3. $15 - 7 = \blacksquare$ 4. $18 - 9 = \blacksquare$

5. $0 + 4 = \blacksquare$ 6. $10 - 7 = \blacksquare$ 7. $8 + 8 = \blacksquare$ 8. $1 + 8 = \blacksquare$

Write the fact family for the set of numbers.

9. 3, 9, 12   10. 1, 6, 7   11. 0, 2, 2   12. 2, 7, 9

Complete the fact. Find the missing number.

13. $7 + \blacksquare = 13$ 14. $8 - \blacksquare = 5$ 15. $\blacksquare + 7 = 10$ 16. $8 + \blacksquare = 16$

17. $6 + \blacksquare = 9$ 18. $16 - 7 = \blacksquare$ 19. $6 + \blacksquare = 6$ 20. $\blacksquare + 4 = 11$

Mental Math: Addition Facts, page 57

Add. Use mental math.

1.	2.	3.	4.	5.	6.
4 + 5	6 + 2	3 + 3	5 + 7	2 + 5	9 + 0

7.	8.	9.	10.	11.	12.
5 + 5	1 + 4	8 + 3	7 + 7	5 + 8	3 + 7

13. $7 + 6$ 14. $8 + 4$ 15. $6 + 6$ 16. $5 + 9$

17. $0 + 3$ 18. $2 + 2$ 19. $3 + 9$ 20. $4 + 7$

Mental Math: Subtraction Facts, page 59 ....

Subtract. Use mental math.

1. 16 − 8	**2.** 14 − 5	**3.** 12 − 3	**4.** 11 − 6	**5.** 13 − 7	**6.** 15 − 9
7. 10 − 3	**8.** 12 − 7	**9.** 14 − 8	**10.** 14 − 7	**11.** 16 − 9	**12.** 18 − 9

13. 14 − 6   **14.** 17 − 9   **15.** 16 − 7   **16.** 11 − 7

17. 10 − 5   **18.** 11 − 5   **19.** 10 − 2   **20.** 17 − 8

Mental Math: Three or More Addends, page 61 ....

Find the sum. Use mental math.

1. 7 2 + 3	**2.** 8 5 + 3	**3.** 6 2 + 4	**4.** 5 7 + 2	**5.** 9 1 + 7	**6.** 2 3 + 9

7. 3 + 4 + 9   **8.** 8 + 3 + 2   **9.** 6 + 4 + 7   **10.** 2 + 1 + 4

11. 1 + 3 + 1 + 5   **12.** 2 + 3 + 5 + 6   **13.** 3 + 4 + 4 + 5

Find the missing addend.

14. (3 + 2) + 5 = 2 + (■ + 5)   **15.** 7 + (3 + ■) = 3 + (6 + ■)

Problem Solving: Finding Needed Information, page 63.........................

Solve. If there is not enough information, write what is needed.

1. Heather saw 4 birds. Her father saw 9 birds. How many more birds did her father see?

2. Garrett bought three postcards. He gave the clerk $10.00. How much change should he get?

3. Valerie went rafting on the Colorado River. The raft traveled 5 miles each hour. How many miles was her trip?

4. Pete hiked 4 miles on Monday and 7 miles on Tuesday. How many miles did he hike in the two days?

EXTRA PRACTICE

Measuring Length: Metric Units, page 67................................

Copy and complete this chart. Estimate. Then use your centimeter ruler to measure.

	Object	Estimate	Measurement
1.	height of a door	■	■
2.	width of your desk	■	■
3.	depth of a shelf	■	■
4.	width of your hand	■	■

Use your centimeter ruler. Draw lines with the following lengths.

5. 12 cm   **6.** 9 cm   **7.** 2 cm   **8.** 7 cm

Estimating Length: Metric Units, page 69..

Tell whether you would use cm, dm, m, or km to measure.

1. the distance across an ocean   **2.** the length of an airplane

3. the length of a ribbon   **4.** the distance from the floor to the ceiling

Write the letter of the best estimate.

5. the height of a spool of thread   **a.** 3 cm   **b.** 3 dm   **c.** 3 m

6. the width of a basket   **a.** 6 cm   **b.** 6 dm   **c.** 6 m

7. the height of a giraffe   **a.** 4 dm   **b.** 4 m   **c.** 4 km

Complete the chart. Use the units given.

Object	Estimate	Measurement
distance around your wrist (cm)	**8.** ■	**9.** ■
length of your leg (dm)	**10.** ■	**11.** ■
length of the chalkboard (m)	**12.** ■	**13.** ■

Problem Solving: Drawing a Diagram, page 71 ....

Draw a diagram to help you solve the problem.

1. Yvette is going to cut a piece of yarn to tie 5 presents. How many cuts will she need to make?

2. Five trucks are parked bumper to bumper. How many bumpers are touching?

3. Jason lives in an apartment building on the ground floor. Jenny lives 8 floors above him. How many floors are there between Jason's and Jenny's apartments?

4. Rob hangs his coat in the middle of a row of hooks. There are 7 hooks to the right of his coat. How many hooks are in the row?

Making Bar Graphs, page 73 ....

Solve. Use the table and graph.

1. Ben's Bread Bakery sells different kinds of bread. Ben wants to compare the sales of four of his most popular types of bread. Complete the bar graph.

BREAD SOLD AT BEN'S BREAD BAKERY

Type	Loaves (to the nearest hundred)
White	300
Wheat	800
Rye	700
Raisin	200

2. Which type of bread did Ben sell the least of? the most of?

3. Ben also sold 118 loaves of banana bread. Show this data on the graph. What do you have to do before you can graph the data?

4. Ben did not sell any bran bread. How would you show this on the graph?

PRACTICE *PLUS*

KEY SKILL: MENTAL MATH: ADDITION FACTS (Use after page 57.)

Level A

Add. Use mental math.

1. 3	**2.** 4	**3.** 1	**4.** 4	**5.** 9	**6.** 5
+ 2	+ 3	+ 5	+ 4	+ 1	+ 2

Write the letter of the correct answer.

7. 8 + 2
a. 10 **b.** 2 **c.** 82

8. 7 + 5
a. 75 **b.** 11 **c.** 12

9. 9 + 4
a. 14 **b.** 94 **c.** 13

10. Alfonso has 4 baseball caps from American League teams and 5 baseball caps from National League teams. How many baseball caps does he have?

Level B

Add. Use mental math.

11. 5	**12.** 8	**13.** 6	**14.** 3	**15.** 9	**16.** 2
+ 3	+ 1	+ 5	+ 8	+ 5	+ 9

17. 7 + 1   **18.** 6 + 3   **19.** 4 + 6   **20.** 9 + 6   **21.** 8 + 9

22. 8 plus 8 is what number?   **23.** What is the sum of 7 plus 9?

24. The school band purchased 8 new flutes and 5 new violins. How many instruments did the band buy?

Level C

Add. Use mental math.

25. 6 + 7   **26.** 9 + 3   **27.** 7 + 4   **28.** 2 + 8   **29.** 6 + 8

Complete the fact. Find the missing number.

30. 9 + ■ = 11   **31.** 7 + ■ = 13   **32.** 8 + ■ = 15   **33.** 5 + ■ = 14

34. Janice takes 9 books from the shelf. There are 7 books left. How many books were there to start with?

KEY SKILL: MENTAL MATH: SUBTRACTION FACTS (Use after page 59.)

Level A

Subtract. Use mental math.

1. 7
− 6

2. 4
− 1

3. 5
− 4

4. 9
− 3

5. 10
− 9

6. 8
− 6

Write the letter of the correct answer.

7. 3 − 1
   a. 4 b. 2 c. 1

8. 11 − 3
   a. 14 b. 7 c. 8

9. 9 − 4
   a. 5 b. 13 c. 6

10. Jessica found 9 starfish. She gave 5 of them to Mark. How many starfish does Jessica have left?

Level B

Subtract. Use mental math.

11. 12
− 5

12. 7
− 2

13. 15
− 6

14. 8
− 5

15. 10
− 8

16. 9
− 6

17. 6 − 1   18. 10 − 4   19. 13 − 8   20. 8 − 4   21. 12 − 3

22. the difference between 10 and 1   23. 6 less than 13

24. Tracie has 9 dimes in her bank. Bo has 17 dimes in his. How many fewer coins does Tracie have than Bo?

Level C

Subtract. Use mental math.

25. 11 − 2   26. 13 − 9   27. 7 − 4   28. 14 − 9   29. 10 − 8

Complete the fact. Find the missing number.

30. 13 − ■ = 8 31. 10 − ■ = 4   32. 9 − ■ = 8   33. 14 − ■ = 8

34. The Snake River is 8 miles longer than the Gekko River. The Snake River is 17 miles long. How long is the Gekko River?

CHAPTER REVIEW

LANGUAGE AND MATHEMATICS

Complete the sentences. Use the words in the chart.

1. The total of two addends is the ■. *(page 53)*

2. The centimeter is a ■ unit of length. *(page 66)*

3. The ■ is used to measure length and distance. *(page 68)*

4. A bar graph is used to organize ■. *(page 72)*

5. *Write a definition* or give an example of the words you did not use from the chart.

> **VOCABULARY**
> difference
> kilometer
> subtraction
> data
> metric
> sum

CONCEPTS AND SKILLS

Add or subtract. *(pages 52, 56–61)*

6. $\begin{array}{r} 7 \\ +5 \\ \hline \end{array}$
7. $\begin{array}{r} 13 \\ -9 \\ \hline \end{array}$
8. $\begin{array}{r} 12 \\ -5 \\ \hline \end{array}$
9. $\begin{array}{r} 9 \\ +2 \\ \hline \end{array}$
10. $\begin{array}{r} 6 \\ 4 \\ +1 \\ \hline \end{array}$
11. $\begin{array}{r} 1 \\ 5 \\ +9 \\ \hline \end{array}$

12. $9 - 3$
13. $8 + 4$
14. $5 + 6$
15. $10 - 3$

16. $8 + 8$
17. $4 + 3$
18. $15 - 9$
19. $14 - 6$

20. $9 + 2 + 3$
21. $8 + 1 + 7$
22. $1 + 3 + 2 + 5$

Can you write a fact family for each set of numbers? *(page 54)*

23. 4, 6, 10
24. 9, 9, 9
25. 3, 7, 8
26. 0, 8, 8

Complete the fact. Find the missing number. *(pages 52–55, 60)*

27. $6 + ■ = 6$
28. $0 + ■ = 8$
29. $4 - ■ = 0$
30. $9 + ■ = 9$

31. $7 + 2 = ■ + 7$
32. $9 + ■ = 17$
33. $7 + (3 + ■) = 3 + (6 + 7)$

Use your centimeter ruler. Draw lines with the following lengths. *(page 66)*

34. 8 cm
35. 5 cm
36. 12 cm
37. 2 cm

Write the letter of the best estimate. *(page 68)*

38. the length of a boat   **a.** 5 dm   **b.** 5 m   **c.** 5 km

39. the width of a television screen   **a.** 50 cm   **b.** 50 m   **c.** 50 dm

40. the distance an airplane travels in an hour   **a.** 700 cm   **b.** 700 m   **c.** 700 km

Use the bar graph on the right to answer the questions. *(page 72)*

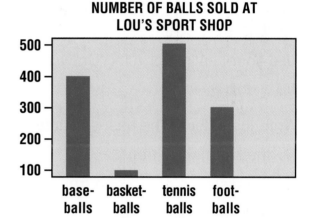

NUMBER OF BALLS SOLD AT LOU'S SPORT SHOP

41. Which type of balls sold the most?

42. Which type sold the least

43. Lou also sold 200 soccer balls. How would you show this data on the graph?

CRITICAL THINKING

44. Write an addition sentence that uses the same digit for both addends and the sum. *(page 52)*

45. Tara says her book is 2 decimeters long. Herb says it is 19 centimeters long. Who is correct? Why? *(page 68)*

MIXED APPLICATIONS

Solve. If there is not enough information, write what is needed.

46. Gabrielle is having a party. She wants to invite 10 boys and 10 girls. So far, she has invited 15 people. Has she invited enough girls? *(page 62)*

47. For homework, Jared received 3 assignments in math, none in science, and 2 in English. How many assignments was he given in all? *(page 60)*

48. Katie has $9. She wants to buy a swimming cap. Her change will be $6. How much is the swimming cap? *(page 58)*

49. Kara wants to cut a board into 6 pieces. How many cuts does she need to make? *(page 70)*

CHAPTER TEST

Complete. Find the missing number.

1. 4 − ■ = 4

2. ■ + 0 = 7

3. 4 + ■ = 2 + 4

4. 3 + 8 = ■ + 3

Add or subtract.

5. 7 + 5

6. 15 − 8

7. 18 − 9

8. 9 + 4

9. 8 + 7

10. 16 − 7

11. 9 + 9

12. 6 + 7

13. 4 + 2 + 5

14. 2 + 7 + 3

Use your centimeter ruler. Draw a line.

15. 1 cm

16. 9 cm

Write the letter of the best estimate.

17. the height of a tree   **a.** 5 dm   **b.** 5 m   **c.** 5 km

18. the distance of a hiking trail   **a.** 6 dm   **b.** 6 m   **c.** 6 km

19. the length of a table   **a.** 2 dm   **b.** 2 m   **c.** 2 km

Use the graph for Questions 20–22.

20. How many more violets than tulips were sold in June?

21. Which type of flower sold the least?

22. Dan sold 35 lilacs in June. How would you show this?

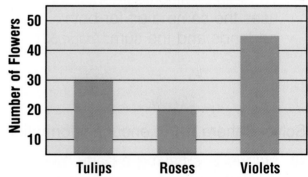

FLOWER SALES IN JUNE

Solve.

23. Nine cars are parked in a row. There is one stripe between each car and at each end. How many stripes are there?

24. Martin sold 9 roses and 8 lilacs. Each was $1 off. What is not needed to find how many flowers he sold?

   a. Martin sold 9 roses.

   b. He sold 8 lilacs.

   c. Each was $1 off.

25. Bill has a piece of rope that he wants to cut into 10 pieces. How many cuts does he make?

ROMAN NUMERALS

The ancient Romans used letters to name numbers. Today we still use Roman numerals to mark the building date on monuments and public buildings. Here are the letters used to name numbers from 1 to 1,000.

I	V	X	L	C	D	M
1	5	10	50	100	500	1000

You can name other numbers by combining letters. When a letter is repeated, add.

III	XX	CCC
1 + 1 + 1 = 3	10 + 10 = 20	100 + 100 + 100 = 300

A letter is never repeated more than 3 times.

When a smaller number is to the *right* of a larger number, add.

LX
50 + 10 = 60

When a smaller number is to the *left* of a larger number, subtract.

XL
50 − 10 = 40

1. Is the Roman numeral system a place-value system?

What does MCMLV equal?

Think: M = 1,000
CM = 1,000 − 100 = 900
LV = 50 + 5 = 55
1,000 + 900 + 55 = 1,955

Match the building with the year it was built.

2. White House (1817)
3. Empire State Building (1931)
4. Roman Colosseum (70 A.D.)

a. MCMXXXI
b. LXX
c. MDCCCXVII

CUMULATIVE REVIEW

Choose the letter of the correct answer.

1. $8 + 3 + 6$

    a. 9   **c.** 17

    b. 11   **d.** not given

2. Round $.26 to the nearest ten cents.

    a. $.10   **c.** $.25

    b. $.20   **d.** $.30

3. $16 - 9$

    a. 5   **c.** 7

    b. 6   **d.** not given

4. Estimate the length of an eraser.

    a. 3 cm   **c.** 3 m

    b. 3 dm   **d.** 3 km

5. Compare: 350 ● 503

    a. <   **c.** =

    b. >   **d.** not given

6. Choose the number for fifty thousand.

    a. 5,000   **c.** 50,000

    b. 500,000   **d.** not given

7. $2 + 6 = $ ■

    a. 4   **c.** 8

    b. 5   **d.** not given

8. Round 5,426 to the nearest hundred.

    a. 5,000   **c.** 5,430

    b. 5,400   **d.** 5,500

9. How much more than 5 is 14?

    a. 7   **c.** 9

    b. 8   **d.** not given

10. Cost: 45¢   Given: 50¢
Change?

    a. 1 penny   **c.** 1 dime

    b. 1 nickel   **d.** not given

11. 4 and 8 more equals what?

    a. 11   **c.** 13

    b. 12   **d.** not given

12. You measure the length of a room in:

    a. cm.   **c.** m.

    b. dm.   **d.** not given

13. Compare: $41.95 ● $14.95

    a. <   **c.** =

    b. >   **d.** not given

14. $4 + 7 + 3 + 1$

    a. 17   **c.** 16

    b. 18   **d.** not given

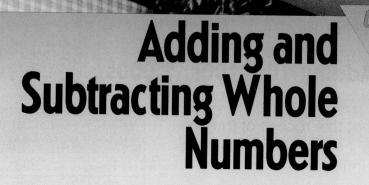

Adding and Subtracting Whole Numbers

CHAPTER 3

MATH CONNECTIONS: PERIMETER • PROBLEM SOLVING

Jim's Deli

Cold Cuts

	one pound	half pound
Turkey	$5.00	$2.50
Roast Beef	$8.00	$4.00
Cheddar Cheese	$4.00	$2.00
Sandwich rolls	$ .40 each	

Sandwiches

Turkey	$3.00
Roast Beef	$3.50
Cheddar Cheese	$2.00
Lettuce & Tomato	$ .50 extra
On a roll	$ .50 extra

1. What information do you see in this picture of a lunch counter?

2. How can you use this information?

3. Write a problem about buying lunch.

UNDERSTANDING A CONCEPT

Mental Math: Adding 10s; 100s; 1,000s

A. The Corner Deli kept this record of the number of sandwiches it sold. How many sandwiches did it sell in the two weeks?

Week	Number of Sandwiches Sold
1	300
2	400

Add: 300 + 400

You can use basic facts to find the sum mentally.

$$\begin{array}{r} 300 \\ + 400 \end{array}$$
Think:
$$\begin{array}{r} 3 \\ + 4 \\ \hline 7 \end{array}$$
3 hundreds
+ 4 hundreds
7 hundreds = 700

300 + 400 = 700
The Corner Deli sold 700 sandwiches.

1. Use basic facts to find 3,000 + 9,000.

2. How can you use this method to find 20 + 60 + 50?

B. You can adjust some numbers so you can add mentally.

Add: 43 + 29

Look to make one of the addends end in zero.

$$\begin{array}{r} 43 \longrightarrow 42 \\ + 29 \longrightarrow + 30 \\ \hline 72 \end{array}$$
Think: Take 1 away from 43 to make 42.
Add 1 to 29 to make 30.

So 43 + 29 = 42 + 30 = 72.

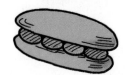

3. Use this method to find 337 + 198.

TRY OUT Write the letter of the correct answer. Add mentally.

4. 4,000 + 7,000   **a.** 1,100   **b.** 11   **c.** 11,000   **d.** 110

5. 600 + 40   **a.** 1,000   **b.** 6,400   **c.** 100   **d.** 640

6. 57 + 89   **a.** 146   **b.** 147   **c.** 148   **d.** 137

7. 98 + 125   **a.** 225   **b.** 223   **c.** 126   **d.** 209

PRACTICE

Add. Use mental math.

8. 20 + 70	**9.** 100 + 500	**10.** 2,000 + 4,000	**11.** 4,000 + 300	**12.** 600 + 900

13. 20 40 + 20	**14.** 30 40 + 70	**15.** 900 60 + 20	**16.** 200 100 + 500	**17.** 600 400 + 700

18. 19 + 31	**19.** 23 + 59	**20.** 65 + 69	**21.** 322 + 499	**22.** 799 + 626

23. 101 + 89	**24.** 299 + 52	**25.** 151 + 799	**26.** 999 + 507	**27.** 561 + 429

28. 10 + 40 + 20 **29.** 5,000 + 5,000 **30.** 71 + 29 **31.** 49 + 36

32. 400 + 800 **33.** 199 + 714 **34.** 74 + 68 **35.** 800 + 30 + 70

36. 99 + 123 **37.** 50 + 30 + 60 **38.** 98 + 142 **39.** 498 + 213

Mixed Applications

40. The Corner Deli sold 400 cans of juice last month and 200 cans this month. How many cans of juice did it sell in the two months?

41. A group of customers orders 15 bowls of soup. The Corner Deli can fill all but 6 bowls. How many bowls of soup can the group get?

42. On Monday Ben made 30 sandwiches. Ray made 20 more sandwiches than Ben. Rachel made 30 more than Ray. How many sandwiches did Rachel make?

43. The Corner Deli used 59 paper bags on Saturday and 44 paper bags on Sunday. How many paper bags did it use on the weekend?

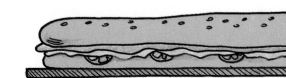

UNDERSTANDING A CONCEPT

Estimating Sums by Rounding

A. Martin uses the number of cartons of milk sold this month to order for next month. About how many cartons were sold?

Sometimes you do not need to know the exact answer. You can round to find an estimate.

Estimate: 489 + 378 + 265

MILK MART

Flavor	Number of Cartons
Vanilla	489
Chocolate	378
Strawberry	265

Step 1

Round each number to the greatest place of the greatest number.

$$
\begin{aligned}
489 &\longrightarrow 500 \\
378 &\longrightarrow 400 \\
+\ 265 &\longrightarrow +\ 300
\end{aligned}
$$

Think: Round to the nearest hundred.

Step 2

Add the rounded numbers.

$$
\begin{aligned}
500 \\
400 \\
+\ 300 \\
\hline
1{,}200
\end{aligned}
$$

About 1,200 cartons of milk were sold.

1. How does this estimate compare with the exact answer? Why?

B. You can also estimate sums when adding money.

Estimate: $5.47 + $.24 + $8.35

Step 1

Round each number to the greatest place of the greatest number.

$$
\begin{aligned}
\$5.47 &\longrightarrow \$5.00 \\
.24 &\longrightarrow 0.00 \\
+\ 8.35 &\longrightarrow +\ 8.00
\end{aligned}
$$

Think: Round to the nearest dollar.

Step 2

$$
\begin{aligned}
\$5.00 \\
0.00 \\
+\ 8.00 \\
\hline
\$13.00
\end{aligned}
$$

Think: Remember to put in the dollar sign and decimal point.

2. How does this estimate compare with the exact answer? Why?

TRY OUT Estimate by rounding.

3. 1,898 + 7,701   **4.** $3.45 + $6.19 + $.82   **5.** 43 + 108 + 279

PRACTICE

Estimate by rounding.

6. 532
  + 290

7. 905
  + 185

8. 5,775
  + 2,680

9. $5.75
  + 3.25

10. $63.80
  + 22.15

11. 160
   308
  + 150

12. 4,050
  2,030
  + 3,200

13. $5.60
  6.25
  + 2.20

14. 2,050
  3,500
  + 350

15. $23.96
  8.14
  + 54.05

16. 788
  350
  + 90

17. $65.00
  15.00
  + 30.00

18. 3,608
  650
  + 4,025

19. 7,604
  1,925
  830
  + 342

20. $30.86
  24.99
  51.09
  + 6.30

21. 329 + 25 + 620

22. 199 + 399 + 299

23. 6,300 + 1,360 + 720

24. $62.20 + $14.80 + $3.10

25. $22.70 + $37.90 + $11.80 + $5.00

Critical Thinking

26. How can you easily tell that this sum is incorrect?

856 + 742 + 699 = 3,297

Mixed Applications

27. Denise bought a quart of fudge ice cream for $2.69. Lee bought a quart of vanilla ice cream for $1.98. Did Lee and Denise spend more than $5.00?

28. Milk Mart gets the milk it sells from 700 cows. When sales go up, it buys 400 more cows. How many cows does Milk Mart use now?

29. The clerk at Milk Mart drops a box of 144 eggs and breaks them. In the next two days, he breaks 276 eggs and 192 eggs. About how many eggs does he break in three days?

30. Dudley's Dairy produces 3,964 gallons of milk in a week. Fern's Cow Farm produces 3,694 gallons of milk in a week. Which place produces more milk in a week?

UNDERSTANDING A CONCEPT

Front-End Estimation

A. Nancy has $5.28, Tom has $4.98, and Kathy has $7.75. Do they have enough money to buy a large deluxe pizza?

PIZZA PALACE		
Pizza	Small	Large
Plain	$9.50	$11.75
Deluxe	$12.55	$15.00
Extra Toppings $2.50 each		

Nancy uses **front-end estimation**.

Estimate: $5.28 + $4.98 + $7.75

Add the front digits.
Write zeros for the other digits.

$$\begin{array}{r} \$5.28 \\ 4.98 \\ + 7.75 \\ \hline \$16.00 \end{array}$$

She thinks:
$5 + $4 + $7 = $16

We have enough to buy a large deluxe pizza.

1. Is Nancy's estimate greater than or less than the exact sum? How do you know?

B. You can adjust a front-end estimate to get closer to the exact answer.

Estimate: 343 + 62 + 409

Step 1

Add the front digits.
Write zeros for the other digits.

$$\begin{array}{r} 343 \\ 62 \\ + 409 \\ \hline 700 \end{array}$$

Step 2

Adjust the estimate.

$$\begin{array}{r} 3\underline{43} \\ \underline{62} \\ + 409 \\ \hline 700 + 100 = 800 \end{array}$$

Think: 43 and 62 is about 100.

TRY OUT
Write the letter of the correct answer.
Estimate. Use the front digits and adjust.

2. 618 + 284   **a.** 400   **b.** 800   **c.** 900   **d.** 1,000

3. $8.65 + $6.55   **a.** $2.00   **b.** $14.00   **c.** $15.00   **d.** $16.00

PRACTICE

Estimate. Use only the front digits.

4.	369 + 936	**5.**	4,986 + 6,559	**6.**	54 2 + 81	**7.**	$2.85 6.81 + .35	**8.**	7,559 1,750 + 2,735

9. $49.74 + $99.37   **10.** 6,873 + 220   **11.** $2.64 + $20.66 + $15.99

Estimate. Use the front digits and adjust.

12.	392 + 317	**13.**	$2.39 + 7.75	**14.**	$.42 .48 + .70	**15.**	4,330 4,489 + 301	**16.**	4,923 104 + 9,015

17. $41.88 + $28.93   **18.** 74 + 99 + 17   **19.** $1.45 + $34.23 + $55.50

Write the letter of the best estimate.

20. $4.48   **a.** less than $12.00   **21.**   8,374   **a.** less than 15,000
       5.29   **b.** greater than $12.00   820   **b.** greater than 16,000
 + 3.43   + 7,011

Critical Thinking   Write *always, sometimes,* or *never* to complete the sentence.

22. A front-end estimate that is not adjusted is _____ less than the exact sum.

23. A front-end estimate that is adjusted is _____ greater than the exact sum.

Mixed Applications   You may need to use the table on page 94.

24. Pizza Palace sold 1,200 pizzas in May, 889 in June, and 2,076 in July. It had 3,000 pizza boxes. Did it have enough boxes for all the pizzas it sold?

25. Kathy and her brother Joe buy a small deluxe pizza. They have a ten-dollar bill and a five-dollar bill. How much change should they get?

26. *Write a problem* about buying a pizza. Solve it. Trade problems with other students and solve them.

PROBLEM SOLVING

Checking for a Reasonable Answer

Mrs. Jackson is the head cook at Lincoln School. She prepares 99 meals for one lunch period and 84 meals for the other lunch period. How many meals does she prepare for the two lunch periods.

Mrs. Jackson estimates by rounding.

$$\begin{array}{r} 99 \rightarrow 100 \\ + 84 \rightarrow + 80 \\ \hline 180 \end{array}$$

Then she uses a calculator to find the sum.

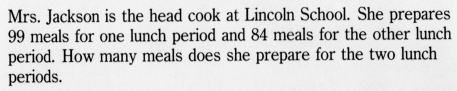

She compares her answer of 111 with the estimated answer of 180. She sees that they are not close. She realizes that she must have hit a wrong key because her answer is not reasonable.

She again uses the calculator to find the sum.

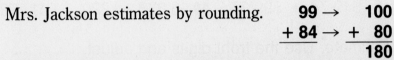

1. Is her new answer reasonable? Why?

2. How can you use estimation to tell if an answer is reasonable?

3. Can an estimate check that the answer is correct? Why or why not?

4. How can you check that your answer is correct?

5. How many meals does Mrs. Jackson prepare?

6. Why is it important to check that an answer is reasonable?

PRACTICE

Use estimation to decide which answer is reasonable.
Write the letter of the correct answer.

7. Scott is buying apples for $1.21 and grapes for $1.53. What is the total cost of the fruit?
   a. $.32   **b.** $1.73   **c.** $2.74

8. Latroy is buying milk for $.73. She gives the cashier $1.00. How much change should she get?
   a. $.27   **b.** $.33   **c.** $.37

9. Sunbeam Raisins cost $.92 a box. Golden Raisins cost $.76 a box. How much will you save if you buy Golden Raisins?
   a. $.04   **b.** $.16   **c.** $.26

10. Fast Lane Supermarket has 794 cans of beans and 425 cans of corn. How many more cans of beans are there?
    a. 369   **b.** 892   **c.** 1,219

Strategies and Skills Review

Solve. Use mental math or paper and pencil.

11. Mrs. Jackson used over 200 pounds of vegetables in this week's school lunches. She used 107 pounds of peas and 85 pounds of carrots. Are these the only vegetables she used? How do you know?

12. At the store, rice is stored on a shelf above the soups. Juice is on a shelf below the spaghetti. The soup is on a shelf between the juice and the spaghetti. Juice is between beans and soup. Which item is on the lowest shelf?

13. Ray has $1.25. He wants to buy 3 pounds of oranges. What else does he need to know to decide whether he has enough money?

14. *Write a problem* that uses addition or subtraction. Ask others to solve it. Then check that the answer is reasonable.

UNDERSTANDING A CONCEPT

Adding Whole Numbers

Last year 267 people went to the Wesley Picnic. This year there were 155 more people. How many people went to the picnic this year?

Add: 267 + 155

Estimate first.

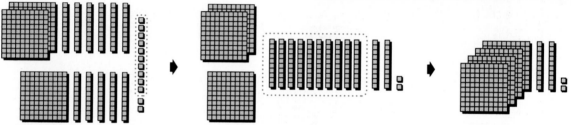

Step 1	Step 2	Step 3
Add the ones. Regroup if necessary.	**Add the tens. Regroup if necessary.**	**Add the hundreds. Regroup if necessary.**
1 **267** **+ 155** ——— 2	1 1 **267** **+ 155** ——— 22	1 1 **267** **+ 155** ——— 422

Think: 12 ones = 1 ten 2 ones   ***Think:*** 12 tens = 1 hundred 2 tens

Check: 155 + 267 = 422

This year 422 people went to the Wesley Picnic.

1. Is this sum reasonable? How do you know?

2. Why can you check addition by changing the order of the addends?

You add money amounts the same way you add whole numbers.

1 1
$25.73
+   9.75
————
$35.48

Think: Write the dollar sign and the decimal point in the answer.

Write the letter of the correct answer.

3. 456 + 385   **a.** 731   **b.** 741   **c.** 841   **d.** 851

4. 3,079 + 960   **a.** 4,049   **b.** 4,039   **c.** 3,939   **d.** 3,039

5. $5.64 + $.29   **a.** $7.54   **b.** $5.83   **c.** $8.54   **d.** $5.93

6. $4.56 + $9.25   **a.** $137.11   **b.** $13.71   **c.** $14.71   **d.** $13.81

PRACTICE

Add. Solve only for sums that are greater than 900 or $9.00.

7.   492
      + 465

8.   439
      + 346

9.   968
      +   42

10.   5,268
       +   85

11.   653
       + 234

12.   772
       + 644

13.   1,058
       +   934

14.   7,586
       +   851

15.   9,931
       +   68

16.   5,486
       + 2,497

17. $4.79
       +   4.42

18. $8.28
       +   3.14

19. $5.00
       +   3.50

20. $72.96
       +   20.22

21. $89.55
       +   40.05

22. 4,656 + 315

23. $6.84 + $.97

24. $40.89 + $40.07

25. $8.55 + $48.25

26. 389 + 951

27. $19.25 + $84.75

Critical Thinking

This is how Ruth added 423 and 214 mentally.

423	4 hundreds	423	2 tens	423	3 ones
+ 214	+ 2 hundreds	+ 214	+ 1 ten	+ 214	+ 4 ones
6	6 hundreds	63	3 tens	637	7 ones

28. How did Ruth add? Why was it easy to add this way mentally?

Mixed Applications

29. At the picnic, 123 tacos and 68 tamales are made. How many tacos and tamales are made?

30. Rudy spends $4.05 on food and $1.25 on drink. Does he spend more than $5.00?

UNDERSTANDING A CONCEPT

Column Addition

A. Sarah kept a record of the number of pies she sold in four days. How many pies did she sell in the four days?

Add: 35 + 29 + 45 + 78

SARAH'S SALES

Day	Pies Sold
Monday	35
Wednesday	29
Friday	45
Saturday	78

Step 1

Add the ones.
Regroup if necessary.

```
  2
  35   Think:  27 ones =
  29            2 tens 7 ones
  45
+ 78
─────
   7
```

Step 2

Add the tens.
Regroup if necessary.

```
  2
  35   Think:  18 tens =
  29            1 hundred 8 tens
  45
+ 78
─────
 187
```

Sarah sold 187 apple pies in the four days.

1. How can you be sure this answer is reasonable?

2. How can you check the addition?

3. What is the sum of $16.75, $23.86, and $6.52?

B. You can use mental math to help you find sums quickly.

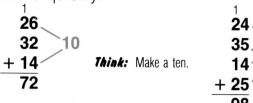

```
  1
  26 ⌐
  32  ⟩10
+ 14 ⌐
─────
  72
```
Think: Make a ten.

```
  1
  24 ⌐  ⟩8
  35  ⟩
  14  ⟩
+ 25 ⌐  ⟩10
─────
  98
```
Think: Use doubles.

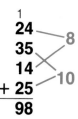

TRY OUT Write the letter of the correct answer.

4. 35 + 12 + 8 + 72   **a.** 117   **b.** 127   **c.** 137   **d.** 199

5. 309 + 90 + 454   **a.** 754   **b.** 843   **c.** 853   **d.** 1,744

6. $5.00 + $.98 + $6.79   **a.** $11.78 **b.** $12.77 **c.** $12.78 **d.** $109.70

7. $129 + $76 + $32 + $46   **a.** $273   **b.** $283   **c.** $284   **d.** $1,669

PRACTICE

Add.

8.	9.	10.	11.	12.
59	88	110	$2.08	$2.25
26	75	65	6.24	1.35
+ 15	+ 16	+ 15	+ .89	+ 6.62

13.	14.	15.	16.	17.
89	68	$.25	$2.03	658
58	41	.36	6.95	247
17	16	.75	1.55	103
+ 52	+ 23	+ .57	+ .93	+ 32

18. 62 + 99 + 127   **19.** $1.55 + $4.38 + $.80   **20.** 55 + 83 + 227

21. 98 + 140 + 663 + 99   **22.** $5.27 + $2.06 + $1.16 + $.21

Critical Thinking

23. What is the greatest sum you can get when you add three 2-digit numbers?

Mixed Applications

24. Juan buys two rolls for $.62, two cupcakes for $.70, and a loaf of bread for $1.35. How much does he spend?

25. David buys 16 cherry tarts and 7 peach tarts. How many fewer peach tarts than cherry tarts does David buy?

26. Last month Sarah made 462 apple pies, 325 cherry pies, 135 peach pies, and 415 banana cream pies. How many pies did she make in all?

27. Sarah used 4,788 bags of flour last year to bake pies. To the nearest hundred, how many bags did she use?

Mixed Review

Add or subtract.

28.	29.	30.	31.	32.	33.
6	9	7	8	16	9
+ 5	+ 3	− 5	+ 2	− 9	+ 5

34. 12 − 9   **35.** 7 + 8   **36.** 12 − 5   **37.** 11 − 7   **38.** 8 + 8

RINGS AROUND THINGS

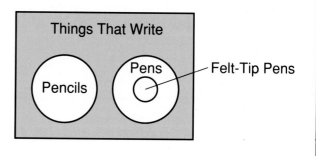

Logical Reasoning

A. You can use circle pictures to sort different groups.

1. The outside rectangle is labeled with the name of the largest group. What is its name? What other groups are included in the largest group?

2. A circle within a circle means that a group is entirely a part of another group. Which group includes one other group?

Things That Write

- Pencils
- Pens
- Felt-Tip Pens

3. If two circles overlap like the circles to the right, it means they share some members. Why don't the Pencils and Pens circles overlap?

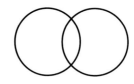

Copy the circle pictures. Write each label where you think it belongs.

4. Living Things
Flowers
Tulips
Trees

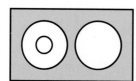

5. Zebras
Black and
White Animals
Pandas
Animals

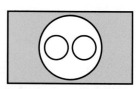

6. Foods
Vegetables
Dairy Products
Desserts

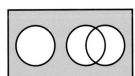

7. Television
Programs
Comedies
News
Programs
Half-hour Shows

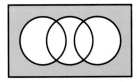

B. Now try to draw some circle pictures of your own.
Here are some group lists to use.

8. Books
Mysteries
Biographies

9. Musical
Instruments
Violins
Stringed
Instruments
Drums

10. Money
Dimes
Coins
Dollar Bills

11. Vehicles
Automobiles
Station
Wagons
Red Vehicles

12. Create some group lists of your own. Ask a partner to draw
the circle pictures.

Mental Math: Subtract 10s; 100s; 1,000s

A. The Golden Dragon Chinese Restaurant usually keeps a stock of 50 boxes of chopsticks. The owner, Mr. Wong, notices that they have only 20 boxes of chopsticks. How many boxes of chopsticks should he order?

Subtract: 50 − 20
You can use basic facts to find the difference mentally.

$$\begin{array}{r} 50 \\ -\ 20 \\ \end{array}$$ ***Think:*** $\begin{array}{r} 5 \\ -\ 2 \\ \hline 3 \end{array}$ $\begin{array}{l} 5 \text{ tens} \\ -\ 2 \text{ tens} \\ \hline 3 \text{ tens} = 30 \end{array}$

50 − 20 = 30

Mr. Wong should order 30 boxes of chopsticks.

1. Use basic facts to find 1,200 − 700.

B. You can adjust some numbers so you can subtract mentally.

Subtract: 48 − 19
Look to make the number that is subtracted end in zero.

$$\begin{array}{r} 48 \\ -\ 19 \\ \end{array} \longrightarrow \begin{array}{r} 49 \\ -\ 20 \\ \hline 29 \end{array}$$ ***Think:*** I can add 1 to both numbers without changing the difference.

So 48 − 19 = 49 − 20 = 29.

2. Use this method to find 356 − 299.

3. How can you find 71 − 38 using this method? What is the difference?

TRY **OUT** Subtract. Use mental math.

4. 8,000 − 5,000   **5.** 1,500 − 600   **6.** 388 − 129   **7.** 67 − 38

PRACTICE

Subtract. Use mental math.

8. $\begin{array}{r} 80 \\ -\ 30 \\ \hline \end{array}$ 9. $\begin{array}{r} 900 \\ -\ 700 \\ \hline \end{array}$ 10. $\begin{array}{r} 1{,}100 \\ -\ 600 \\ \hline \end{array}$ 11. $\begin{array}{r} 1{,}600 \\ -\ 800 \\ \hline \end{array}$ 12. $\begin{array}{r} 7{,}000 \\ -\ 2{,}000 \\ \hline \end{array}$

13. $\begin{array}{r} 60 \\ -\ 10 \\ \hline \end{array}$ 14. $\begin{array}{r} 800 \\ -\ 300 \\ \hline \end{array}$ 15. $\begin{array}{r} 140 \\ -\ 70 \\ \hline \end{array}$ 16. $\begin{array}{r} 1{,}700 \\ -\ 800 \\ \hline \end{array}$ 17. $\begin{array}{r} 1{,}300 \\ -\ 900 \\ \hline \end{array}$

18. $\begin{array}{r} 68 \\ -\ 29 \\ \hline \end{array}$ 19. $\begin{array}{r} 45 \\ -\ 19 \\ \hline \end{array}$ 20. $\begin{array}{r} 82 \\ -\ 38 \\ \hline \end{array}$ 21. $\begin{array}{r} 60 \\ -\ 48 \\ \hline \end{array}$ 22. $\begin{array}{r} 100 \\ -\ 79 \\ \hline \end{array}$

23. $\begin{array}{r} 566 \\ -\ 299 \\ \hline \end{array}$ 24. $\begin{array}{r} 652 \\ -\ 439 \\ \hline \end{array}$ 25. $\begin{array}{r} 781 \\ -\ 198 \\ \hline \end{array}$ 26. $\begin{array}{r} 494 \\ -\ 128 \\ \hline \end{array}$ 27. $\begin{array}{r} 83 \\ -\ 37 \\ \hline \end{array}$

28. $90 - 40$　　29. $57 - 39$　　30. $150 - 60$　　31. $555 - 298$

32. $6{,}000 - 3{,}000$　33. $1{,}800 - 900$　34. $900 - 199$　35. $7{,}435 - 3{,}998$

Mixed Applications

36. Mrs. Wong bought 1,300 containers to pack take-out orders. Last week she used 500 of the containers. How many containers are left?

37. The restaurant gives out 199 fortune cookies at lunch. They are left with 635 fortune cookies. How many fortune cookies did they have to start with?

38. Mr. Wong has 57 bottles of soy sauce and 38 bottles of duck sauce. How many more bottles of soy sauce than duck sauce does he have?

39. **Write a problem** about the Golden Dragon Chinese Restaurant. Solve it. Then ask others to solve your problem.

Mixed Review

Compare. Write >, <, or =.

40. $0 + 8 \bullet 8 - 0$ 41. $\$345.54 \bullet \$345.45$ 42. $\$13.27 \bullet \$6 + \$3 + \$4$

43. $55 - 19 \bullet 46$ 44. $12 - 6 \bullet 3 + 4$ 45. $616 \bullet 589 + 27$

46. $30 + 40 \bullet 10 + 20 + 30$ 47. 289 rounded to the nearest ten $\bullet$ 300

UNDERSTANDING A CONCEPT

Estimating Differences

A. The school lunchroom served 6,382 people last month. If 845 of them were teachers, about how many were students?

You can round to estimate the answer.

Estimate: 6,382 − 845

Step 1

Round to the greatest place of the greater number.

6,382 ⟶ **6,000**   ***Think:*** Round to the
−   845 ⟶ **− 1,000**   nearest thousand.

Step 2

Subtract the rounded numbers.

6,000
− 1,000
5,000

About 5,000 students were served at the school lunchroom.

B. You can also use front-end estimation to estimate differences.

Estimate: 7,439 − 3,705

Step 1

Subtract the front digits.
Write zeros for the other digits.

7,439
− 3,705
4,000

Step 2

Adjust the estimate.

7,439 ***Think:*** 4 < 7
− 3,705 The exact answer is
less than 4,000 less than 4,000.

1. Estimate 8,796 − 459. Will the exact answer be greater than or less than 8,000? How do you know?

TRY OUT

2. Estimate 5,894 − 3,226 by rounding.

3. Estimate 604 − 97 by using the front digits and adjusting.

PRACTICE

Estimate by rounding.

4.   874   **5.**   2,397   **6.** $3.50   **7.**   8,100   **8.** $39.75
   − 396   − 1,049   − 1.26   − 5,724   − 23.99

9.   509   **10.** 8,622   **11.** $4.28   **12.** 3,027   **13.** $14.14
   − 81   − 755   − .49   − 804   − 3.93

14. 9,630 − 3,389   **15.** $81.93 − $1.99   **16.** 7,500 − 629

Estimate. Use the front digits and adjust.

17.   863   **18.** $8.37   **19.** 3,906   **20.** $91.35   **21.** 9,655
   − 425   − 3.62   − 1,057   − 19.88   − 7,726

22. $7.35   **23.**   989   **24.** 1,370   **25.** $93.75   **26.** 6,508
   − .75   − 44   − 785   − 9.06   − 3,585

27. 478 − 83   **28.** $6.61 − $2.54   **29.** 828 − 423

Choose the letter of the best estimate.

30.   5,468 **a.** greater than 7,000   **31.** $7.86 **a.** greater than $5.00
   − 307 **b.** greater than 5,000   − 2.55 **b.** greater than $6.00
          c. less than 5,000   **c.** less than $4.00
          d. less than 3,000   **d.** less than $5.00

Mixed Applications Solve. Which method did you use?

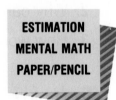

ESTIMATION
MENTAL MATH
PAPER/PENCIL

32. In one week the lunchroom served 436 boxes of raisins with lunch. They had 194 boxes left. How many boxes of raisins did they have to start with?

33. Tony has $4.47 to spend on lunch. He spends $3.65 on a hot turkey sandwich and some juice. He needs at least $1.19 to buy a piece of pie as well. Does he have enough money?

34. About 800 students ate in the lunchroom last week. About 500 students brought their lunch. About how many more students ate in the lunchroom than brought their lunch?

UNDERSTANDING A CONCEPT

Subtracting Whole Numbers

The Yorktown Fair is becoming more popular each year. How many more tickets did the fair sell in 1991 than in 1989?

Estimate. Then subtract to find the answer.

Subtract: 324 − 145

TICKETS SOLD FOR THE YORKTOWN FAIR	
Year	Number Sold
1989	145
1990	263
1991	324

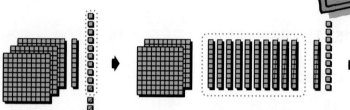

Step 1	Step 2	Step 3
Regroup if necessary. Subtract the ones.	**Regroup if necessary. Subtract the tens.**	**Subtract the hundreds.**
$$\begin{array}{r} {\scriptstyle 1\ 14} \\ 3\,2\,4 \\ -\,1\,4\,5 \\ \hline 9 \end{array}$$	$$\begin{array}{r} {\scriptstyle 11} \\ {\scriptstyle 2\ 1\ 14} \\ 3\,2\,4 \\ -\,1\,4\,5 \\ \hline 7\,9 \end{array}$$	$$\begin{array}{r} {\scriptstyle 11} \\ {\scriptstyle 2\ 1\ 14} \\ 3\,2\,4 \\ -\,1\,4\,5 \\ \hline 1\,7\,9 \end{array}$$
Think: 2 tens 4 ones = 1 ten 14 ones	*Think:* 3 hundreds 1 ten = 2 hundreds 11 tens	

Check: 179 + 145 = 324

In 1991 the fair sold 179 more tickets than it sold in 1989.

1. Is 179 a reasonable answer? How do you know?

2. Why can you check subtraction by adding?

3. How would you subtract $29.36 from $75.43?

TRY OUT   Write the letter of the correct answer.

4. 23 − 17   **a.** 6   **b.** 16   **c.** 30   **d.** 50

5. 567 − 278   **a.** 389   **b.** 299   **c.** 289   **d.** 399

6. $5.15 − $3.25   **a.** $2.30   **b.** $1.90   **c.** $2.90   **d.** $8.90

7. $71.36 − $6.87 **a.** $75.49 **b.** $65.59 **c.** $75.23 **d.** $64.49

PRACTICE

Subtract.

8. 68
− 22

9. 81
− 55

10. 216
− 94

11. 367
− 108

12. $9.25
− 6.50

13. $10.95
− 7.88

14. 760
− 92

15. 1,185
− 815

16. 614
− 195

17. 2,274
− 875

18. $25.35
− 9.98

19. $35.50
− 14.68

20. 3,374
− 2,998

21. $7.81
− 6.98

22. 4,522
− 3,867

23. $.74 − $.25

24. 109 − 86

25. $8.65 − $.70

26. 1,048 − 930

27. $3.25 − $1.85

28. 3,746 − 707

29. 217 − 48

30. 734 − 559

31. $23.44 − $19.37

Mixed Applications

Solve. You may need to use the Databank on page 518.

32. The Plum Tree Café has a food stand at the fair. How much would it cost to buy a broiled fish dinner and some milk?

33. Bobby has $8.50. He buys a chicken salad sandwich from the Plum Tree Café. How much money does he have left?

34. Tina goes on the roller coaster 4 times, the Hurricane 3 times, the Ferris wheel 5 times, and the Whip 4 times. How many times does she ride in all?

35. Letitia brought 473 balloons to the fair. At the end of the fair, she had 196 balloons left. How many balloons did Letitia sell?

Mixed Review

Add or subtract. Which method did you choose?

MENTAL MATH
PAPER/PENCIL

36. $3.65
+ 2.96

37. 220
+ 330

38. 364
− 198

39. $47.58
+ 49.62

40. 850 − 300

41. $7.34 − $3.75

42. 79 + 62

UNDERSTANDING A CONCEPT

Subtracting Across Zeros

There are 300 restaurants in Newton.
Don Figg, the food critic, has eaten in
158 of them. In how many restaurants
has he not eaten?

Subtract: 300 − 158

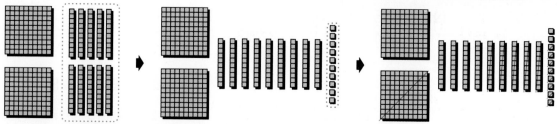

Step 1	Step 2	Step 3
Subtract the ones. **No tens.** **Regroup the hundreds.**	**Regroup the tens.**	**Subtract.**

Step 1:
```
  2 10
  3 0 0
− 1 5 8
```

Step 2:
```
      9
  2 10 10
  3 0 0
− 1 5 8
```

Step 3:
```
      9
  2 10 10
  3 0 0
− 1 5 8
  1 4 2
```

Think: 3 hundreds 0 tens = 2 hundreds 1 ten   *Think:* 10 tens 0 ones = 9 tens 10 ones

So Don Figg has not eaten in 142 of the restaurants in Newton.

1. Is 142 a reasonable answer? How do you know?

2. How can you check your subtraction?

3. Subtract $9.00 − $3.47. What is the difference?

TRY OUT   Write the letter of the correct answer.

4. 306 − 129   **a.** 177   **b.** 187   **c.** 277   **d.** 435

5. 5,000 − 138   **a.** 4,862   **b.** 4,972   **c.** 3,862   **d.** 4,762

6. $7.00 − $.73   **a.** $5.36   **b.** $5.37   **c.** $6.37   **d.** $6.27

7. $9.00 − $6.04   **a.** $2.94   **b.** $2.95   **c.** $2.96   **d.** $3.06

PRACTICE

Subtract.

8. 400
− 109

9. 605
− 586

10. 200
− 68

11. 3,008
− 2,779

12. 8,000
− 673

13. 6,200
− 435

14. $5.00
− 2.89

15. $1.00
− .71

16. $9.06
− 7.38

17. $60.05
− 4.08

18. 2,000 − 84

19. $10.00 − $.42

20. $90.00 − $71.06

21. $7.05 − $6.09

22. $25.05 − $17.99

23. 1,040 − 992

Mixed Applications

Solve. You may need to use the Databank on page 518.

24. Don Figg catches 263 clams for the Newton clambake. His friend Megan catches 400 clams. How many more clams does Megan catch than Don?

25. Stan has $7.00. He wants to buy a tuna salad sandwich, a lettuce and tomato salad, french fries, and a lemonade. Does he have enough money?

26. Carla has $6.00. If she buys a shrimp salad sandwich, how much money will she have left?

27. *Write a problem* about subtraction that uses the menu in the Databank.

MENTAL MATH

You can mentally subtract 600 − 438.

600 $\xrightarrow{-1}$ 599
− 438 $\xrightarrow{-1}$ − 437
162

Think:
If I subtract 1 from both numbers, the difference remains the same.

So 600 − 438 = 162.

Subtract mentally.

1. 700
− 295

2. 400
− 260

3. 80
− 57

4. 1,000
− 496

5. 3,000
− 1,083

Add and Subtract Larger Numbers

The table lists how many meals the Food Court in the mall sold each month. How many meals did the Food Court sell all together?

Month	Number of Meals
April	34,569
May	22,316
June	27,482

WORKING TOGETHER

You can use mental math, paper and pencil, or a calculator to find the answer. Work in a group. Have each person use a different method to solve the problem.

1. Can you easily find the total using mental math? Why or why not?

2. Tell how you can use paper and pencil to find the sum. What is the sum?

3. How can you be sure your sum is reasonable?

4. Which method was fastest in finding the sum?

5. **What if** the numbers had been rounded to 30,000; 20,000; and 30,000? Which method would be fastest?

How many more meals did the Food Court sell in April than in May? Solve the problem using all three methods.

6. Can you find the difference mentally? Why or why not?

7. Tell how you can use paper and pencil to find the difference. What is the difference?

8. Which method was fastest? Why?

Add or subtract. Which method did you use?

9.	10.	11.	12.	13.
40,000 + 20,000	4,037 − 3,859	$53.41 + 78.69	75,000 − 4,935	3,226 4,504 + 321

14. 34 + 46 + 58 + 79   **15.** 27,689 − 16,327   **16.** $79.70 − $59.99

SHARING IDEAS

17. How is adding or subtracting 5-digit numbers like adding or subtracting smaller numbers? How is it different?

18. How did you decide which method to use to find the answers for Exercises 9–16?

19. When do you think it is easier to use mental math? paper and pencil? a calculator? Why?

20. Can you always be sure your answer is correct when you use a calculator? Why or why not?

ON YOUR OWN

Solve. Which method did you use?

**MENTAL/MATH
CALCULATOR
PAPER/PENCIL**

21. Last month the Food Court sold 36,765 soft drinks and 13,689 milk shakes. To the nearest ten thousand, how many drinks did the Food Court sell?

22. The Art Club spends $437.27 on lunch. The Music Club spends $673.13. How much more does the Music Club spend?

23. *Write a problem* that involves addition or subtraction of large numbers. Choose which method you think should be used to solve it. Trade your problem with other students and compare methods.

PROBLEM SOLVING

Strategy: Choosing the Operation

A. Hiko has 1 serving of cereal and 1 serving of orange juice for breakfast. How many calories are in his breakfast?

CALORIES PER SERVING				
Boiled egg 75	Oatmeal 65	Slice of toast 80	Cereal 140	Orange juice 100

1. What information does Hiko need?

To solve the problem he needs to decide which **operation** to use.

2. Hiko is **combining** groups of calories. Which operation should he use?

$$140 + 100 = \blacksquare$$

3. How many calories are in his breakfast?

B. Hiko wants to burn 400 calories by exercising. After riding his bicycle for 1 hour, how many more calories does he need to burn?

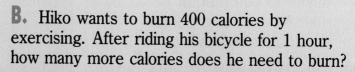

CALORIES BURNED PER HOUR (for a 75-pound person)			
Tennis 200	Bicycle riding 255	Bowling 150	Swimming 335

To solve the problem, he needs to **separate** the 400 calories into two parts.

4. Which operation should he use?

$$400 - 255 = \blacksquare$$

5. How many more calories does he need to burn?

6. How many calories would he burn if he rode his bicycle for 1 hour and swam for 1 hour?

7. Which operation should you use when you combine two groups?

8. Which operation should you use when you separate two groups?

PRACTICE

Decide which operation to use. Then solve the problem.

9. Mrs. Ho weighed 147 pounds before she went on a diet. She lost 19 pounds in six months. How much did she weigh then?

10. Paul took in 695 calories for breakfast and 585 calories for lunch. How many calories was that for both meals?

11. A medium-size peach has 35 calories. A medium-size pear has 100 calories. How many more calories does the pear have than the peach?

12. A serving of raisin cereal has 50 more calories than the 140 calories in a serving of regular cereal. How many calories are in a serving of raisin cereal?

Strategies and Skills Review

Solve. You may need to use the Databank on page 518.

13. The normal heart rate for a fourth grader is about 85 heartbeats a minute. For adults, it is about 72 heartbeats a minute. How many more times does the heart of a fourth grader beat in a minute than the heart of an adult?

14. The recommended daily intake of vitamin A for a fourth grader is 700 units. Phil has a fried egg and a glass of milk. He says he has met the vitamin A requirement for the day. Is he right? How do you know?

15. Lynn had a hamburger and a glass of milk for lunch. Then she exercised for an hour and burned 300 calories. Did she burn all the calories she took in for lunch? How do you know?

16. The average weight of a 9-year-old girl is 62 pounds. At age 14 her average weight is 107 pounds. How much greater is her average weight at age 14 than at age 9?

17. Maya notices that there is a scale between every two rowing machines at the gym. There are 4 rowing machines. How many scales are there?

18. *Write a problem* that can be solved by addition or by subtraction. Solve the problem. Then ask others to solve it.

DEVELOPING A CONCEPT

Finding Perimeter

You can use what you know about measuring length to find the distance around objects.

The distance around an object is its **perimeter**.

WORKING TOGETHER

1. Estimate the perimeter of your desk.

2. Use a centimeter ruler to measure the distance around your desk to the nearest centimeter.

3. Record your estimate and actual measure in a chart like this.

Object	Estimate	Perimeter
desk		

4. Estimate and then measure the perimeter of other objects in your classroom. Record your findings in your chart.

5. Compare your chart with those of others.

6. What other units of measure could you use to measure the perimeter of objects?

You can use addition to find the perimeter of objects.

The perimeter of this rectangle is
3 cm + 3 cm + 4 cm + 4 cm,
or 14 cm.

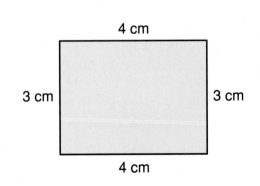

7. What is the perimeter of a triangle with sides 5 m, 8 m, and 10 m?

8. What is the perimeter of a square with sides 4 cm?

9. How would you measure the perimeter of your home? Compare your methods with those of others.

10. What is the least number of sides you can measure to find the perimeter of a rectangle? of a square? How do you know?

11. What must you do before you can find the perimeter of this triangle? Why?

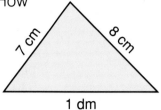

PRACTICE

Find the perimeter.

12.
6 m
6 m 6 m
6 m

13.
5 cm
4 cm
6 cm

14.
5 m
1 m   1 m
5 m

15.
15 cm
8 cm   8 cm
15 cm

16.
4 m
4 m   4 m
4 m   4 m
4 m

17.
12 m
12 m   12 m
12 m

18.
3 m   3 m
3 m

19.
8 m
2 m   4 m
8 m

20.
2 cm
2 cm   4 cm
2 cm
3 cm

Critical Thinking

21. The perimeter of this rectangle is 20 m. What is the length of the rectangle? How do you know?

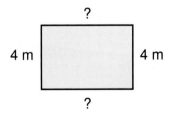

?
4 m   4 m
?

Mixed Applications

22. A rectangle has a width of 22 m and a length of 7 m. What is the difference between its width and length?

23. Tanya runs around a square field that is 100 m on each side. How far does Tanya run?

DECISION MAKING

COOPERATIVE LEARNING

Problem Solving: Planning a Lunch

SITUATION

Emmanuel, Monica, Vincent, and Michele are planning to have lunch in the park on Saturday. They are going to meet at Jim's Deli and buy their lunch to take to the park.

PROBLEM

Should they buy sandwiches or buy the things they need to make their own sandwiches?

DATA

1 loaf of bread – 12 sandwiches
1 pound of cold cuts – 8-10 sandwiches
1 pound of cheese – 12 sandwich-size slices
half pint – 1 drink
half gallon – 8 drinks

JIM'S DELI

Cold Cuts
Sold in half and whole pounds only

	Half Pound	One Pound
Turkey	$2.50	$5.00
Roast Beef	4.00	8.00
Cheddar Cheese	2.00	4.00

Sandwiches

Turkey	$3.00
Roast Beef	3.50
Cheddar Cheese	2.00
Lettuce + Tomato	.50¢ EXTRA
On a roll	.50¢ EXTRA

...ped
...heese
...9¢

Lettuce 75¢ a head

Tomatoes 35¢ each

Sandwich Roll .40¢ each
Whole Wheat Bread $1.40 a loaf

Milk: half-gallon $1.50 half-pint $ .50

Juice: half-gallon $1.40 half-pint $ .40

USING THE DATA

How much will 4 of the following sandwiches cost?

1. roast beef

2. roast beef on a roll

3. roast beef on a roll, lettuce and tomato

4. cheddar cheese

5. cheddar cheese on a roll

6. turkey with lettuce and tomato

How much of each must they buy to make 4 sandwiches?

7. rolls

8. lettuce

9. roast beef

10. tomatoes

11. whole wheat bread

12. cheddar cheese

Find the total cost to make these sandwiches.

13. 4 cheddar cheese on whole wheat with lettuce and tomato

14. 4 roast beef on a roll

MAKING DECISIONS

15. The children all want the same sandwich. Will it cost more to make or to buy the 4 sandwiches? Tell how you made your choice.

16. *What if* only one child wants lettuce and tomato and three want sandwich rolls? Is it cheaper to make or buy 4 cheese sandwiches?

17. What other things will the children need to bring if they want to make sandwiches in the park?

18. Would you buy or make sandwiches for a lunch in the park? Tell how you made your choice.

19. The children must also decide what to drink with their lunches. Should they buy a half gallon or half-pint containers? Why might they not want to buy a half-gallon container?

20. *Write a list* of the things you would buy for a picnic if you had $15.00 to spend.

Math and Social Studies

Hundreds of years ago people measured land by describing it. A farmer might say he or she owned everything from the ridge to the river, with the forest on one side and the waterfall on the other. This was called the **landmark method**.

In the 1300s people started using units such as feet, rods, furlongs, and acres to measure land. But the definitions of these units were still descriptions. For example, an **acre** was first defined as the amount of land two oxen could plow in one day. The **furlong** was defined as the length of a row in a plowed field.

Today we still use these units, but we define them differently. An acre is a square plot of land with sides that each measure 209 feet. A furlong is about 600 feet.

What if Mr. Farmer wants to build a fence around his land? Look at the drawing. The length of each side of the property is given in furlongs. How much fence will he need?

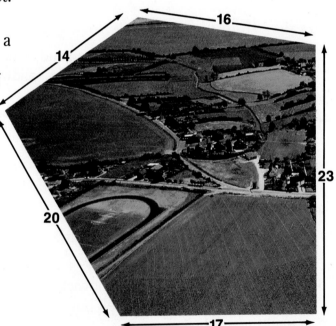

Think: The distance around his land is the perimeter.
23 + 16 + 20 + 17 + 14 = 90

Mr. Farmer will need 90 furlongs of fence.

ACTIVITIES

1. What problems might arise in using the landmark method to measure land? Tell why the old method of defining an acre will not give an exact measure.

2. Use the landmark method to describe your school or another well-known place in your town. Share your description with others.

Computer Graphics

Your class is having bicycle races on three different courses.
You want to study the shape and distance of each racecourse.
You can use a computer to draw the courses quickly and
compute their perimeters. All you need to do is input the
length of each side.

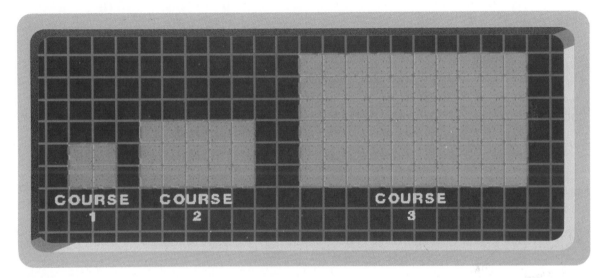

DATA

Using graph paper and a pencil, copy the racecourses shown
above. Let the sides of each square equal 50 meters.

THINKING ABOUT COMPUTERS

1. What is the shape and perimeter of each racecourse?

2. **What if** each side of Course 1 were changed to 300
   meters? What would be its shape and perimeter?

3. **What if** each side of Course 2 were increased by 100
   meters? What would be its shape and perimeter?

4. Why would it be easier to use a computer to solve
   Problems 1–3?

5. Why would a computer be useful for designing
   different racecourses?

EXTRA PRACTICE

Mental Math: Adding 10s; 100s; 1,000s, page 91 . . . .

Add. Use mental math.

1.	30	2.	100	3.	40	4.	900	5.	500
	+ 50		+ 400		30		10		500
					+ 60		+ 70		+ 300

6.	79	7.	399	8.	162	9.	888	10.	452
	+ 201		+ 43		+ 699		+ 309		+ 507

11. 20 + 10 + 50   **12.** 300 + 600   **13.** 39 + 17   **14.** 398 + 143

Estimating Sums by Rounding, page 93 . . . .

Estimate by rounding.

1.	335	2.	809	3.	4,375	4.	$4.55	5.	$52.65
	+ 480		+ 175		+ 2,870		+ 2.17		+ 21.17

6.	130	7.	3,090	8.	$3.52	9.	2,070	10.	$24.89
	209		2,010		5.10		4,500		9.12
	+ 150		+ 2,200		+ 1.17		+ 125		+ 49.87

11. 319 + 10 + 410   **12.** 299 + 499 + 399   **13.** 2,400 + 1,470 + 620

Front-End Estimation, page 95 . . . .

Estimate. Use only the front digits.

1.	478	2.	3,987	3.	34	4.	$2.39	5.	6,452
	+ 218		+ 7,109		3		7.71		1,817
					+ 79		+ .25		+ 2,809

6. $52.73 + $89.15   **7.** 4,372 + 5,872   **8.** $1.89 + $20.05 + $16.89

Estimate. Use the front digits and adjust.

9.	293	10.	$2.17	11.	$.35	12.	3,880	13.	3,817
	+ 117		+ 8.95		.27		2,189		204
					+ .80		+ 207		+ 8,019

EXTRA PRACTICE

Problem Solving: Checking for a Reasonable Answer, page 97 ....

Use estimation to decide which answer is reasonable.

1. Josie spent $3.52 on lunch. She gave the clerk $5.00. How much change should she receive?

    a. $2.52   **b.** $1.48   **c.** $.98

2. David buys 156 mL of tomato juice and 245 mL of grape juice. How much juice does David buy?

    a. 401 mL   **b.** 621 mL   **c.** 261 mL

3. The Bagel Bonanza store has sold 653 bagels so far this month. If they sell 761 more bagels this month, how many bagels will they have sold all together?

    a. 714 bagels   **b.** 1,414 bagels   **c.** 13,114 bagels

Adding Whole Numbers, page 99 ....

Add.

1. 387 + 292	**2.** 420 + 204	**3.** $8.58 + .32	**4.** 3,159 + 88	**5.** $4.57 + 2.10
6. 2,069 + 835	**7.** $66.88 + 7.21	**8.** 9,871 + 58	**9.** 3,673 + 2,417	**10.** $95.52 + 46.68

11. $89.52 + $38.50   **12.** 4,476 + 2,389   **13.** $129.85 + $74.50

Column Addition, page 101 ....

Add.

1. 63 24 + 39	**2.** $3.12 2.19 + 4.07	**3.** 53 29 18 + 37	**4.** $3.07 2.12 4.10 + .93	**5.** 353 217 104 + 22

6. 39 + 89 + 143   **7.** $1.57 + $4.19 + $.70   **8.** 44 + 87 + 212

9. 87 + 130 + 554 + 89   **10.** $5.19 + $2.17 + $1.14 + $.22

EXTRA PRACTICE

Mental Math: Subtract 10s; 100s; 1,000s, page 105.........................

Subtract. Use mental math.

1.	90	**2.**	800	**3.**	1,200	**4.**	1,300	**5.**	6,000
	− 20		− 600		− 600		− 700		− 1,000

6.	57	**7.**	47	**8.**	455	**9.**	387	**10.**	681
	− 28		− 19		− 319		− 128		− 227

11. 80 − 30 **12.** 78 − 19 **13.** 120 − 40 **14.** 700 − 299

Estimating Differences, page 107.................................

Estimate by rounding.

1.	763	**2.**	2,175	**3.**	$4.50	**4.**	7,575	**5.**	$3.27
	− 274		− 1,050		− 1.12		− 895		− .39

6. $6.17 − $2.89 **7.** 8,962 − 4,129 **8.** 8,500 − 529

Estimate. Use the front digits and adjust.

9.	755	**10.**	$7.43	**11.**	2,809	**12.**	$94.12	**13.**	6,907
	− 109		− 2.65		− 1,072		− 8.07		− 2,873

14. 389 − 25 **15.** $7.69 − $2.12 **16.** 812 − 324

Subtracting Whole Numbers, page 109.................................

Subtract.

1.	77	**2.**	89	**3.**	$7.12	**4.**	458	**5.**	$8.15
	− 12		− 27		− .81		− 109		− 4.50

6.	$24.09	**7.**	3,173	**8.**	780	**9.**	$38.12	**10.**	4,808
	− 8.98		− 1,899		− 479		− 27.85		− 725

11. 209 − 89 **12.** 892 − 375 **13.** $7.95 − $5.40

Subtracting Across Zeros, page 111 ....

Subtract.

1. 300
  − 107
	2. 705
	  − 365
		3. 100
		  − 57
			4. 408
			  − 399
				5. 900
				  − 553

6. 600
  − 345
	7. $5.00
	  − 3.89
		8. $1.00
		  − .62
			9. $8.07
			  − 3.20
				10. $4.03
				  − 2.07

11. 600 − 214   **12.** $8.00 − $6.09   **13.** $2.09 − $.50

Problem Solving: Choosing the Operation, page 115 ....

Decide which operation to use. Then solve the problem.

1. Miles wants to burn 350 calories. He burns 220 calories playing basketball. How many more calories does he have to burn?

2. Daphne had 765 calories at lunch and 809 calories at dinner. How many calories did she have at both meals?

3. A serving of raisin bran has 170 calories. A serving of cornflakes has 140 calories. How many more calories does a serving of raisin bran have?

Finding Perimeter, page 117 ....

Find the perimeter.

1.

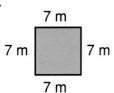

2.

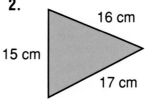

3.

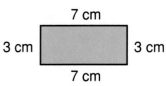

4.

5.

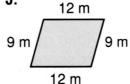

6.

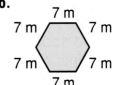

PRACTICE PLUS

KEY SKILL: COLUMN ADDITION (Use after page 101.)

Level A ....

Add.

1.	12	**2.**	31	**3.**	$.56	**4.**	$.16	**5.**	31
	23		28		.22		.62		48
	+ 14		+ 10		+ .10		+ .02		+ 11

6. $.43 + $.22 + $.65   **7.** 29 + 31 + 20   **8.** 83 + 13 + 87

9. Kasha buys three pastel pencils for $.49, two paintbrushes for $.89, and four tubes of oil paint for $1.51. How much does she spend in all?

Level B ....

Add.

10.	123	**11.**	$3.08	**12.**	198	**13.**	$3.67	**14.**	454
	49		2.78		12		1.43		577
	+ 22		+ .32		+ 310		+ .34		+ 40

15. 88 + 34 + 115   **16.** 76 + 117 + 483 **17.** 583 + 297 + 86 + 19

18. Professor Yampa found 179 brontosaurus bones on a dig in Utah last summer. He also found 651 tyrannosaurus bones and 39 stegosaurus bones. How many bones did he find?

Level C ....

Add.

19.	72	**20.**	$6.86	**21.**	529	**22.**	$13.43	**23.**	1,625
	87		2.38		237		.89		2,209
	+ 54		+ 1.77		188		24.63		342
					+ 650		+ 5.06		+ 3,848

24. 435 + 78 + 22 + 169   **25.** 1,167 + 529 + 4,298 + 331

26. Mary Jo scores 651 points on a video game. Li scores 418 points more than Mary Jo. Alix scores 163 more points than Li. How many points does Alix score?

KEY SKILL: SUBTRACTING ACROSS ZEROS (Use after page 111.)

Level A ....

Subtract.

1. 306 − 89	**2.** $4.01 − .66	**3.** 904 − 35	**4.** 702 − 17	**5.** $1.02 − .24

6. $8.07 − .49	**7.** 505 − 326	**8.** 803 − 565	**9.** $2.06 − 1.48	**10.** 603 − 567

11. Terry has 427 calories. Gary has 603 calories.
How many more calories does Gary have than Terry?

Level B ....

Subtract.

12. 600 − 211	**13.** 700 − 418	**14.** 302 − 109	**15.** 908 − 779	**16.** 407 − 289

17. 3,705 − 289   **18.** $11.04 − $8.95   **19.** $24.03 − $5.47

20. Lacrosse sticks are on sale for $45.77. They usually cost
$51.01. How much can you save by buying one now?

Level C ....

Subtract.

21. 2,600 − 1,731	**22.** $70.02 − 46.38	**23.** 4,005 − 1,159	**24.** 8,003 − 6,926	**25.** $103.08 − 74.49

26. $95.07 − 76.78	**27.** 7,005 − 4,369	**28.** 40,702 − 33,884	**29.** $280.01 − 197.34	**30.** 20,803 − 13,919

31. One year the Boston Celtics scored 9,004 points and
the L.A. Clippers scored 5,906 points. What was the
difference in point totals between the two teams?

Chapter Review

LANGUAGE AND MATHEMATICS

Complete the sentences. Use the words in the chart.

1. The distance around an object is its ■. *(page 116)*

2. You can ■ numbers to find an estimate. *(page 106)*

3. Use ■ to check for reasonableness. *(page 96)*

4. ■ is the opposite of subtraction. *(page 108)*

5. **Write a definition** or give an example of the words you did not use from the chart.

VOCABULARY
difference
addition
perimeter
operations
estimation
round

CONCEPTS AND SKILLS

Estimate by rounding. *(pages 92, 106)*

6. 170 204 + 250	**7.** 7,619 − 512	**8.** $71.82 − 2.99	**9.** $6.50 4.23 + 1.50	**10.** 8,032 2,001 + 478

11. 2,450 + 730   **12.** 7,540 − 2,481   **13.** $8.26 − $5.73

Estimate. Use the front digits and adjust. *(pages 94, 106)*

14. 629 + 395	**15.** $8.57 − 3.17	**16.** $.25 .77 + .89	**17.** 7,889 − 2,917	**18.** 798 103 + 99

19. 329 + 988 + 496   **20.** 781 − 217   **21.** $62.21 − $29.72

Add. *(pages 90, 98–101)*

22. 276 + 328	**23.** 47 54 + 12	**24.** 500 300 + 49	**25.** $2.53 .74 + 3.81	**26.** $8.61 9.29 + .50

27. 48 + 95   **28.** 1,317 + 2,562   **29.** 200 + 50 + 90

30. 28 + 166 + 35 + 471   **31.** $3.11 + $5.60 + $.72 + $1.24

Subtract. *(pages 104, 108–111)*

32. 5,000
  −   76

33. $6.00
  − 4.02

34. $80.00
  − 25.03

35.   577
  − 398

36. $15.99
  −   1.51

37. 8,000 − 1,200

38. 724 − 355

39. 6,541 − 2,983

40. 98 − 57

41. $4.38 − $2.58

42. $9.27 − $.86

Find the perimeter. *(page 116)*

43.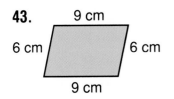
9 cm
6 cm   6 cm
9 cm

44.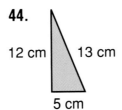
12 cm   13 cm
5 cm

45.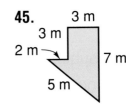
3 m
3 m
2 m   7 m
5 m

CRITICAL THINKING *(pages 98, 106–109, 116)*

Write *always*, *sometimes*, or *never* to complete the sentence.

46. The sum of two 2-digit numbers is ■ a 4-digit number.

47. The difference between two 3-digit numbers is ■ a 3-digit number.

48. A front-end estimate is ■ less than the exact difference.

49. The perimeter of a square is ■ four times the length of a side.

MIXED APPLICATIONS

50. Jeff is buying juice for 59¢. He gives the cashier 75¢. How much change should he get? *(page 96)*

 a. $1.04   **b.** 14¢   **c.** 16¢

51. A large fish tank can hold 150 fish. It contains 75 minnows. Is there enough room in the tank to add 25 guppies and 40 mollies? *(page 94)*

52. The aquarium has 1,458 plants and 3,412 fish. How many more fish than plants does it have? Which operation did you use to find the answer? *(page 114)*

53. Sandy is painting a line on a square frame that is 22 cm on each side. How long will the line be? *(page 116)*

CHAPTER TEST

Estimate by rounding.

1.   368
      + 812

2.   5,298
      − 3,712

3. $7.95
        .92
      + 2.26

4.   2,675
      − 619

Estimate. Use the front digits and adjust.

5.   293
      + 314

6.   285
      − 107

7.   3,411
        3,322
      + 201

8. $8.95
      − 4.25

Add or subtract.

9.   800
      + 40

10.   700
       − 40

11.   625
       + 594

12. $9.03
       − 4.09

13.   620
          90
       + 40

14.   5,623
       −   89

15. $59.05
       + 22.89

16.   9,000
       − 2,987

17. 3,599 + 625

18. $36.00 − $17.24

19. 729 + 42 + 697

20. 2,004 − 876

Find the perimeter.

21.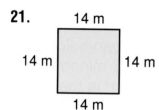

14 m (top), 14 m (left), 14 m (right), 14 m (bottom)

22.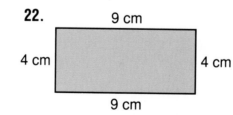

9 cm (top), 4 cm (left), 4 cm (right), 9 cm (bottom)

23.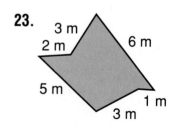

3 m, 2 m, 6 m, 5 m, 1 m, 3 m

Solve.

24. Mr. Watt baked 412 muffins in the morning and 228 in the afternoon. How many more muffins did he bake in the morning?

25. Delia bought a pen for $1.24 and a pencil for $.69. What is her total bill?

a. $4.09   **b.** $1.93   **c.** $19.23

PALINDROMES

Reverse the order of the letters in the words in the box.

What do you notice about each word?

Reverse the order of the digits in the numbers in the box.

What do you notice about each number?

Words and numbers that can be read the same in either direction are called **palindromes.**

Is the number 42 a palindrome? Why or why not?

You can use the number 42 to find a palindrome.

```
   42
 + 24  ⟵  Reverse the digits and add.
   66  ⟵  Palindrome
```

1. How would you use the number 36 to find a palindrome?

Tell if the number is a palindrome.
Write *yes* or *no.*

2. 26   **3.** 88   **4.** 414   **5.** 123   **6.** 707

Use the number to find a palindrome.

7. 15   **8.** 63   **9.** 71   **10.** 53   **11.** 13

Use these numbers to find palindromes. You may need to reverse the digits and add more than once.

11. 82   **12.** 76   **13.** 84   **14.** 59   **15.** 164

16. Can you think of more words that are palindromes?

CUMULATIVE REVIEW

Choose the letter of the correct answer.

1. $5.89 − $3.99

   a. $1.90 **c.** $2.09
   b. $1.99 **d.** not given

2. Estimate by rounding:
   892 + 435

   a. 1,100 **c.** 1,300
   b. 1,230 **d.** 1,400

3. What is 17 take away 8?

   a. 7 **c.** 9
   b. 8 **d.** not given

4. $72.91 + $.85

   a. $73.76 **c.** $72.06
   b. $72.76 **d.** not given

5. Choose the number for ten million.

   a. 1,000,000 **c.** 10,000
   b. 10,000,000 **d.** not given

6. 5,300 + 900

   a. 5,200 **c.** 6,200
   b. 5,900 **d.** not given

7. Estimate by rounding:
   $7.85 − $2.31

   a. $9.00 **c.** $5.00
   b. $6.00 **d.** $4.00

8. What is 6 and 5 more?

   a. 1 **c.** 16
   b. 11 **d.** not given

9. Estimate the width of a pillow.

   a. 4 cm **c.** 4 m
   b. 4 dm **d.** 4 km

10. 72 + 263 + 56

    a. 154 **c.** 381
    b. 291 **d.** not given

11. Compare: 598 ● 589

    a. < **c.** =
    b. > **d.** not given

12. 600 − 132

    a. 468 **c.** 532
    b. 458 **d.** not given

13. Round 287 to the nearest ten.

    a. 200 **c.** 290
    b. 280 **d.** 300

14. Find the perimeter of a square with 7-cm sides.

    a. 14 cm **c.** 28 cm
    b. 21 cm **d.** not given

Measuring Time, Capacity, and Mass

CHAPTER 4

MATH CONNECTIONS: GRAPHING
• TEMPERATURE • PROBLEM SOLVING

EXIT

The Brontosaurus could have measured over
30 meters in length and 6 meters in height
and weighed up to 37,000 kilograms.

1. What information do you see in this picture of a museum exhibit?

2. How can you use the information?

3. Write a problem about the dinosaur in the picture.

133

Estimating Time

Brian studies the exhibit on the history of travel.

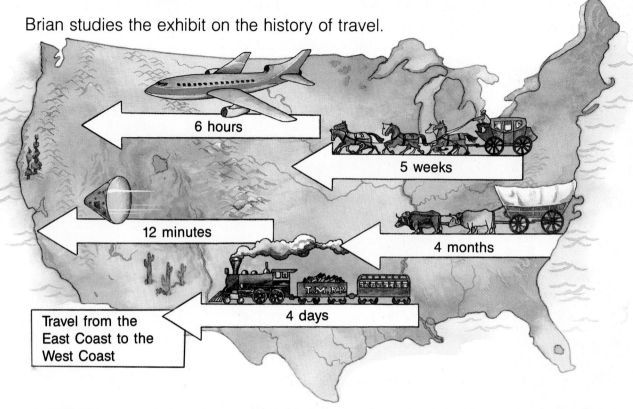

6 hours

5 weeks

12 minutes

4 months

4 days

Travel from the East Coast to the West Coast

1. What happened to the amount of time it takes to travel from coast to coast? Why do you think this happened?

This table shows different units used to measure time.

2. Which units measure long periods of time?

3. Which units measure short periods of time?

UNITS OF TIME

60 minutes (min)	=	1 hour (h)
24 hours	=	1 day (d)
7 days	=	1 week (wk)
12 months (mo)	=	1 year (y)
about 52 weeks	=	1 year
365 days	=	1 year
366 days	=	1 leap year

Choose the best unit of time to measure the activity.

4. sleep at night

5. take a shower

6. count by fives to 100

7. become a doctor

8. go to summer camp

9. run 100 yards

10. drive from Los Angeles, California to Washington, D.C.

11. Name three activities that take about one minute to complete; about one hour to complete.

12. Name three activities that take about one day to complete; about one month to complete.

Practice

Which is the more reasonable estimate for the activity?

13. see a movie
120 seconds, 120 minutes

14. go from home to school
30 minutes, 130 minutes

15. make a sand castle
60 minutes, 600 minutes

16. learn to play a guitar
24 weeks, 240 weeks

17. build a tree house
2 hours, 2 days

18. grow an inch
230 days, 230 weeks

19. take a test
45 minutes, 245 minutes

20. walk a mile
25 minutes, 125 minutes

Mixed Applications

Solve. You may need to use the Databank on page 519.

21. Shara spent 2 hours at the science museum and 100 minutes at the art museum. Where did she spend more time?

22. One museum display showed inventions made in the last 100 years. Would the telephone be in the display?

23. How many years apart were the invention of television and the invention of radio?

24. *Write a story* about an adventure at a museum. Be sure it involves time.

CALCULATOR

A fourth-grader's heart beats about 64 times each minute. Complete the table to find out how many times your heart beats in 1 hour, 1 day, or longer. (*Hint:* Multiply to find the answer.)

Time	1 minute	1 hour	1 day	1 week	1 year	your lifetime
Total	64	■	■	■	■	■

UNDERSTANDING A CONCEPT

Telling Time

A. Lauren and her friend Shara are going to the museum at 9:00.

They see a movie.	The movie ends.	They go on a tour.

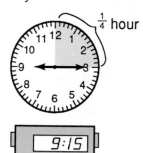 ¼ hour

½ hour

¾ hour

```
9:15          9:30          9:45
```

Read: nine-fifteen, fifteen minutes after nine, or a quarter after nine

Read: nine-thirty, thirty minutes after nine, or half past nine

Read: nine forty-five, forty-five minutes after nine, or a quarter to ten

1. As the clock goes from 9:00 to 10:00, how far does the hour hand go? how far does the minute hand go?

2. What number is the hour hand closest to at 9:15? at 9:30? at 9:45? How do you know?

B. The hour hand goes around the clock two times each day. It goes around once for A.M. hours and once for P.M. hours. A.M. means the time between midnight and noon. P.M. means the time between noon and midnight.

3. List three things that you might do during A.M. hours.

4. List three things that you might do during P.M. hours.

TRY OUT   Write the letter of the correct time.

5.
   a. 2:25
   b. 5:10
   c. 6:10
   d. 5:12

6. 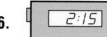 `2:15`
   a. two-fifteen
   b. two-thirty
   c. three forty-five
   d. three-fifteen

PRACTICE

Write the time using numbers and A.M. or P.M. Then write the time in words.

7. Jeff eats lunch at:

8. The sun rose at:

9. Lois makes her bed at:

10. The birthday party begins at:

Mixed Applications

11. Morrie went to the museum cafeteria at 4:35 P.M. Tara went to the cafeteria at 4:15 P.M. Who went to the cafeteria first?

12. Nancy arrives for the noon movie about dinosaurs at 11:50 A.M. Is she early or late? By how much?

13. Mrs. Katz does volunteer work at the museum 5 hours a day on Sunday, Tuesday, and Friday. How many hours does she volunteer each week?

14. Museum tickets are $1.50 for adults and $.85 for children. How much more does an adult's ticket cost than a child's ticket?

Mixed Review

Find the answer. Which method did you use?

15. 60 + 70
16. 3,917 + 9,496
17. 74 − 38
18. 1,000 − 305
19. 172 − 61
20. 218 + 35 + 9

MENTAL MATH
CALCULATOR
PAPER/PENCIL

UNDERSTANDING A CONCEPT

Elapsed Time

A. A class spends 2 hours and 15 minutes at the nature preserve. The class arrives at 1:45 P.M. What time will it leave?

Here is a way you can find the time after 2 hours and 15 minutes have **elapsed.**

Step 1

Count the hours.

Begin at: 1:45 P.M.   *Think:*
            2:45 P.M.   1 hour
Stop at: 3:45 P.M.   2 hours

Step 2

Count the minutes.

Begin at: 3:45 P.M.   *Think:*
Stop at: 4:00 P.M.   15 minutes

The class will leave at 4:00 P.M.

1. How could you find the starting time of the trip if you were given the end time?

B. A guided tour of the Wildlife Trail begins at 2:30 P.M. and ends at 3:42 P.M. How long is the tour?

Step 1

Count the hours.

Begin at: 2:30 P.M.
Stop at: 3:30 P.M.
Think: 1 hour

Step 2

Count the minutes.

Begin at: 3:30 P.M.
Stop at: 3:42 P.M.
Think: 12 minutes

Step 3

Write the hours and minutes.

1 hour and 12 minutes

The tour is 1 hour and 12 minutes long.

TRY OUT   Find the start time, end time, or elapsed time.

2. Start time: 6:53 P.M.   End time: ■   Elapsed time: 3 h 35 min

3. Start time: ■   End time: 3:40 P.M.   Elapsed time: 1 h 18 min

4. Start time: 9:45 A.M.   End time: 11:30 A.M.   Elapsed time: ■

PRACTICE

What time will it be:

5. in 15 minutes?

6. in 2 hours 35 minutes?

7. in 1 hour 56 minutes?

What time was it:

8. a half hour ago?

9. 1 hour and 15 minutes ago?

10. 50 minutes ago?

What time will it be:

11. 20 minutes after 10:40 A.M.?

12. 30 minutes before 3:15 P.M.?

13. 1 hour and 12 minutes before 5:00 P.M.?

14. 50 minutes after 6:15 A.M.?

How much time passes between:

15. two-thirty and two-fifty?

16. half past twelve and 1:08?

17. 3:35 and a quarter to four?

18. midnight and noon?

Critical Thinking Solve. Tell how you answered each question.

19. How much time elapses between 11:30 A.M. and 3:40 P.M.?

20. What time will it be 40 minutes before 12:20 P.M.?

Mixed Applications

21. Ramón and Lucia went to the park at 1:14 P.M. They stayed for 3 hours and 20 minutes. At what time did they leave?

22. It takes two days to walk through the nature preserve. What metric unit would best measure its length?

EXTRA Practice, page 164; Practice *PLUS*, page 168   Measuring Time, Capacity, and Mass   **139**

PROBLEM SOLVING

Identifying Extra Information

The Sanchez family is looking forward to Thanksgiving weekend. Today is Election Day. Carmen sees on the calendar that Veterans Day is November 11th and Thanksgiving Day is November 23rd. How many days is it until Thanksgiving?

NOVEMBER

S	M	T	W	T	F	S
			1	2	3	4
5	6	7 Election Day	8	9	10	11 Veteran's Day
12	13	14	15	16	17	18
19	20	21	22	23 Thanksgiving Day	24	25
26	27	28	29	30		

1. What information does Carmen know?

2. What does she need to find out?

The calendar has more information than Carmen needs to solve the problem.

3. List the information Carmen needs to solve the problem.

4. What information is not needed?

5. How many days is it until Thanksgiving?

6. Why is it important to know what information you need to solve a problem?

PRACTICE

**List the extra information. Then solve the problem.
Use mental math, a calculator, or paper and pencil.**

7. There were 27 boys and 19 girls in the Columbus Day pageant. Another 48 students watched the pageant. How many students were in the pageant?

8. There are 198 school days and 9 holidays this year. If there are 365 days this year, how many days are there with no school?

9. There are 7 people running for mayor. Another 4 people are running for sheriff. There are 24 people running for some office this election. How many are not running for mayor?

10. The class members used 46 sheets of green construction paper and 61 sheets of red paper to make 158 holiday decorations. How many sheets did they use?

11. The Veterans Day parade has 8 floats with a total of 24 people on them. There are 3 people on each float. There are 21 people in the marching band. How many more people rode on the floats than marched in the band?

Strategies and Skills Review

12. A string of Christmas tree lights is made of red and green bulbs. A red bulb always follows a green bulb. If the first bulb is green, what color is the seventh bulb? the tenth bulb?

13. It is 29 days from Columbus Day to Election Day, 16 days from Election Day to Thanksgiving, and 32 days from Thanksgiving to Christmas. How many days is it from Columbus Day to Christmas?

14. Mindy made macaroons for Hanukkah. She ate 6. Then she had 21 left. How many macaroons did Mindy make?

15. ***Write a problem*** that has extra information. Solve your problem. Ask others to solve it.

UNDERSTANDING A CONCEPT

Ordered Pairs

A. Beth and Mike use a map to find their way to the Art Museum. Where is the Art Museum on the map?

An **ordered pair** of numbers can be used to describe the location of a point.

Beth and Mike start at 0. On the map they move 4 spaces to the right and 3 spaces up. So the ordered pair (4, 3) gives the location of the Art Museum.

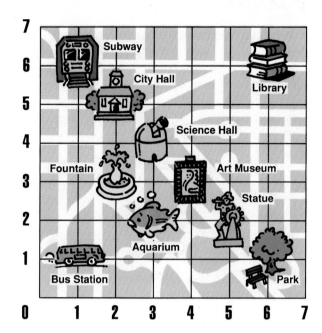

1. Write the ordered pair that tells where the Science Hall is on the map.
   Think: First move to the right; then move up.

2. Compare these two ordered pairs. Why is the order of the numbers important when you name a point?

B. You can find a point given an ordered pair.

What is at the point named by the ordered pair (2, 5)?

Start at 0. Move 2 spaces to the right. Move 5 spaces up. City Hall is at (2, 5).

3. What is at the point named by (5, 2)? Tell how you found the point.

TRY OUT   Write the letter of the correct answer.

4. Give the ordered pair that names the location of the Fountain.

   a. (2, 2)   **b.** (3, 2)   **c.** (3, 3)   **d.** (2, 3)

5. Tell what is at the point named by (6, 1).

   a. Aquarium   **b.** Subway   **c.** Library   **d.** Park

PRACTICE

Write the ordered pair that gives the location.

6. C **7.** F **8.** K **9.** N

10. A **11.** G **12.** D **13.** E

Tell what letter is at the point named by the ordered pair.

14. (5, 6) **15.** (2, 3) **16.** (3, 2)

17. (4, 5) **18.** (0, 1) **19.** (6, 5)

20. Make a grid map of your classroom. List some ordered pairs for your map. Challenge another student to name what is at each location.

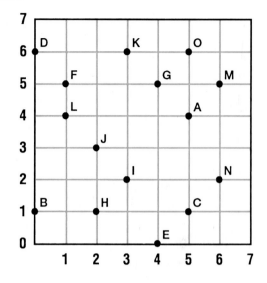

Mixed Applications Use the map on page 142.

21. A parking lot is being built between the Science Hall and the Aquarium. What ordered pair names this location?

22. The museum had 134 visitors on Saturday. This was 16 more than on Sunday. How many visitors came on Sunday?

23. Beth and Mike met at the point named by (1, 6). Where in the city did they meet?

24. Janine walks from the Park to the Library. How many spaces on the map does she walk?

CHALLENGE

Hidden Treasure
Make a treasure map on a grid that is 5 squares long and 5 squares wide. Draw treasures such as gold coins, silver coins, and diamonds. Let your partner try to find each treasure by naming ordered pairs. You can answer "hot," "cold," or "warm," while your partner marks your answers on a blank grid.

Making Line Graphs

The rangers at Grandee State Park keep track of how many people visit the park each week. They want to see how the numbers change during the summer.

A **line graph** shows changes over time.

VISITORS DURING THE SUMMER

Week in July	Number of People (rounded to the nearest ten)	Week in August	Number of People (rounded to the nearest ten)
1	150	1	190
2	170	2	210
3	180	3	210
4	200	4	170

A ranger used these steps to show the data on a line graph.

Step 1 Choose a scale for the vertical side.

Step 2 Draw and label the horizontal and vertical sides.

Step 3 Place the points on the graph.

Step 4 Connect the dots in order.

Step 5 Write a title above the graph.

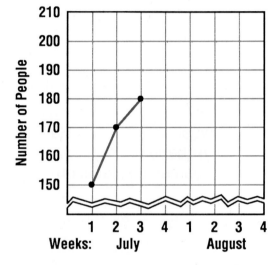

WORKING TOGETHER

1. Copy and complete the graph.

2. Why do you think 150 to 210 was used for the scale?

3. Why do you think there is a broken line below 150?

4. **What if** 154 people visited the park in Week 1 of September? How would you show this on the graph? What do you need to do to the data first?

SHARING IDEAS

5. Compare your graph with those of other students. How are they similar? How are they different?

6. Where do you see the greatest changes? Are they increases or decreases?

7. Why is a line graph useful for showing how things change over time?

PRACTICE

CARS PASSING SCHOOL EACH HOUR

Time	8–9	9–10	10–11	11–12	12–1	1–2	2–3
Cars (to the nearest ten)	60	40	20	50	80	70	30

8. Complete the line graph, using the data in the table. Then answer the questions below.

9. In which hour did the most cars pass? the fewest cars pass?

10. How would you describe the changes as the day went on?

11. Between 3:00 and 4:00 41 cars passed and between 4:00 and 5:00, 76 cars passed. Show this data on your graph.

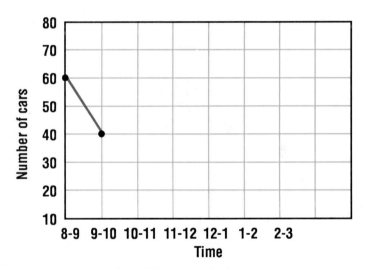

12. On Saturday Kevin and Andrea counted people going into Kerry's Korner Market. The table below shows their data. Make a line graph of their data. Tell what the graph shows.

From To	10:00 10:15	10:15 10:30	10:30 10:45	10:45 11:00	11:00 11:15	11:15 11:30	11:30 11:45	11:45 12:00
People	10	12	9	15	18	20	25	28

13. Record the outdoor temperature each day for one week. Make a line graph of the data.

GO WITH THE PUNCHES

Visual Reasoning

A. You can use a hole puncher to conduct an experiment. Take a sheet of paper. Fold it in half like this:

Then punch two holes in this pattern:

What will the paper look like when you open it up? How many holes will there be? How do you know?

This is what the opened paper will look like:

Were you correct?

1. Now fold and punch another square sheet this way:

What will it look like unfolded?

2. Fold a square sheet of paper in half. Then fold it in half again. Now punch it in the middle. What will the unfolded square look like? How many holes will there be?

4 holes

3. Fold another square in half twice. Punch it like this:

4. What will the unfolded square look like? How do you know?

B. Work on these puzzles. Try to guess what the unfolded square will look like before opening the paper.

5.

6.

7.

Now try to guess how each paper was folded and punched to get the design. Experiment with your own paper. Try to make the fewest punches possible.

8.

9.

10.

11. Design three hole-punching puzzles of your own. Have another student try to solve your puzzles.

DEVELOPING A CONCEPT

Capacity

The **capacity** of a container is the amount of liquid it can hold.

WORKING TOGETHER

1. Collect three cups of different sizes. About how many of each size cup will it take to fill a large juice carton? Make a table to record your findings.

2. Compare the sizes and the numbers of cups needed to fill the juice carton. What do you notice?

The **milliliter (mL)** is a metric unit for measuring small amounts of liquid. A teacup holds about 100 mL. The **liter (L)** is a metric unit for measuring larger amounts of liquid.

3. Collect large containers. Use a liter container. Estimate and measure the capacities of your large containers.

CAPACITY TABLE

Size of juice carton:	
Size of cup	Number of cups
small	■
medium	■
large	■

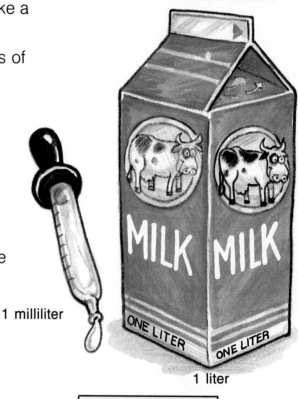

1 milliliter

1 liter

1,000 mL = 1 L

CAPACITY IN LITERS

Container	Estimate	Measure
juice carton	■	■
milk carton	■	■
pail	■	■

SHARING IDEAS

4. What other kinds of containers have large capacities? small capacities? Which units would you use to measure their capacities?

5. What would happen if people used different-size containers to measure the same capacity?

6. What would be a better way to measure the same capacity?

PRACTICE

Choose the best estimate of capacity. Write the letter of the answer.

7. a vase
    a. 1 mL   **b.** 1 L   **c.** 10 mL

8. a teaspoon
    a. 5 mL   **b.** 500 mL   **c.** 5 L

9. a water pitcher
    a. 10 mL   **b.** 1 L   **c.** 10 L

10. a sink
    a. 7 mL   **b.** 70 mL   **c.** 7 L

11. a soup bowl
    a. 12 mL   **b.** 120 mL   **c.** 12 L

12. a tablespoon
    a. 15 mL   **b.** 150 mL   **c.** 15 L

Tell why the sentence does not make sense.

13. Jack drank 250 L of orange juice for breakfast.

14. The Murphys bought 20 mL of paint to paint their whole house.

Critical Thinking

15. Look for some 1 liter containers. Are they all the same shape? Why or why not? How can you show that they have the same capacity?

Mixed Applications

16. A rare plant needs to be fed every 3 hours. The last feeding was at 1:15 P.M. When is the next feeding?

17. Roy puts a few tablespoons of liquid plant food in each can of water. Would he put in about 30 mL or 30 L?

18. Roy is a Botanical Gardens gardener. Would his watering can hold about 4 mL or 4 L?

19. Lisa buys a plant for $4.50 and a book for $7.79. How much does she spend all together?

Mixed Review

MENTAL MATH
CALCULATOR
PAPER/PENCIL

Find the answer. Which method did you use?

20.	**21.**	**22.**	**23.**
803	97,216	2,002	575
+ 288	+ 38,965	− 795	− 200

DEVELOPING A CONCEPT

Mass

The **mass** of an object tells how much of it there is.

WORKING TOGETHER

1. Hold a calculator in one hand and a book in the other. Which has the greater mass?

2. Compare a calculator and a pencil. Which has the greater mass?

3. Find two different books that have the same mass. See if others agree.

The **gram (g)** is a metric unit for measuring the mass of light objects.

A shoelace has a mass of about 1 gram.
A nickel has a mass of about 5 grams.
An apple has a mass of about 200 grams

4. Name other things that could be measured using grams.

The **kilogram (kg)** is a metric unit for measuring heavier things.

| 1,000 g = 1 kg |

A baseball bat has a mass of about 1 kilogram.

A bowling ball has a mass of about 7 kilograms.

A television set has a mass of about 20 kilograms.

5. Name some other things that could be measured using kilograms.

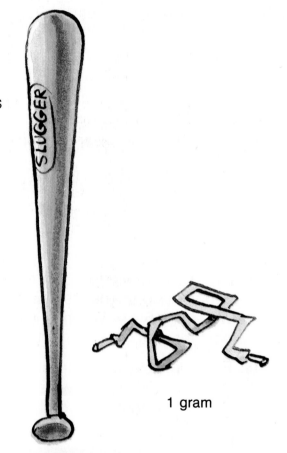

1 gram

1 kilogram

SHARING IDEAS

6. Why do you use small units to measure light things and large units to measure heavy things?

PRACTICE

Choose the best estimate of mass. Write the letter
of the answer.

7. calculator		**a.** 2 g		**b.** 130 g		**c.** 20 kg
8. can of tomatoes		**a.** 370 g		**b.** 37 kg		**c.** 370 kg
9. bicycle		**a.** 25 g		**b.** 510 g		**c.** 15 kg
10. dictionary		**a.** 3 g		**b.** 3 kg		**c.** 30 kg
11. dog		**a.** 18 kg		**b.** 18 g		**c.** 180 kg
12. ice skates		**a.** 2 g		**b.** 2 kg		**c.** 20 kg

Mixed Applications Solve. Which method did you use?

13. The Gold Rush Museum has 3 gold nuggets on
display. They have masses of 350 g, 29 g, and
128 g. What is the total mass of these nuggets?

14. A gold miner's pack has a pan for sifting gold. The
pan has a mass of 682 g. The pack has a mass of
399 g. Does the gold miner carry more than 1 kg?

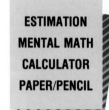

ESTIMATION
MENTAL MATH
CALCULATOR
PAPER/PENCIL

15. Suppose a miner found
750 g of gold. How much
more gold does he need
to have 1 kg of gold?

16. Nelle's class arrived at
the museum at 1:30 P.M.
The museum closes at
5:00 P.M. How much
time can the class
spend there?

LOGICAL REASONING

Apples in the Bag
You need to fill a bag with 14 kg of apples. You
have a balance scale and these weights: 1 kg, 5 kg,
and 10 kg. How can you measure out 14 kg of apples?

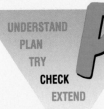

PROBLEM SOLVING

Strategy: Solving a Two-Step Problem

A. The Princetown Science Museum charges $3.50 admission for adults and $1.25 for children. Mrs. Parker takes her young nieces Kata and Chris to the museum. She gives the cashier a $10 bill. How much change should she get?

1. What information do you know?

2. What do you need to find out?

When it takes more than one step to solve a problem, it is a good idea to make a **plan.**

3. What do you need to find out *before* you can decide how much change Mrs. Parker should get? What do you need to do next?

PLAN	
Step 1	**Step 2**
$3.50	$10.00
1.25	− ■
+ 1.25	■
■	

Try your plan.

4. How much change should she get?

B. There were 12 precious gems on display. On Monday 4 gems were removed for polishing. On Tuesday 6 gems were placed in the display case. How many gems were on display then?

5. Make a plan.
   What do you need to do first?
   What do you need to do next?

PLAN	
Step 1	**Step 2**

6. How many gems were on display at the end of Tuesday?

7. Why is it important to make a plan when solving a two-step problem?

PRACTICE

**Make a two-step plan. Then solve the problem.
Use mental math, a calculator, or paper and pencil.**

8. The Princetown Museum has a collection of 147 shells. Recently it received another 12 shells. The Riverside Museum has 203 shells. How many more shells does the Riverside Museum have than the Princetown Museum?

9. Each child is given $2.50 to spend for lunch in the cafeteria. Alma buys a grilled cheese sandwich for $1.95, a drink for $.85, and an apple for $.60. How much of her own money does Alma have to spend to pay for lunch?

10. There are 78 children and 3 adults going to the museum on two buses. One bus leaves with 45 riders. How many riders are on the second bus?

11. While browsing in the museum shop, Kevin chooses 15 animal cards. He puts 7 cards back and buys 6 wildflower cards. How many cards does he buy?

Strategies and Skills Review

Solve. Make a two-step plan if you need one.

12. A total of 4,783 people came to see the coal mine exhibit at the museum last summer. There were 1,364 visitors in June and 1,807 visitors in July. The rest came in August. How many people saw the exhibit in August?

13. The Museum of Modern Art has a collection of 168 paintings and 84 sculptures. When Carl visited the museum, he saw 37 of the paintings in the collection. How many paintings did he not see?

14. Tamla walked through the entire museum. Did she walk about 2 miles? 20 miles? 200 miles?

15. **_Write a problem_** that can be solved in two steps. Solve it. Then ask others to solve it.

Renaming Metric Measures

You can find and use patterns among the metric units of measurement.

WORKING TOGETHER

1. Look at a meter stick. How many decimeters are equal to 1 meter? How many centimeters are equal to 1 meter?

2. Use the meter stick to measure. Cut off a piece of string 2 m long. Use your meter stick to find its length in decimeters and centimeters.

3. Suppose you had a piece of string 3 m long. What would be its length in decimeters? in centimeters?

4. Complete the table.

m	1	2	3	4
dm	■	■	■	■
cm	■	■	■	■

5. You know that 1,000 meters is equal to 1 kilometer. How many meters are equal to 2 kilometers? How do you know?

6. Complete the table.

km	1	2	3	4
m	■	■	■	■

SHARING IDEAS

7. As the size of the unit decreases, what happens to the number of units in the measure of the same length?

8. If you know a length in meters, how can you find the length in decimeters? in centimeters?

ON YOUR OWN

9. You know that 1,000 milliliters is equal to 1 liter. If you know a capacity in liters, how can you find the capacity in milliliters?

10. If you know a mass in kilograms, how can you find the mass in grams?

Complete.

11. 6 m = ■ cm   **12.** 900 cm = ■ m   **13.** 7 dm = ■ cm

14. 80 dm = ■ m   **15.** 5 m = ■ dm   **16.** 5 L = ■ mL

17. 3,000 mL = ■ L   **18.** 4 kg = ■ g   **19.** 2,000 g = ■ kg

Solve.

20. Lucy's class walked 3 km to the aquarium. How far did the class walk in meters?

21. Sam saw an eel that was 90 cm long. What was its length in decimeters?

22. At home or at a grocery store, find three bottles that show their capacities in liters or milliliters.
Write a problem about the capacity of the three bottles. Have another student solve your problem.

Temperature

To measure temperature in metric units, use a **Celsius thermometer.**

To read the temperature on a thermometer, look at the mark or number next to the top of the red column. Each mark on the thermometer at the right shows one **degree Celsius (°C).**

A warm bath is about 50°C.

A glass of iced tea is about 3°C.

Water boils — 100

Room temperature — 20

Water freezes — 0

Cold day — ⁻10

°C

1. What is room temperature?

2. At what temperature does water boil?

3. At what temperature does water freeze?

On the thermometer, the temperature mark for a cold day is at ⁻10°. Read this as "minus 10 degrees Celsius" or "10 degrees below 0, Celsius."

4. Write the temperature for 15 degrees below 0, Celsius.

TRY OUT Write the letter of the correct answer.
Choose the most reasonable temperature.

5. hot oatmeal	a. 0°C	b. 23°C	c. 40°C	d. 72°C
6. snowball	a. 1°C	b. 35°C	c. 55°C	d. 72°C
7. winter day	a. 22°C	b. 56°C	c. ⁻13°C	d. 90°C
8. hot desert	a. 12°C	b. 40°C	c. 100°C	d. 200°C

PRACTICE

Write the temperature in degrees Celsius.

9.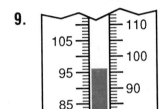

10.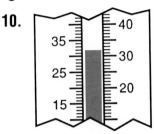

11.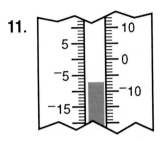

Choose the most reasonable temperature.

12. swimming at the beach   **a.** ⁻3°C   **b.** 10°C   **c.** 35°C   **d.** 115°C

13. building a snow sculpture   **a.** ⁻10°C   **b.** 5°C   **c.** 30°C   **d.** 100°C

14. raking leaves on a cool day   **a.** 0°C   **b.** 12°C   **c.** 50°C   **d.** 90°C

Choose the object closest to the given temperature.

15. 50°C   **a.** glass of cold water   **b.** pan of warm water   **c.** ice cube

16. 85°C   **a.** cup of hot tea   **b.** campfire   **c.** warm bath

17. ⁻20°C   **a.** glass of lemonade   **b.** frozen yogurt   **c.** food freezer

Mixed Applications

18. Carl wants to wear shorts today. The temperature outside is 10°C. Should Carl wear shorts? Why or why not?

19. Gail left the house at 7:30 A.M. She arrived at school at 8:10 A.M. How long did it take Gail to get to school?

20. *Write a problem* about temperature. Solve it. Then trade your problem with others.

Mixed Review

Compare. Use >, <, or =.

21. 1,046 − 368 ● 728   22. 4 + 8 + 6 ● 23 − 5   23. 389 ● 97 + 281

24. 650 + 250 ● 850   25. $42.88 − $36.52 ● $9.37

26. 3 dm ● 30 cm   27. the perimeter of a square with 5 m sides ● 25 m

EXPLORING A CONCEPT

What Do You Measure?

How you measure an object depends on what you need to know about the object.

WORKING TOGETHER

1. Suppose you have a large watering can. What would you measure to find out if it fits on a shelf?

2. What would you measure to find out how much water the can will hold?

3. What would you measure to see if someone can lift the can if it is filled with water?

4. Most fish cannot take very cold water. What should you measure before putting fish in an aquarium?

5. Fill a container with water. Estimate all of its measures. Find every measure that you can.

SHARING IDEAS

6. Name some situations at school or at home when you might need to measure length, capacity, mass, or temperature.

ON YOUR OWN

Write *capacity*, *length*, *mass*, or *temperature* to tell what should be measured in each of the following situations. Then tell what tool should be used to measure it.

7. You need to know if a table will fit in your room.

8. You want to know if you should wear a sweater outside today.

9. You need to know how much punch to make to fill a punch bowl.

10. A mail clerk wants to know how much postage to put on a large package.

Match the descriptions to the objects at the right.

11. Temperature: 20°C
    Width: 7.5 cm
    Length: 12 cm
    Mass: 90 g

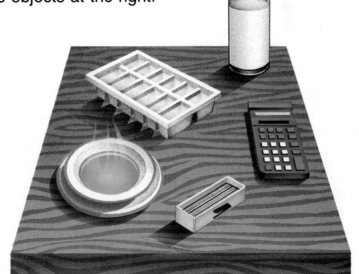

12. Temperature: 0°C
    Length: 23 cm
    Width: 15 cm
    Mass: 500 g

13. Temperature: 75°C
    Mass: 350 g
    Capacity: 250 mL

14. Find some common object in your classroom or at home. Describe it with different measurements. Then ask another student to guess the object.

DECISION MAKING

Problem Solving: Planning a Trip to the Museum

SITUATION

The fourth-grade students in Mr. Levinsky's class are going to the Natural History Museum. The travel time between school and the museum is 30 minutes. The class will leave at 9:30 A.M. and must be back in school by 3:00 P.M.

PROBLEM

Which tours, exhibits, and films should the students see? How should they arrange their day's schedule?

DATA

Natural History Museum

Exhibits open from 10:00 A.M. until 5:30 P.M.

Museum Tours (No charge)

Tours	Time	Length	Comments
Touch a Dinosaur (Hands-On Tour)	10:30 A.M.	$1\frac{1}{2}$ hours	Tour of the Dinosaur exhibit: Includes a classroom program during which students can touch fossils of dinosaur bones.
Dinosaurs	1:00 P.M.	1 hour	Tour of the Dinosaur exhibit.
Birds and Reptiles	12 noon	1 hour	Tour of the Bird and Reptile exhibits.
Marvelous Sea Mammals (Hands-On Tour)	12:30 P.M.	$1\frac{1}{2}$ hours	Tour of the Ocean Animal exhibit: Includes a classroom program during which students can touch a whale's skeleton.
Ocean Animals	2:00 P.M.	1 hour	Tour of the Ocean Animal exhibit.
Gems and Minerals	1:00 P.M.	20 minutes	Tour of the Gems and Minerals exhibit.
Ancient Peoples	10:30 A.M.	45 minutes	Tour of the exhibits of ancient cultures.

Museum Theater Schedule

Films	Times	Length
Meet Mother Nature	10:00 A.M. and 1:30 P.M.	20 minutes
Animals of the Ocean	11:00 A.M. and 2:15 P.M.	60 minutes
Life in Space	12:45 P.M. and 4:30 P.M.	30 minutes

The museum has a lunchroom on the ground floor. It is open from 11:00 A.M. until 2:00 P.M.

USING THE DATA

1. When will the students arrive at the museum? When must they leave? How many hours will they be at the museum?

What time will the following end?

Tours:

2. Ocean Animals   3. Ancient Peoples   4. Birds and Reptiles

Films:

5. 12:45 P.M. *Life in Space*   6. 2:15 P.M. *Animals of the Ocean*

The students need 25 minutes to eat lunch. Should they eat lunch before or after they take these tours?

7. Touch a Dinosaur

8. Marvellous Sea Mammals

9. Dinosaurs, Birds, and Reptiles

MAKING DECISIONS

10. *Write a list* of the things that the students should consider when deciding which museum tours to see.

11. The students want to take both Hands-On tours. Will they have time to take any other tours? To see any film?

12. *What if* the students are studying about ocean animals? Will they have time for both the film and the tour on the subject?

13. *What if* the students want to take the Dinosaur Tour and also want to learn more about ocean animals? Decide how you would arrange a schedule.

14. *Write a schedule* for the students that interests you. Be sure to include a break for lunch. Give reasons for your choices.

STEGOSAURUS

CURRICULUM CONNECTION

Math and Social Studies

There are 24 hours in a day, but we use a 12-hour clock. We show that a time is before or after noon by writing A.M. or P.M.

A **24-hour clock** is used in space travel and for other science projects, when it is important to be very precise.

On a 24-hour clock, the first two digits tell the hour. For example, 1100 is 11:00 A.M. Add 100 to find each hour that follows. So 1200 is 12:00 noon, 1300 is 1:00 P.M., and so on to 2400, or 12:00 midnight.

The second two digits on a 24-hour clock tell the minutes after the hour. So 0130 is 1:30 A.M. and 1745 is 5:45 P.M.

What if you flew to the moon? Lift-off is at 1530. What time would you tell your friends to watch you take off if they have a 12-hour clock only?

Think: Subtract. 1530 − 1200 = 330 1530 is after noon, so it is P.M.

Lift-off time is at 3:30 P.M.

ACTIVITIES

1. Make a set of flash cards. Show a time in the 12-hour system on one side and the same time in the 24-hour system on the other. Work with a partner to practice reading the time both ways.

2. Write a schedule of your day using the 24-hour system. List everything you do from the time you get up until the time you go to bed.

Computer Applications: Scheduling

Your teacher wants you to create a two-month calendar for May and June. She also wants you to enter certain activities for those months. You can use a computer program to create a calendar for two months. You can also use it to enter activities on the calendar.

			MAY			
SUNDAY	MONDAY	TUESDAY	WEDNESDAY	THURSDAY	FRIDAY	SATURDAY
		1	2	3	4	5
6	7	8	9	10	11	12
13	14	15	16	17	18	19
20	21	22	23	24	25	26
27	28	29	30	31		

DATA

Use paper, pencil, and a ruler to draw a calendar for the months of May and June. Let May 1 begin on a Tuesday. Then fill in the calendar with the following activities.

Activities

- Class trip to the park: May 10
- French Club meetings: Every Tuesday
- See the school play: Third Wednesday in June
- Math tests: May 11, 25; June 8, 22
- School holiday: May 28
- Last day of school: June 30

THINKING ABOUT COMPUTERS

1. On what day of the week is your class trip?

2. How many days are there between the first two math tests?

3. **What if** the French Club meetings were moved to Thursdays? How could this affect your schedule?

4. Why would a computer be more useful than paper and pencil if you needed to update your calendar often?

5. **What if** you want to create monthly schedules for the entire year? How can a computer help you?

Measuring Time, Capacity, and Mass **163**

EXTRA PRACTICE

Estimating Time, page 135..

Choose the most reasonable unit.

1. see a play (hours or days)

2. learn to play the piano (minutes or months)

3. read a book (minutes or days)

4. graduate from college (months or years)

Telling Time, page 137..

Write the time using numbers and A.M. or P.M. Then write
the time in words.

1. Jim does his homework at:

2. Zoe eats breakfast at:

7:25

Elapsed Time, page 139..

What time will it be:

1. in 15 minutes?

2. in 1 hour and 25
minutes?

6:05

3. in 2 hours and 46
minutes?

4. 15 minutes after 11:35 A.M.?

5. 1 hour and 6 minutes before
3:00 P.M.?

How much time passes between:

6. half past six and 7:08?

7. noon and midnight?

EXTRA PRACTICE

Problem Solving: Identifying Extra Information, page 141 ....

Solve. Then list the extra information.

1. The fourth and fifth grades made floats for the parade. The fourth grade has 48 students and the fifth grade has 54. There were 86 students at the parade. How many students worked on floats?

2. Lia needs 8 yards of fabric for her costume. Andy needs 12 yards and Mia needs 3 yards. How many yards of fabric will Andy and Mia need?

3. The Adams County Courthouse has 4 flags flying. The LaPlata County Courthouse has 16 flags and the Fremont County Courthouse has 3 banners. How many flags are flying?

4. Nan and Pete are having a cookout. They have 20 hamburgers and 68 hot dogs. They bought 40 plates. How many hamburgers and hot dogs do they have?

Ordered Pairs, page 143 ....

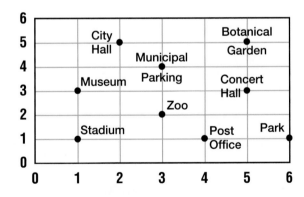

Name the ordered pair for each location.

1. City Hall   2. Park   3. Zoo   4. Museum

What is located at each ordered pair?

5. (1, 1)   6. (3, 4)   7. (4, 1)   8. (5, 3)

EXTRA PRACTICE

Making Line Graphs, page 145 ....

1. Complete the line graph using the data in the table. Then answer Questions 2–4.

People Getting on the City Bus							
Time	8–9	9–10	10–11	11–12	12–1	1–2	2–3
People	52	75	68	24	17	29	59

2. Which hours had the most people getting on the bus?

3. Which had the fewest?

4. How would you describe the changes as the day went on?

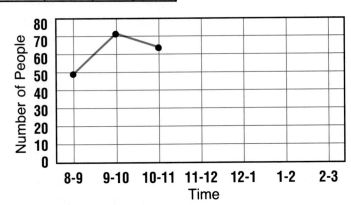

Capacity, page 149 ....

Choose the best estimate of the capacity of the containers.

1. thermos   1 mL or 1 L

2. coffee cup   150 mL or 150 L

3. tablespoon   7 mL or 7 L

4. baby's bottle   50 mL or 50 L

5. watering can   2 mL or 2 L

Mass, page 151 ....

Choose the best estimate of the mass of the object.

1. kitten   **a.** 35 g   **b.** 3 g   **c.** 3 kg

2. box of cereal   **a.** 380 g   **b.** 10 g   **c.** 630 kg

3. man   **a.** 180 kg   **b.** 100 g   **c.** 80 kg

4. moving van   **a.** 100 kg   **b.** 3,000 kg   **c.** 300 g

Problem Solving Strategy: Solving a Two-Step Problem, page 153...........

Use a two-step plan to solve the problem.

1. There are 89 students and 4 teachers going to the museum. One bus leaves school with 48 passengers. How many passengers are on the other bus?

2. In the museum shop, Ginger buys 12 Old West keychains. She returns 6 Old West keychains and buys 5 leather keychains instead. How many keychains did she buy?

3. The Art Museum has a collection of 18 clay pots. Another 11 clay pots are donated to the museum. The Indian Museum has 35 clay pots. How many more clay pots does the Indian Museum have?

4. Mr. Gilmore has 14 Navajo rugs. He buys 11 more. He donates 15 of them to the museum. How many rugs does he have left?

Temperature, page 157 ....

Choose the most reasonable temperature.

1. cross-country skiing   **a.** $^-5°C$ **b.** 10°C   **c.** 45°C   **d.** 100°C

2. sunbathing   **a.** $^-15°C$ **b.** 10°C   **c.** 40°C   **d.** 80°C

3. mowing the lawn   **a.** 30°C   **b.** 3°C   **c.** $^-3°C$ **d.** 80°C

4. sled riding   **a.** $^-8°C$ **b.** 8°C   **c.** 18°C   **d.** 80°C

Choose the object closest to the given temperature.

5. 1°C   **a.** hot tea   **b.** icicles   **c.** glass of milk

6. 10°C   **a.** cold water   **b.** room temperature   **c.** hot soup

7. 100°C   **a.** lemonade   **b.** room temperature   **c.** hot tea

PRACTICE *PLUS*

KEY SKILL: Elapsed Time (Use after page 139.)

Level A

Give the time:

1. in 5 minutes.

2. in 10 minutes.

3. in 15 minutes.

4. 5 minutes ago.

5. 1 hour ago.

6. 30 minutes ago.

Level B

Give the time:

7. in 20 minutes.

8. in 1 hour 45 minutes.

9. in 1 hour 38 minutes.

10. 11 minutes ago.

11. 1 hour 45 minutes ago.

12. 33 minutes ago.

What time will it be:

13. 15 minutes after 9:50 A.M.?

14. 1 hour 14 minutes before 6:00 P.M.?

Level C

What time will it be:

15. 36 minutes after 8:50?

16. 42 minutes before 12:20 P.M.?

How much time passes between:

17. seven thirty-one and 8:09?

18. 12:09 and half past one?

KEY SKILL: Mass (Use after page 151.)

Level A . . . .

Choose the best estimate of the mass of the object.

1. bird   **a.** 3 g   **b.** 4 kg   **c.** 50 g

2. girl   **a.** 300 g   **b.** 30 kg   **c.** 10 g

Choose the correct unit.

3. Janice's book has a mass of 50 (g or kg).

4. Matthew's basketball has a mass of 2 (g or kg).

Level B . . . .

Choose the best estimate of the mass of the object.

5. truck   **a.** 1,800 g   **b.** 180 kg   **c.** 1,800 kg

6. watermelon **a.** 2 g   **b.** 2 kg   **c.** 20 kg

Match the object with a possible mass.

7. man   **a.** 11 kg

8. 50 nickels **b.** 75 kg

9. bicycle   **c.** 1 kg

Level C . . . .

Choose the best estimate of the mass of the object.

10. airplane   **a.** 6,000 kg   **b.** 400 kg   **c.** 4,000 g

11. computer   **a.** 55 kg   **b.** 15 g   **c.** 1 kg

12. medicine bottle **a.** 79 kg   **b.** 7 g   **c.** 150 g

Match the unit with a possible mass.

13. motorcycle   **a.** 70 kg

14. air conditioner **b.** 190 kg

15. desk   **c.**   9 kg

CHAPTER REVIEW

LANGUAGE AND MATHEMATICS

Complete the sentences. Use the words in the chart.

1. There are 12 ■ in the year. *(page 134)*

2. ■ is the time between noon and midnight. *(page 136)*

3. (3, 5) is called an ordered ■. *(page 142)*

4. The ■ of a container tells how much it holds. *(page 146)*

5. ***Write a definition*** or give an example of the words you did not use from the chart.

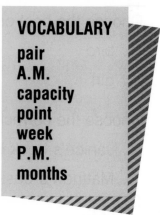

VOCABULARY
pair
A.M.
capacity
point
week
P.M.
months

CONCEPTS AND SKILLS

Choose the more reasonable estimate for the activity. *(page 134)*

6. swim 5 laps: minutes or hours

7. fly cross-country: days or hours

8. listen to a cassette tape

    a. 45 seconds   **b.** 45 minutes

9. bake cookies

    a. 1 hour   **b.** 1 minute

Write the time, using numbers and A.M. or P.M. *(page 136)*

10. Karen goes to the library at

11. Dina eats lunch at

Find the start time, end time, or elapsed time. *(page 138)*

12. Start time: 7:21 A.M.   End time: 8:36 A.M.   Elapsed time: ■

13. Start time: 11:47 A.M.   End time: ■   Elapsed time: 5 h, 15 min

14. Start time: ■   End time: 3:07 A.M.   Elapsed time: 3 h, 3 min

Use the graph to find the ordered pair, or name the point. *(page 142)*

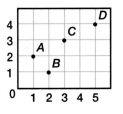

15. *A*   **16.** (2, 1)

17. (3, 3)   **18.** *D*

19. Use the data in the table to make a line graph. *(page 144)*

NUMBER OF BIRDS SEEN DURING MIGRATION FOR A WEEK

Sunday	Monday	Tuesday	Wednesday	Thursday	Friday	Saturday
12	20	38	41	59	51	40

Choose the better estimate. *(pages 148–151, 156)*

20. pitcher of cream: 1 L or 1 mL

21. cup of cider: 250 L or 250 mL

22. jar of jam: 120 g or 12 kg

23. bag of potatoes: 5 g or 5 kg

24. temperature inside a refrigerator:

    a. 5°C   **b.** 25°C

25. hiking on a hot day:

    a. 3°C   **b.** 32°C

CRITICAL THINKING

26. How much time passes between 11:43 A.M. and 12:17 P.M.? Tell how you found your answer. *(page 138)*

27. Which has more mass: a kilogram of feathers or a kilogram of lead? How do you know? *(page 150)*

MIXED APPLICATIONS

Solve. If there is extra information, list it.

28. Jon went to the planetarium at 11:45 A.M. Pat arrived at 11:20 A.M. Who got to the planetarium first? *(page 136)*

29. Mr. Hou bought a cup of cocoa for $.85 and a bagel for $1.29. How much change does he get from $5.00? *(page 152)*

30. Halloween is on October 31. Thanksgiving is on November 23. New Year's Day is on January 1. How many days are between Halloween and New Year's Day? *(page 140)*

CHAPTER TEST

Choose the most reasonable unit.

1. read a book

    seconds   hours   years

2. length of vacation

    minutes   days   years

What time will it be:

3. ten after six at night?

4. 15 minutes before 11:50 P.M.?

Choose the best estimate.

5. paint in a can: 4 L or 4 mL

6. water in a sink: 6 L or 6 mL

7. canary: 85 g, 850 g, or 8 kg

8. van: 190 g; 190 kg; or 1,900 kg

Choose the most reasonable temperature.

9. cup of hot cocoa   **a.** 20°C   **b.** 85°C   **c.** 150°C

10. skiing on a cold day   **a.** ⁻10°C   **b.** 20°C   **c.** 50°C

Use this grid to name or locate the ordered pair.

11. B   **12.** E   **13.** F

14. (1, 3)   **15.** (5, 5)   **16.** (6, 1)

Use the graph for Questions 17 and 18.

17. During which month were the most magazines sold?

18. Dan sold 70 magazines in the 5th month. How would you show this on the graph?

Solve.

19. There are 238 students at lunch and 174 students in class. If there are 510 students in the school, how many are not at lunch or in class?

20. On the field trip, there were 3 vans with 9 people in each van. There were 18 people who did not go. How many people went on the field trip?

TIME ZONES

The world is divided into different time zones.
This map shows the six time zones in the United States.
The clocks show you what time it is in each time zone
when it is 12 noon in the Pacific Time Zone.

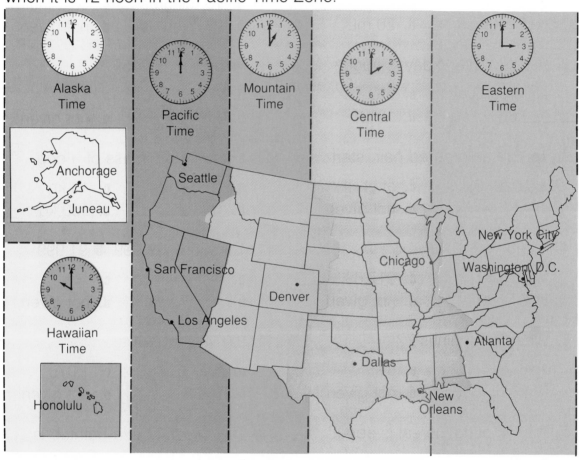

1. You are on the East Coast and telephone the West Coast.
What is the time difference?

2. How much earlier or later is the time in these cities than in Seattle?
   a. Denver   **b.** Dallas   **c.** Anchorage   **d.** Atlanta

3. When it is 7:00 A.M. in the Eastern Time Zone, what time
is it in these time zones?
   a. Central   **b.** Mountain   **c.** Pacific   **d.** Hawaiian

4. It is 10:30 P.M. in Denver. What time is it in these cities?
   a. San Francisco   **b.** Chicago   **c.** New York   **d.** Honolulu

CUMULATIVE REVIEW

Choose the letter of the correct answer.

1. Estimate the amount of water in a small spoon.

   a. 2 L   c. 20 L
   b. 2 mL   d. 20 mL

2. A cold winter's day is about:

   a. ⁻100°C.   c. 6°C.
   b. ⁻6°C.   d. 10°C.

3. To plot an ordered pair, start

   a. at (0, 0)   c. at (1, 1)
   b. at (0, 1)   d. not given

4. $4.98 + $12.42 + $.29

   a. $16.69   c. $17.69
   b. $17.59   d. not given

5. What is 16 minus 7?

   a. 6   c. 8
   b. 7   d. not given

6. A bowl of hot cereal is about:

   a. 10°C.   c. 80°C.
   b. 40°C.   d. 100°C.

7. Round 542 to the nearest hundred.

   a. 500   c. 550
   b. 540   d. 600

8. How much time passes between 7:55 and 8:23?

   a. 18 minutes   c. 33 minutes
   b. 28 minutes   d. not given

9. What is 7 plus 5?

   a. 2   c. 12
   b. 11   d. not given

10. Estimate the mass of a dog.

    a. 10 g   c. 10 kg
    b. 100 g   d. 100 kg

11. Compare: 99,999 ● 91,999

    a. <   c. =
    b. >   d. not given

12. 3,974 − 1,999

    a. 5,973   c. 1,075
    b. 1,975   d. not given

13. The capacity of a bathtub is:

    a. 10 L.   c. 100 L.
    b. 10 mL.   d. 100 mL.

14. What time is 8 minutes before 1 P.M.?

    a. 12:52 A.M.   c. 1:08 A.M.
    b. 12:52 P.M.   d. not given

Multiplication and Division Facts

MATH CONNECTIONS: AREA
• GRAPHING • PROBLEM SOLVING

1. What is happening in this picture?
2. What types of groups do you see?
3. How many of each group are there?
4. Write a problem about this picture.

DEVELOPING A CONCEPT

The Meaning of Multiplication

There are 5 groups of dogs at the pet show. There are 3 dogs in each group. How many dogs are there?

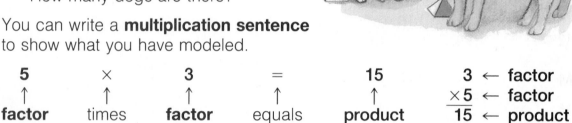

WORKING TOGETHER

Use counters to model the problem.

1. Write an addition sentence to show what you have modeled.

2. How many counters are there in all? How many dogs are there?

You can write a **multiplication sentence** to show what you have modeled.

5	×	3	=	15
↑	↑	↑	↑	↑
factor	times	**factor**	equals	**product**

$$3 \leftarrow \text{factor}$$
$$\underline{\times 5} \leftarrow \text{factor}$$
$$15 \leftarrow \text{product}$$

Compare your two number sentences to your model.

3. Which factor tells the number of addends? Which factor tells the number of groups?

4. Which factor names the addend? Which factor tells how many in each group?

5. What does the sum or the product tell you?

When a multiplication fact contains two factors that are the same, it is called a **square.**

6. **What if** there are 3 groups of 3 dogs? How many dogs are there? Use counters to model the problem. Why do you think 3 × 3 is called a square?

7. **What if** one group has 7 dogs, another has 3 dogs, and another has 4 dogs? How many dogs will there be? Can you write a multiplication sentence? Why or why not?

8. When can you multiply to find how many in all? When must you add?

PRACTICE

Complete the multiplication sentence.

9.  **10.**  **11.**

$4 \times 3 = \blacksquare$ $3 \times \blacksquare = \blacksquare$ $\blacksquare \times \blacksquare = \blacksquare$

Write a multiplication sentence.

12. $5 + 5 + 5 + 5$ **13.** $9 + 9$ **14.** $8 + 8 + 8 + 8 + 8$

Multiply. Use counters if needed.

15. 3×6 **16.** 8×3 **17.** 2×5 **18.** 3×9

19. 4×3 **20.** 6×8 **21.** 8×6 **22.** 3×4

23. 9×4 **24.** 5×8 **25.** 5×6 **26.** 5×1

27. $\begin{array}{r} 7 \\ \times 6 \\ \hline \end{array}$ **28.** $\begin{array}{r} 9 \\ \times 9 \\ \hline \end{array}$ **29.** $\begin{array}{r} 4 \\ \times 2 \\ \hline \end{array}$ **30.** $\begin{array}{r} 5 \\ \times 9 \\ \hline \end{array}$ **31.** $\begin{array}{r} 2 \\ \times 7 \\ \hline \end{array}$ **32.** $\begin{array}{r} 2 \\ \times 6 \\ \hline \end{array}$

33. How many are in 2 groups of 8?   **34.** Multiply 3 by 2.

35. What is 4 times 6?   **36.** Find the product of 1×8.

Mixed Applications

37. The pet show winners get 7 blue, 7 red, and 7 gold ribbons. How many ribbons is this?

38. There are 80 pets in the show. Only 21 will win a prize. How many pets will not win a prize?

39. A group picture is taken of the top 2 winners in each of the 7 showings. How many winners are in the picture?

40. *Write a problem* that can be solved using multiplication. Solve your problem. Ask others to solve it.

DEVELOPING A CONCEPT

The Meaning of Division

There are 24 horses grazing on the range. The horses are standing in groups of 3. How many groups of horses are there?

WORKING TOGETHER

Use counters to model the problem Start with 24 counters and take away groups of 3.

1. How many times can you take away 3 from 24? How many groups of horses are there?

You can write a division sentence to show what you have modeled.

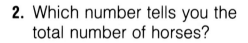

24	÷	3	=	8
↑	↑	↑	↑	↑
dividend	divided by	**divisor**	equals	**quotient**

$$\overset{8}{3\overline{)24}}$$ ← **quotient**
← **dividend**
divisor

2. Which number tells you the total number of horses?

3. Which number tells you the number in each group?

4. Which number tells you the number of equal groups?

The 24 horses regrouped into 3 equal groups. How many horses were in each group?

Use counters to model this problem. Start with 24 counters. Put them in 3 groups, one by one, until you run out of counters.

5. How many counters are in each group? How many horses are in each group?

6. Write a division sentence to show what you modeled.

7. Find other groups you can make with 24 counters. Record your results in a table like this.

Number in All	÷	Number in Each Group	=	Number of Equal Groups
24	÷	3	=	8
24	÷	8	=	■

8. Find other ways to separate 24 counters into equal groups. Record them in a table like this.

Number in All	÷	Number of Equal Groups	=	Number in Each Group
24	÷	3	=	■
24	÷	8	=	■

SHARING IDEAS

9. Compare the division sentences in each table. What do you notice?

10. What two questions does a division sentence answer?

PRACTICE

Complete the division sentence.

11.

$10 \div ■ = 2$

12.

$■ \div ■ = 4$

13.

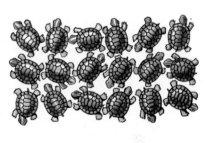

$■ \div ■ = ■$

Divide. Use counters if needed.

14. $16 \div 8$ **15.** $8 \div 2$ **16.** $18 \div 6$ **17.** $15 \div 5$

18. $9\overline{)27}$ **19.** $6\overline{)36}$ **20.** $5\overline{)30}$ **21.** $4\overline{)32}$ **22.** $3\overline{)18}$

23. How many groups of 5 are in 20?

24. What is 27 divided by 3?

Mixed Applications

25. Alexis has 16 horses lined up in groups of 4. How many groups do the horses make?

26. A magician made 2 groups of 6 pigeons disappear. How many pigeons disappeared?

27. There are 21 monkeys living in equal groups in 7 trees. How many monkeys live in each tree?

28. *Write a problem* that can be solved using division. Have others solve it.

DEVELOPING A CONCEPT

Multiplication and Division Properties

A. You can use counters and diagrams to explore the properties of multiplication.

Here are two ways to show a group of 6 counters.

← 3 groups of 2   ← 2 groups of 3

3×2 = 2×3 = 6

1. What do you notice about the products?

2. What if one of the factors is 1? What do you notice about the product?

$3 \times 1 = 3$ $1 \times 3 = 3$

3. What if one of the factors is 0? What do you notice about the product?

$3 \times 0 = 0$ You cannot picture 0 groups of 3. $0 \times 3 = 0$

4. What if you multiply 3 numbers? What happens to their product if you group them in different ways?

$(3 \times 2) \times 4 = 6 \times 4 = 24$
$3 \times (2 \times 4) = 3 \times 8 = 24$

B. You can also explore the properties of division.

5. What is the quotient when you divide a number by 1?

← 1 group of 4
$4 \div 1 = 4$ *Think:* $1 \times 4 = 4$

6. What is the quotient when you divide a number by itself?

← 4 groups of 1
$4 \div 4 = 1$ *Think:* $4 \times 1 = 4$

7. What if you divide zero by a number? What is the quotient?

← 4 groups of 0
$0 \div 4 = 0$ *Think:* $4 \times 0 = 0$

8. What if you divide a number by 0? What do you notice?

You cannot show 0 groups of 4.
$4 \div 0 = $ ■   *Think:* $0 \times $ ■ $= 4$
No number works.

9. Use counters or diagrams to model each of the multiplication and division examples using different numbers. What do you notice?

SHARING IDEAS

With which statements do you agree? disagree? If you disagree, tell why.

10. If you multiply a number by 1, the product is 1.

11. When you divide a number by itself, the quotient is 1.

12. Changing the way factors are grouped changes the product.

13. Any number divided by 0 equals 0.

14. When you divide a number by 1, the quotient is 1.

15. When 0 is divided by a number, the quotient is the number.

16. If you change the order of the factors, the product remains the same.

PRACTICE

Find the missing number.

17. $2 \times 0 = \blacksquare$ **18.** $1 \times 4 = \blacksquare$ **19.** $0 \div 8 = \blacksquare$ **20.** $1 \div 1 = \blacksquare$

21. $8 \times 1 = \blacksquare$ **22.** $5 \div 1 = \blacksquare$ **23.** $\blacksquare \times 2 = 0$ **24.** $6 \div 0 = \blacksquare$

25. $7 \div \blacksquare = 1$ **26.** $\blacksquare \times 2 = 2$ **27.** $3 \times 5 = \blacksquare \times 3$

28. $(2 \times 3) \times 6 = \blacksquare \times (3 \times 6)$ **29.** $(4 \times \blacksquare) \times 2 = 4 \times (2 \times 2)$

Find the rule.

30. Multiply by $\blacksquare$.

0	1	2	3	4	5	6	7	8	9
0	1	2	3	4	5	6	7	8	9

31. Divide by $\blacksquare$.

1	2	3	4	5	6	7	8	9
1	1	1	1	1	1	1	1	1

Mixed Applications

32. Len takes 10 pictures of deer and 2 pictures of elk. How many more pictures of deer does he take?

33. There are 3 tigers. Each tiger has 3 cubs. How many tiger cubs are there?

DEVELOPING A CONCEPT
Fact Families

You can use number cubes to explore fact families. Suppose that you roll two number cubes and get a 3 and a 5.

You can write related multiplication sentences.

Factor	×	Factor	=	Product
3	×	5	=	15
5	×	3	=	15

You can then write related division sentences.

Dividend	÷	Divisor	=	Quotient
15	÷	3	=	5
15	÷	5	=	3

WORKING TOGETHER

Copy the charts above.

1. The number sentences in both charts belong to the same **fact family.** Tell how they are related.

2. Use two number cubes and repeat this activity four more times. See how many related sentences you can write. Record them on your charts.

SHARING IDEAS

3. Did you write the same number of related sentences for every fact family? Why or why not?

4. Compare your sentences with those of other students. How are they alike? How are they different?

5. How do the multiplication and division properties help you write fact families?

6. How does the relationship between multiplication and division help you write fact families?

PRACTICE

Find the product or quotient. Write the related
number sentences that make up the fact family.

7. 3)21   **8.** 1)6   **9.** 4 × 5   **10.** 2)4   **11.** 2 × 8

12. 9 ÷ 3   **13.** 5 × 5   **14.** 5 × 3   **15.** 18 ÷ 9   **16.** 3 ÷ 3

Write the fact family for the set of numbers.

17. 3, 4, 12   **18.** 4, 1, 4   **19.** 2, 4, 8   **20.** 8, 8, 64   **21.** 2, 5, 10

Complete the fact. Find the missing number.

22. 7 × ■ = 28   **23.** 27 ÷ ■ = 9   **24.** ■ × 4 = 16   **25.** 5 ÷ ■ = 1

26. ■ × 8 = 40   **27.** ■ ÷ 3 = 6   **28.** 6 × ■ = 48   **29.** ■ ÷ 6 = 0

Mixed Applications

30. Lisa walks dogs in groups of 4. She walks 5 groups a day. How many dogs does she walk a day?

31. Mark has 15 tropical fish. He puts 5 in each fishbowl. How many fishbowls does he use?

32. Nona bought a book about the care of gerbils. It cost $4.79. She paid with a $10 bill. What was her change?

33. Harry started riding his horse at 3:30 P.M. He rode for 1 hour 30 minutes. At what time did he finish riding?

Mixed Review

Find the answer. Which method did you use?

34. 1   **35.**   247   **36.** 4)32   **37.**   126   **38.** $768.32
     × 2   − 168   − 10   + 205.19

MENTAL MATH
CALCULATOR
PAPER/PENCIL

UNDERSTANDING A CONCEPT

2 Through 9 as Factors

COLUMNS

×	0	1	2	3	4	5	6	7	8	9
0	0	0	0	0	0	0	0			
1	0	1	2	3	4	5				
2	0	2	4			10				
3	0	3	6			15				
4			8			20				
5										
6										
7										
8										
9										

ROWS

Mario and Jill are making a table so that they will be sure not to skip any multiplication facts. When the table is complete, they will use it for a fact race.

1. Make a table like this for yourself.

Mario says that it is easy to fill in the first two rows and columns.

2. What does Mario know about multiplying by 0 and 1 that makes this easy?

Jill uses skip-counting to fill in column 2 and column 5.

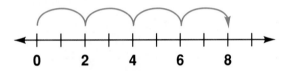

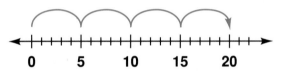

3. Why can Jill fill in row 2 and row 5 the same way?

Jill uses a table to remember the squares.

Rule: Multiply the Number by Itself.

1	2	3	4	5	6	7	8	9
1	4	9	■	■	■	■	■	■

4. Use the rule to complete Jill's table.

Think: 2 × 2

Mario says that he can use squares to help him multiply mentally.

$3 \times 3 = 9$, so $4 \times 3 = 9 + 3 = 12$ and $2 \times 3 = 9 - 3 = 6$.

5. What other facts can you find from $4 \times 4 = 16$?

6. Complete the multiplication table. You can use counters to check your work.

TRY OUT   Multiply.

7. 8×4

8. 3×9

9. 0×6

10. 9×1

11. 6×2

12. 5×5

13. 2×8

14. 7×7

PRACTICE

Multiply.

15. 2 × 5	**16.** 1 × 1	**17.** 5 × 0	**18.** 4 × 3	**19.** 4 × 7	**20.** 3 × 9
21. 9 × 5	**22.** 2 × 4	**23.** 3 × 5	**24.** 6 × 2	**25.** 8 × 7	**26.** 3 × 6
27. 7 × 8	**28.** 1 × 7	**29.** 6 × 6	**30.** 7 × 3	**31.** 0 × 9	**32.** 7 × 9

33. 0 × 4 **34.** 9 × 2 **35.** 4 × 8 **36.** 8 × 5 **37.** 7 × 0

38. 8 × 2 **39.** 8 × 8 **40.** 1 × 9 **41.** 9 × 5 **42.** 9 × 8

43. 7 × 5 **44.** 6 × 9 **45.** 8 × 3 **46.** 9 × 9 **47.** 6 × 7

48. Multiply 5 by 7.   **49.** What is 6 times 4?

50. Find the product of 0 × 3.   **51.** How many are in 9 groups of 3?

Give the rule. Complete the table.

52.

0	1	2	3	4	5	6	7	8	9
0	4	8	■	■	■	■	■	■	■

53.

0	1	2	3	4	5	6	7	8	9
0	7	14	■	■	■	■	■	■	■

Mixed Applications

54. There are 48 swans living on the lake. If there are 6 swans in each family, how many families of swans live on the lake?

55. Ramon has 4 sets of books about animals. There are 2 books in each set. How many books are there in all?

UNDERSTANDING A CONCEPT

Dividing by 2 through 9

Joy and Rick are helping each other learn division facts. They use all the models, properties, and relationships they have learned.

A. To find the quotient 18 ÷ 3, Joy draws a picture. She makes rows of 3 until she has 18 items.

1. What is the quotient?

2. What is the related division fact?

To find the quotient 20 ÷ 4, Rick draws circles. He makes groups of 4 until he has 20 items.

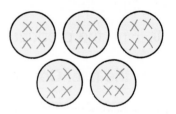

3. What is the quotient?

4. What is the related division fact?

B. Joy uses a property to find the quotient 9 ÷ 9.

5. Which property does Joy use? What is the quotient?

6. Write the related division fact.

To find the quotient 45 ÷ 5, Rick thinks of the related multiplication sentence.

7. What is the related multiplication sentence? What is the quotient?

8. What is the related division fact?

TRY OUT   Divide.

9. 3)‾24‾   **10.** 5)‾35‾   **11.** 6 ÷ 3   **12.** 36 ÷ 9

PRACTICE

Divide.

13. $2\overline{)16}$ **14.** $7\overline{)14}$ **15.** $9\overline{)27}$ **16.** $5\overline{)25}$ **17.** $8\overline{)0}$ **18.** $8\overline{)24}$

19. $5\overline{)30}$ **20.** $8\overline{)40}$ **21.** $6\overline{)42}$ **22.** $9\overline{)45}$ **23.** $4\overline{)20}$ **24.** $7\overline{)28}$

25. $7\overline{)49}$ **26.** $8\overline{)72}$ **27.** $2\overline{)18}$ **28.** $1\overline{)9}$ **29.** $6\overline{)30}$ **30.** $4\overline{)36}$

31. $4\overline{)28}$ **32.** $9\overline{)81}$ **33.** $6\overline{)24}$ **34.** $8\overline{)8}$ **35.** $6\overline{)54}$ **36.** $7\overline{)56}$

37. $0 \div 5$ **38.** $6 \div 2$ **39.** $32 \div 8$ **40.** $12 \div 6$ **41.** $42 \div 7$

42. $36 \div 6$ **43.** $21 \div 3$ **44.** $72 \div 9$ **45.** $40 \div 5$ **46.** $0 \div 1$

47. $32 \div 4$ **48.** $7 \div 1$ **49.** $24 \div 4$ **50.** $2 \div 2$ **51.** $64 \div 8$

52. $14 \div 2$ **53.** $35 \div 7$ **54.** $63 \div 7$ **55.** $54 \div 9$ **56.** $48 \div 8$

57. Divide 8 by 1.　　　　　　　**58.** How many 6s are in 42?

59. Find the quotient of $56 \div 7$.　　**60.** What is 0 divided by 9?

Use the rule to complete.

61. Rule: Divide by 5.

0	5	10	15	20	25	30	35	40	45
0	1	2	■	■	■	■	■	■	■

62. Rule: Divide by 8.

0	8	16	24	32	40	48	56	64	72
0	1	2	■	■	■	■	■	■	■

Mixed Applications

63. Joy keeps her 36 glass animals on 4 shelves. She puts the same number of animals on each shelf. How many animals are on each shelf?

64. Rick is going to buy 5 sets of animal posters. There are 6 posters in each set. How many posters will Rick buy?

65. *Write a problem* involving multiplication or division. Solve it. Then ask others to solve it.

EXTRA Practice, page 209; Practice Plus, page 212

UNDERSTAND
PLAN
TRY
CHECK
EXTEND

PROBLEM SOLVING

Strategy: Choosing the Operation

The New England Aquarium has 7 sea turtles in each of 6 tanks. How many sea turtles does the Aquarium have? You can use the five-step process to solve this problem.

UNDERSTAND	
What do I know?	I know there are 6 tanks. There are 7 sea turtles in each tank.
What do I need to find out?	I need to find out how many sea turtles there are in all.
PLAN	
What can I do?	Since each tank has the same number of turtles, I can multiply the number of tanks by the number of sea turtles in each tank.
TRY	
Let me try my plan.	Number of turtles in each tank ↓ $6 \times 7 = 42$ ↑ Number of tanks So there are 42 sea turtles at the New England Aquarium.
CHECK	
Have I answered the question?	Yes. There are 42 sea turtles at the New England Aquarium.
Does my answer make sense?	Yes. I get the same answer if I add. $7 + 7 + 7 + 7 + 7 + 7 = 42$
EXTEND	
What have I learned?	I have learned that I can use multiplication to solve problems that combine groups of the same size. I can check my answer by adding.

PRACTICE

Decide which operation to use. Then solve the problem.
Use mental math, a calculator, or paper and pencil.

1. A farmer has 63 rabbits. During the next month 9 rabbits are born. How many rabbits does the farmer have now?

2. Susan feeds her pet lizard 2 times a day. How many times does she feed her lizard in 1 week?

3. Beth buys 72 vitamins for her guinea pig. She puts them in 8 packets, each with the same number of vitamins. How many vitamins does she put in each packet?

4. Ramón's dog weighs 56 pounds. Tamar's dog weighs 8 pounds less. How many pounds does Tamar's dog weigh?

Strategies and Skills Review

5. Angie and Wayne had a turtle race. After one minute Angie's turtle had gone a distance of 36 inches. This was 4 times the distance covered by Wayne's turtle. How far had Wayne's turtle gone?

6. There are 2 fish tanks in the pet store. One tank has 349 fish swimming in it, and the other has 307 fish in it. Does the pet store have at least 600 fish? How can you tell?

7. Tonio's parrot knows 15 words. He knows 8 one-syllable words and 7 two-syllable words. How many syllables does Tonio's parrot know?

8. ***Write a problem*** using one of the four operations. Solve your problem. Ask others to solve it.

Experimenting and Predicting

A. Here's a game you can play to test your luck and skill. It's called STRIKE. You will need two number cubes. Each cube should be numbered 1–6. You will also need a sheet of paper to keep track of your score.

Take turns. On your turn, roll the cubes and add the numbers shown on the cubes to your score. Then decide if you want to roll again. You can keep rolling but watch out for STRIKE and STRIKE OUT!

If **one** of the cubes shows a 1, you have rolled STRIKE. You lose all of the points you scored that turn. Your turn is over as well.

Roll		Score
6 3		9
5 2		16
3 1		0
	↑ STRIKE	

If **both** of the cubes show a 1, you have rolled STRIKE OUT. You lose all of the points you have so far. It is also the end of your turn.

The first player to get 100 points wins the game!

Play several games of STRIKE with another student. Think about the best way to play.

B. How do you decide when to stop rolling the cubes?

1. Suppose you have already rolled the cubes 5 times and have not rolled STRIKE or STRIKE OUT. Will you roll again?

You can change the game by adding new rules. Try some of these:

2. BIG STRIKE: All points are doubled. How does this change the way you play?

3. DOUBLE-TROUBLE STRIKE: If you roll double sixes, you lose 20 points. You do not lose your turn.

4. Invent your own STRIKE variations and try them out. Discuss how the new rules change your method of play.

Missing Factors

Martin and Peggy are buying dog food. The cans are packed into 6-packs, and there are 48 cans in a large case. How many 6-packs are there in a large case?

You need to find the missing factor: ■ × 6 = 48.

Martin and Peggy use related facts.

Martin thinks:
I know that 48 ÷ 6 = 8, so
8 × 6 = 48.

Peggy thinks:
I know that 6 × 8 = 48, so
8 × 6 = 48.

There are eight 6-packs of cans in a large case.

1. Use a related fact to find the missing factor in ■ × 7 = 63.

2. Find each missing factor. Tell how you know.

    a. 3 × ■ = 3   **b.** 6 × ■ = 0   **c.** 1 × ■ = 7

TRY OUT Write the letter of the correct answer. Find the missing factor.

3. ■ × 6 = 54   **a.** 6   **b.** 9   **c.** 48   **d.** 60

4. 7 × ■ = 35   **a.** 5   **b.** 7   **c.** 28   **d.** 42

5. ■ × 9 = 0   **a.** 10   **b.** 9   **c.** 1   **d.** 0

6. 8 × ■ = 64   **a.** 72   **b.** 9   **c.** 8   **d.** 7

PRACTICE

Find the missing factor.

7. ■ × 9 = 27   **8.** 8 × ■ = 32   **9.** 6 × ■ = 30   **10.** ■ × 6 = 42

11. ■ × 7 = 28   **12.** 1 × ■ = 5   **13.** 4 × ■ = 36   **14.** ■ × 5 = 40

15. 8 × ■ = 0   **16.** ■ × 6 = 24   **17.** 8 × ■ = 56   **18.** ■ × 6 = 54

19. 3 times what number is 15?   **20.** What number times 4 is 16?

Critical Thinking

Find the missing number. Tell how you found it.

21. 49 ÷ ■ = 7   **22.** 24 ÷ ■ = 3   **23.** 35 ÷ ■ = 7   **24.** 72 ÷ ■ = 8

Mixed Applications Solve. Which method did you use?

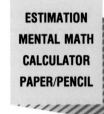

ESTIMATION
MENTAL MATH
CALCULATOR
PAPER/PENCIL

25. Emmy Lou buys a bag containing 652 g of bird food and a box containing 391 g of fish food. Does she buy more than 1 kg of pet food?

26. There are 36 bags of cat food on sale. There are 9 bags on each shelf. How many shelves have bags of cat food?

27. During a sale the pet store gets 2,142 customers. For the rest of the month, it gets 978 customers. How many more customers came during the sale?

28. The pet store gets a shipment of 6 boxes of cat toys. Each box contains 9 cat toys. The store sells 50 of the toys in a week. How many toys are left?

CHALLENGE

1. Pick two even numbers. Multiply them.

2. Pick two odd numbers. Multiply them.

3. Pick an even number and an odd number. Multiply them.

4. Repeat Steps 1–3 with different numbers.

5. Compare your products. What does this tell you about multiplying even and odd numbers?

UNDERSTANDING A CONCEPT

Other Fact Strategies

A. During summer vacation Vida and Vera take care of pets and plants for people who are away from home. They charge $7 a week. This week they have 8 customers. How much will they earn?

Vida and Vera use **squares** to find 7×8.

Vida thinks: $7 \times 7 = 49$.
I can count up from 7×7.
$7 \times 8 = (7 \times 7) + 7$, or 56

Vera thinks: $8 \times 8 = 64$.
I can count down from 8×8.
$7 \times 8 = (8 \times 8) - 8$, or 56

Vida and Vera will earn $56 this week.

1. Use counting up and then use counting down to find 7×5.

B. You can use multiplication facts you know to help you remember other facts.

Think of 4×7 as $(2 \times 7) + (2 \times 7)$.
Then add: $14 + 14 = 28$.

Think of 7×6 as $(7 \times 3) + (7 \times 3)$.
Then add: $21 + 21 = 42$.

2. Use this method to find 8×6.

C. One week Vida and Vera earned $42. How many customers did they have that week?

You need to find $42 \div 7$.

Think of finding the missing factor: $\blacksquare \times 7 = 42$.

$6 \times 7 = 42$, so $42 \div 7 = 6$.

They had 6 customers that week.

3. Use the missing factor to find $72 \div 9$.

TRY OUT Multiply or divide.

4. 8×6
5. 9×8
6. $63 \div 9$
7. $40 \div 5$

PRACTICE

Multiply.

8. 7
  ×5

9. 6
  ×9

10. 8
  ×6

11. 6
  ×7

12. 7
  ×7

13. 9
  ×8

14. 6 × 7   **15.** 8 × 6   **16.** 6 × 9   **17.** 7 × 5   **18.** 9 × 9

Divide.

19. 7)‾63‾   **20.** 6)‾54‾   **21.** 9)‾72‾   **22.** 5)‾45‾   **23.** 9)‾81‾   **24.** 7)‾56‾

25. 48 ÷ 6   **26.** 35 ÷ 7   **27.** 72 ÷ 8   **28.** 56 ÷ 8   **29.** 54 ÷ 9

Complete the table. Use the rule.

30. Rule: Multiply by 8.

4	5	6	7	8	9
■	■	■	■	■	■

31. Rule: Multiply by 9.

4	5	6	7	8	9
■	■	■	■	■	■

32. Rule: Divide by 6.

24	30	36	42	48	54
■	■	■	■	■	■

33. Rule: Divide by 7.

28	35	42	49	56	63
■	■	■	■	■	■

Mixed Applications

34. The Millers paid $7 each week for pet and plant service. How much did it cost them for 6 weeks?

35. One week Vida and Vera earned $63. If each customer paid $7, how many customers did they have that week?

36. Mrs. Garcia has 42 African violets and 7 ferns. How many plants does she have?

Mixed Review

Find the answer. Which method did you use?

37. 6 + 8 + 4 + 3   **38.** $704.23 − $68.44   **39.** 45 + 786

40. 5 × 8   **41.** 504 − 167   **42.** 12 ÷ 2

| MENTAL MATH |
| CALCULATOR |
| PAPER/PENCIL |

UNDERSTANDING A CONCEPT

Factors and Multiples

A. The **multiples** of a number are the products of that number and other numbers.

Complete the tables of multiples. Multiply or skip-count.

MULTIPLES OF 2

1.
×	0	1	2	3	4	5	6	7	8	9
	0	2	4	■	■	■	■	■	■	■

MULTIPLES OF 3

2.
×	0	1	2	3	4	5	6	7	8	9
	0	3	6	■	■	■	■	■	■	■

Multiples of 2 are called **even numbers.** All other counting numbers are called **odd numbers.**

If two or more numbers have the same number as a multiple, that number is called a **common multiple.** The number 6 is a common multiple of 2 and 3.

3. Name another common multiple of 2 and 3.

B. The **factors** of a number are the numbers that can be multiplied to produce that number.

The factors of 3 are 1 and 3. $3 \times 1 = 3$ and $1 \times 3 = 3$

The factors of 6 are 1, 2, 3, and 6. $6 \times 1 = 6$, $3 \times 2 = 6$,
$2 \times 3 = 6$, and $1 \times 6 = 6$

Two or more numbers can have some of the same factors. These factors are called **common factors.**

The common factors of 3 and 6 are 1 and 3.

4. What are the factors of 8 and 16? What are their common factors?

C. If a number has only two factors, itself and 1, it is called a **prime number.** If a number has more than those two factors, it is called a **composite number.**

5. Show that 5 is a prime number.

6. Show that 10 is a composite number.

Write the letter of the correct answer.

7. Which number is not a multiple of 8?
   a. 56   **b.** 18   **c.** 48   **d.** 8

8. The number 4 is a factor of both:
   a. 4 and 14. **b.** 4 and 10. **c.** 14 and 24. **d.** 8 and 12.

PRACTICE

Copy and complete to find the multiples.

	×	0	1	2	3	4	5	6	7	8	9
9.	Multiples of 6	■	■	■	■	■	■	■	■	■	■
10.	Multiples of 7	■	■	■	■	■	■	■	■	■	■
11.	Multiples of 8	■	■	■	■	■	■	■	■	■	■
12.	Multiples of 9	■	■	■	■	■	■	■	■	■	■

Write *even* or *odd.*

13. 4   **14.** 12   **15.** 7   **16.** 15   **17.** 24   **18.** 35

Name a common multiple of the pair of numbers.

19. 2 and 6   **20.** 4 and 5   **21.** 3 and 4   **22.** 3 and 9

Name the factors of each pair of numbers. What are the common factors?

23. 7, 14   **24.** 8, 24   **25.** 11, 15   **26.** 21, 35   **27.** 18, 27

28. Which numbers in Exercises 23–27 are prime? Which are composite?

Critical Thinking

29. What number is a multiple of all other numbers? a factor of all other numbers?

Mixed Applications

Solve. You may need to use the Databank on page 519.

30. How many times faster than a chicken can a human being run?

31. What is the difference between the speed of a cat and the speed of a cheetah?

PROBLEM SOLVING

Strategy: Finding a Pattern

A. During a sale at the Peppy Pet store, Bertha gives away 2 cat toys for every 6 cans of cat food sold, 4 toys for every 12 cans, and 6 toys for every 18 cans. She gives away 12 cat toys on Tuesday. How many cans of cat food did she sell?

Bertha plans to make a table to help her find a pattern to solve the problem.

She knows that for every 6 cans of cat food sold she gives away 2 cat toys.

1. What can she do to continue the pattern for the number of cans sold?

2. What can she do to find the number of cat toys sold for each group of 6 cans?

3. Complete the table. How many cans of cat food did she sell?

Cans	Toys
6	2
12	4
18	6
24	8

B. When a dog collar first goes on sale the price is $12.00. After a month the price is $10.00. After 2 months it costs $8.00. How much does the collar cost after 4 months?

Sometimes you do not need a table to find a pattern.

4. By how much does the cost of the collar decrease each month?

5. What can you do to find what the cost of the collar will be the next month?

6. Continue the pattern. How much does the collar cost after 4 months?

7. How does finding a pattern help you solve problems like these?

PRACTICE

Find a pattern to solve the problem. Use a table if needed.
Use mental math, a calculator, or paper and pencil.

8. Bertha gives away 3 free cans of fish food for every 5 fish sold, 6 cans for every 10 fish, and 9 cans for every 15 fish. She gives Bert 15 cans of fish food. How many fish did Bert buy?

9. Gary is buying dog food for a kennel. He sees that 6 cans cost $3.00, 12 cans cost $6.00, and 18 cans cost $9.00. Gary spends $15.00 on dog food. How many cans does he buy?

10. Tom sells 8 guppies on Monday, 16 guppies on Tuesday, and 24 guppies on Wednesday. If this pattern continues, how many guppies will he sell on Friday?

11. Pat stacks 8 rows of dog dishes in a pattern. There are 16 dishes in the bottom row, 14 dishes in the row above that, and 12 dishes in the next row. How many dishes will be in the top row?

Strategies and Skills Review

12. Chang has a 34-minute video about horses. He watches it for 12 minutes. Then he takes an 8-minute break before watching the rest of the video. How much time is left on the video?

13. The pet store has a sale on the 1st, 8th, and 15th of the month. If this pattern continues, on what day will the fifth sale fall?

14. When Tamla walks her dog, she has to stop for a red light every other block. At the end of 5 blocks, is the light red or green?

15. *Write a problem* that can be solved by finding a pattern. Solve your problem. Ask others to solve it.

DEVELOPING A CONCEPT

Area

The **area** of a figure is the number of square units it takes to cover the figure. A metric unit for measuring area is the **square centimeter.**

1 cm
1 cm

WORKING TOGETHER

Draw shapes like these on heavy paper. Cut them out and trace around them on centimeter graph paper.

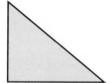

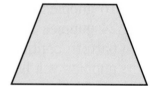

1. Count squares inside each figure to find the area. Combine incomplete squares in any reasonable way.

2. Have another student find the area of each of your figures. Compare results. Are they the same? Why or why not?

3. Draw some rectangles and squares on the graph paper. Count to find the area. Make a table.

4. What pattern do you see in the table?

5. How could you find the area without counting each square?

Number of Rows	Number of Squares in Each Row	Total Number of Squares (Area)
8	2	■

6. Which method is easier for finding the area of a rectangle or a square?

SHARING IDEAS

7. When you measured your figures on graph paper, which areas were easier to measure? Why?

8. What metric unit would you use to measure the area of your classroom? Why?

PRACTICE

Find the area in square centimeters

9.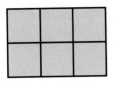

10.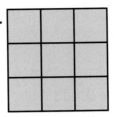

11.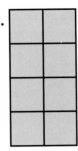

12.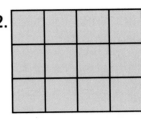

Measure the figure to the nearest centimeter. Find the area.

13.

14.

15.

Find the area.

16. rectangle with length 7 cm and width 5 cm

17. rectangle with length 6 dm and width 4 dm

18. square with sides 8 cm

19. square with sides 5 m

Critical Thinking

20. The area of a rectangular rug is 18 square m. The length is 6 m. How can you find the width? What is the width?

Mixed Applications

21. The school library is 30 m long and 20 m wide. What is the perimeter of the room?

22. A rug in the library is 9 m long and 6 m wide. What is the area of the rug?

Mixed Review

Find the answer. Which method did you use?

MENTAL MATH
CALCULATOR
PAPER/PENCIL

23. 63 ÷ 9

24. 22,003 − 1,679

25. 3,400 + 200

26. 4 × 8

27. $24.33 + $19.99

28. 75 − 20

DEVELOPING A CONCEPT

Making Pictographs

Alicia has found data about rare animals.

She uses these steps to show the data in a pictograph.

Step 1 List each kind of animal.

Step 2 Choose a picture to represent the number of living animals.

Step 3 Draw pictures for each animal.
Let 🐢 = 50 animals

Step 4 Write the title above the graph.

Step 5 Write the number of animals each picture represents at the bottom of the graph.

WORLD POPULATION OF SOME RARE ANIMALS

Kind of Animal	Number Living
Golden lion tamarin monkey	350
Pere David's deer	350
Spurred tortoise	100
Whooping crane	150

Kind of Animal	Number Living
Golden lion tamarin monkey	🐢 🐢 🐢 🐢 🐢 🐢 🐢
Pere David's deer	
Spurred tortoise	
Whooping crane	
Each 🐢 = 50 animals	

WORKING TOGETHER

1. Complete the pictograph.

2. How do you find the number of pictures for each animal?

3. What title did you write above the graph?

SHARING IDEAS

4. Why do pictographs use one picture to show many items?

5. Why is a pictograph easier to read than a table of the same data?

6. There are between 700 and 800 snow leopards in the world. If Alicia had included the snow leopard in her graph, would she use 50 for each picture? Why or why not?

PRACTICE

Use this pictograph.

7. How many years is each picture worth?

8. Which animal has the greatest life span? the least life span?

9. How much greater is the life span of the box turtle than that of the horse?

10. List the life spans from the greatest to the least.

11. *Write a problem* using the information in the pictograph. Solve the problem. Then ask others to solve it.

AVERAGE LIFE SPAN OF ANIMALS

Animal	Life Span
Box turtle	🐰🐰🐰🐰🐰🐰
Goat	🐰🐰
Horse	🐰🐰🐰
Pig	🐰🐰
Rabbit	🐰
Each 🐰 = 5 years	

Use the data to make a pictograph.

12. **CATS IN THE CAT SHOW**

Kind of Cat	Number
American shorthair	45
Manx	15
Persian	20
Siamese	25

13. **ANIMALS BOUGHT LAST YEAR AT ANN'S ANIMAL HOUSE**

Kind of Animal	Number Bought
Cat	70
Dog	60
Goldfish	90
Hamster	50
Parrot	40

14. Make a survey among classes in your school to find the students' favorite pet. Make a pictograph to show the results.

DECISION MAKING

Problem Solving: Adopting a Pet

SITUATION

Kim and Dan and their parents are adopting a dog. At the Adopt-A-Pet center, they find two dogs they would like.

PROBLEM

Which dog should they adopt?

DATA

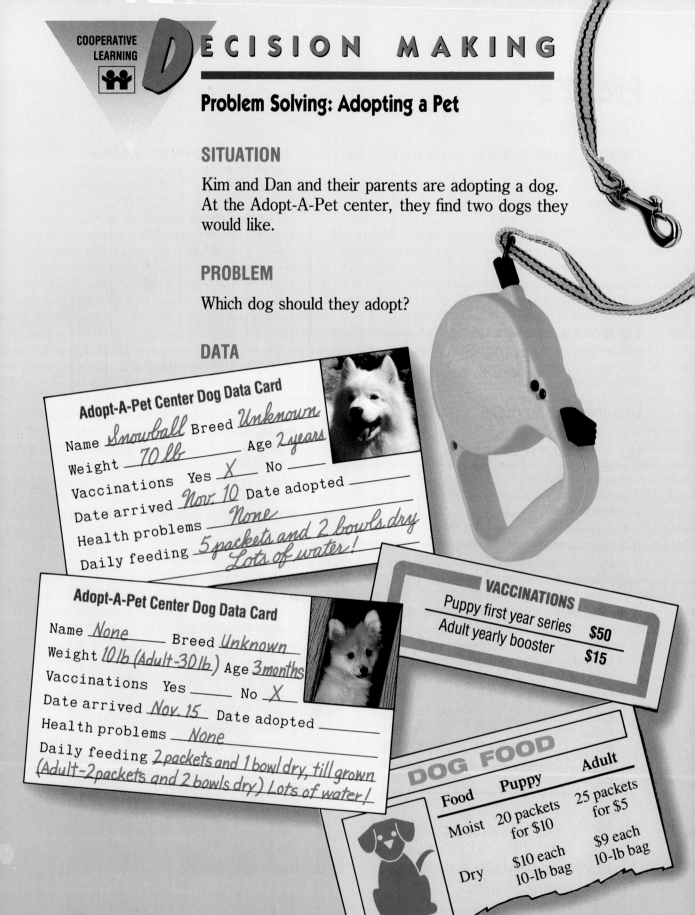

Adopt-A-Pet Center Dog Data Card

Name _Snowball_ Breed _Unknown_
Weight _70 lb_ Age _2 years_
Vaccinations Yes _X_ No ___
Date arrived _Nov. 10_ Date adopted ___
Health problems _None_
Daily feeding _5 packets and 2 bowls dry_
Lots of water!

Adopt-A-Pet Center Dog Data Card

Name _None_ Breed _Unknown_
Weight _10 lb (Adult–30 lb)_ Age _3 months_
Vaccinations Yes ___ No _X_
Date arrived _Nov. 15_ Date adopted ___
Health problems _None_
Daily feeding _2 packets and 1 bowl dry, till grown_
(Adult–2 packets and 2 bowls dry) Lots of water!

VACCINATIONS

Puppy first year series	$50
Adult yearly booster	$15

DOG FOOD

Food	Puppy	Adult
Moist	20 packets for $10	25 packets for $5
Dry	$10 each 10-lb bag	$9 each 10-lb bag

USING THE DATA

How many packets of moist food will each dog eat in one week?

1. puppy   **2.** Snowball

Each bag holds 18 bowls of dry food. How long will it take each dog to finish a bag?

3. the puppy   **4.** the dog

5. How long will 25 packets feed Snowball? How much will it cost each day?

6. How long will 20 packets feed the puppy? How much will it cost each day?

7. Will the puppy cost more or less to feed as an adult than as a puppy? How much more or less?

MAKING DECISIONS

8. *Write a list* of the things Kim and Dan should think about before deciding which pet to adopt.

9. What will happen to Snowball's fluffy white coat in the summer? What extra care will Snowball need?

10. What are some reasons why Kim and Dan might prefer to adopt a dog that is already full-grown?

11. What are some reasons why Dan and Kim might prefer to adopt a puppy?

12. Would it make a difference in Kim's and Dan's decision whether they live in the country or in a city apartment? Why or why not?

13. *Write a list* of the things you would care about the most if you were adopting a dog.

Math and Literature

Tall tales are folk tales about heroes or heroines who do things no real person can do. These amazing deeds make us laugh. One popular tall-tale hero is Paul Bunyan.

The tales about Paul Bunyan say that he was the greatest lumberjack in history. He was so tall he towered over the trees. He crossed rivers in one step. His ax was as wide as a barn door, and the handle was made out of a whole oak tree.

When Paul was a baby, it took 14 cows to supply him with milk. Each day, Paul's parents found their baby son was 2 feet taller than he was the day before!

What if Paul grows 2 feet every day for a week? How much would Paul grow?

Think: Multiply. $7 \times 2 = 14$

Paul would grow 14 feet in a week.

ACTIVITIES

1. The pancakes Paul ate were so big that it took 5 ordinary men to eat one. Write a word problem about this that involves multiplication.

2. Write a tall tale about a hero or heroine you make up.

Calculator: Discover the Patterns

There are six different patterns on this page. You can use your calculator to discover them.

For each exercise, use your calculator to find the first few answers. When you discover a pattern, write the remaining answers without using a calculator. Then check your answers.

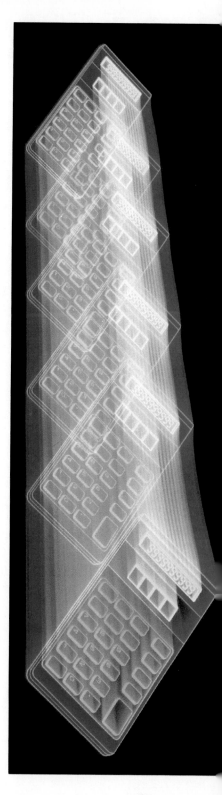

1. 4 × 9
   4 × 99
   4 × 999
   4 × 9,999
   4 × 99,999
   4 × 999,999

2. 1 × 999
   2 × 999
   3 × 999
   4 × 999
   5 × 999
   6 × 999

3. 9 × 9
   9 × 99
   9 × 999
   9 × 9,999
   9 × 99,999

4. 1 × 1
   11 × 11
   111 × 111
   1,111 × 1,111
   11,111 × 11,111

5. 1 × 2 × 101
   2 × 2 × 101
   3 × 2 × 101
   4 × 2 × 101
   5 × 2 × 101
   6 × 2 × 101

6. 1 × 12 × 101
   2 × 12 × 101
   3 × 12 × 101
   4 × 12 × 101
   5 × 12 × 101
   6 × 12 × 101

USING THE CALCULATOR

7. Make up two patterns of your own for others to find and complete.

EXTRA PRACTICE

The Meaning of Multiplication, page 177 ....

Write as a multiplication sentence and solve.

1. 3 + 3 + 3   **2.** 7 + 7 + 7 + 7 + 7   **3.** 6 + 6

Multiply.

4. 2 × 7   **5.** 4 × 6   **6.** 5 × 5   **7.** 3 × 9

8. How many in 4 groups of 9?   **9.** Find the product of 9 × 7.

10.	**11.**	**12.**	**13.**	**14.**	**15.**
6	8	5	2	8	9
× 7	× 8	× 6	× 9	× 1	× 6

16.	**17.**	**18.**	**19.**	**20.**	**21.**
5	3	7	6	3	4
× 2	× 6	× 8	× 6	× 4	× 5

The Meaning of Division, page 179 ....

Divide.

1. 16 ÷ 4   **2.** 28 ÷ 7   **3.** 14 ÷ 7   **4.** 10 ÷ 2

5. 3)18   **6.** 6)30   **7.** 8)32   **8.** 4)28   **9.** 7)49

10. 5)20   **11.** 7)21   **12.** 9)18   **13.** 2)16   **14.** 8)64

15. How many 4s are in 20?   **16.** What is 42 divided by 7?

Multiplication and Division Properties, page 181 ....

Find the missing number.

1. 3 × 0 = ■   **2.** 7 × 1 = ■   **3.** 6 × 2 × 3 = 2 × ■ × 6

4. 16 ÷ 4 = ■   **5.** 14 ÷ 2 = ■   **6.** 18 ÷ 6 = ■   **7.** 4 ÷ 4 = ■

Fact Families, page 183 ....

Write the fact family for the set of numbers.

1. 2, 6, 12   **2.** 5, 1, 5   **3.** 5, 4, 20   **4.** 5, 7, 35

2 Through 9 as Factors, page 185...

Multiply.

1. 7 ×1	**2.** 4 ×4	**3.** 6 ×7	**4.** 2 ×6	**5.** 4 ×7	**6.** 8 ×8

7. 6 × 7   **8.** 5 × 8   **9.** 9 × 8   **10.** 7 × 8   **11.** 4 × 6

12. 7 × 7   **13.** 6 × 5   **14.** 9 × 8   **15.** 6 × 8   **16.** 9 × 9

17. Multiply 7 × 7.   **18.** Find the product of 0 × 9.

Dividing by 2 Through 9, page 187...

Divide.

1. $2\overline{)18}$ **2.** $4\overline{)36}$ **3.** $5\overline{)15}$ **4.** $8\overline{)48}$ **5.** $9\overline{)81}$ **6.** $6\overline{)36}$

7. $4\overline{)24}$ **8.** $8\overline{)40}$ **9.** $5\overline{)25}$ **10.** $7\overline{)42}$ **11.** $9\overline{)72}$ **12.** $3\overline{)12}$

13. $5\overline{)40}$ **14.** $2\overline{)18}$ **15.** $6\overline{)18}$ **16.** $3\overline{)21}$ **17.** $7\overline{)35}$ **18.** $8\overline{)32}$

19. 14 ÷ 2   **20.** 54 ÷ 6   **21.** 8 ÷ 8   **22.** 30 ÷ 6   **23.** 45 ÷ 5

24. Find the quotient of 63 ÷ 7.   **25.** How many 4s are in 28?

Problem Solving Strategy: Choosing the Operation, page 189 ....

Decide which operation to use. Then solve the problem.

1. Brian changes his cat's water 3 times a day. How many times does he change the water in 1 week?

2. Chuck's turtle weighs 12 pounds. Tom's turtle weighs 7 pounds less. How many pounds does Tom's turtle weigh?

3. Cassie bought 63 dog treats for her dog. She put an equal number of treats into 9 bags. How many treats are in each bag?

4. The pet store owner has 45 gerbils. In one month, 8 more gerbils are born. How many gerbils are there in all?

EXTRA PRACTICE

Missing Factors, page 193 ....

Find the missing factor.

1. ■ × 3 = 27   **2.** 6 × ■ = 12   **3.** 4 × ■ = 12   **4.** ■ × 5 = 35

5. 6 × ■ = 48   **6.** ■ × 7 = 49   **7.** ■ × 9 = 45   **8.** 7 × ■ = 63

Find the missing number.

9. ■ ÷ 3 = 9   **10.** ■ ÷ 6 = 6   **11.** 36 ÷ ■ = 4   **12.** 18 ÷ ■ = 9

Other Fact Strategies, page 195 ....

Multiply or divide.

1.   8   **2.**   5   **3.** 9)$\overline{54}$ **4.** 8)$\overline{56}$ **5.**   3   **6.** 6)$\overline{36}$
       × 9   × 8   × 8

Complete the table. Use the rule.

7. Rule: Multiply by 7.

3	4	5	6	7
■	■	■	■	■

8. Rule: Divide by 4.

16	20	24	28	32	36
■	■	■	■	■	■

Factors and Multiples, page 197 ....

Write *even* or *odd*.

1. 6   **2.** 14   **3.** 5   **4.** 17   **5.** 30   **6.** 45

Name a common multiple of the pair of numbers.

7. 3 and 6   **8.** 5 and 6   **9.** 4 and 8

Name the factors of the pair of numbers.
What are the common factors?

10. 5, 9   **11.** 6, 8   **12.** 13, 14   **13.** 20, 35   **14.** 7, 28

15. Which numbers in Questions 10–14 are prime? Which are composite?

EXTRA PRACTICE

Problem Solving Strategy: Finding a Pattern, page 199 ....

Look for a pattern. Then solve the problem.

1. Tina is stacking cans of cat food in a pattern. She puts 4 cans in the first row, 9 cans in the second row, and 14 cans in the third row. How many cans will she put in the next 3 rows?

2. The pet store gives away 8 covers with 2 cases of food, 12 covers with 3 cases, and 16 covers with 4 cases. How many covers do they give away with 5 cases of food?

Area, page 201 ....

Find the area in square centimeters.

1.

2.

3.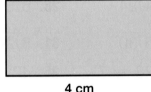

2 cm

4 cm

4.

3 cm

Making Pictographs, page 203 ....

Use this pictograph.

1. How many votes is each picture worth?

2. Which kind of dog had the most votes? the least?

3. How many pictures do you need to show 54 dogs?

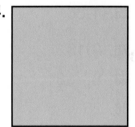

FAVORITE DOGS

Kind of Dog	Number of Votes
Poodle	🐕🐕 🐕
German shepherd	🐕 🐕 🐕 🐕
Irish setter	🐕 🐕 🐕 🐕 🐕 🐕
Labrador	🐕 🐕 🐕 🐕 🐕 🐕 🐕 🐕
Each 🐕 = 6 votes	

Practice PLUS

KEY SKILL: Dividing by 2 Through 9 (Use after page 187.)

Level A

Divide.

1. $2\overline{)6}$ **2.** $3\overline{)3}$ **3.** $5\overline{)15}$ **4.** $4\overline{)16}$ **5.** $6\overline{)18}$

6. $7\overline{)14}$ **7.** $9\overline{)18}$ **8.** $3\overline{)9}$ **9.** $2\overline{)14}$ **10.** $5\overline{)30}$

11. $20 \div 4$ **12.** $0 \div 3$ **13.** $8 \div 2$ **14.** $12 \div 6$ **15.** $32 \div 4$

16. $7 \div 1$ **17.** $12 \div 3$ **18.** $8 \div 4$ **19.** $10 \div 5$ **20.** $30 \div 6$

21. Nan has 24 video games in 4 boxes. She has the same number in each box. How many games are in each box?

Level B

Divide.

22. $6\overline{)36}$ **23.** $4\overline{)36}$ **24.** $5\overline{)40}$ **25.** $4\overline{)24}$ **26.** $8\overline{)32}$

27. $3\overline{)27}$ **28.** $2\overline{)18}$ **29.** $4\overline{)0}$ **30.** $6\overline{)30}$ **31.** $8\overline{)24}$

32. $28 \div 4$ **33.** $8 \div 8$ **34.** $0 \div 5$ **35.** $28 \div 7$ **36.** $42 \div 7$

37. $18 \div 3$ **38.** $25 \div 5$ **39.** $16 \div 2$ **40.** $54 \div 6$ **41.** $40 \div 8$

42. Ned has 45 model planes. He puts 9 planes on each shelf. How many shelves does he use?

Level C

Divide.

43. $9\overline{)36}$ **44.** $8\overline{)64}$ **45.** $7\overline{)63}$ **46.** $6\overline{)42}$ **47.** $1\overline{)8}$

48. $7\overline{)49}$ **49.** $8\overline{)48}$ **50.** $9\overline{)45}$ **51.** $5\overline{)35}$ **52.** $8\overline{)56}$

53. $21 \div 7$ **54.** $6 \div 1$ **55.** $54 \div 9$ **56.** $72 \div 8$ **57.** $81 \div 9$

58. $48 \div 6$ **59.** $56 \div 7$ **60.** $45 \div 5$ **61.** $0 \div 9$ **62.** $72 \div 9$

63. Ann has 63 tickets to give away evenly to 9 friends. How many tickets does each friend get?

KEY SKILL: Factors and Multiples (Use after page 197.)

Level A ....

Write *even* or *odd*.

1. 6　　　　**2.** 11　　　　**3.** 13　　　　**4.** 8　　　　**5.** 16　　　　**6.** 21

Name a common multiple of the pair of numbers.

7. 3 and 4　　　　**8.** 2 and 5　　　　**9.** 2 and 3　　　　**10.** 4 and 5

Name the factors of the number.

11. 6　　　　**12.** 9　　　　**13.** 13　　　　**14.** 15　　　　**15.** 17

16. Name the common factors of 2 and 6.

Level B ....

Write *even* or *odd*.

17. 20　　　**18.** 25　　　**19.** 31　　　**20.** 49　　　**21.** 70　　　**22.** 104

Name a common multiple of the pair of numbers.

23. 5 and 6　　　　**24.** 4 and 6　　　　**25.** 7 and 8　　　　**26.** 3 and 9

Name the factors of each pair of numbers. What are the common factors?

27. 8, 9　　　**28.** 15, 19　　　**29.** 10, 25　　　**30.** 12, 18　　　**31.** 14, 21

Level C ....

Name a common multiple of the pair of numbers.

32. 6 and 8　　　　**33.** 5 and 7　　　　**34.** 8 and 9　　　　**35.** 6 and 9

Name the factors of each pair of numbers. What are the common factors?

36. 10, 20　　　**37.** 5, 35　　　**38.** 16, 24　　　**39.** 19, 28　　　**40.** 11, 22

41. Which numbers in Exercises 36–40 are prime numbers? Which are composite numbers?

CHAPTER REVIEW

LANGUAGE AND MATHEMATICS

Complete the sentences. Use the words in the chart.

1. The numbers being multiplied are called ■.
   (page 176)

2. The ■ of a number are the products of that
   number. *(page 196)*

3. The ■ of a figure is the number of square units
   it takes to cover the figure. *(page 200)*

4. In 56 ÷ 4, 4 is the ■. *(page 178)*

5. ***Write a definition*** or give an example of the words
   you did not use from the chart.

VOCABULARY

area
multiples
product
divisor
pictograph
factors

CONCEPTS AND SKILLS

Multiply. *(pages 176, 184)*

6.	7.	8.	9. 6 × 5	10. 7 × 8

6. 3
   × 9

7. 5
   × 4

8. 6
   × 7

9. 6 × 5

10. 7 × 8

Divide. *(pages 178, 186)*

11. 7)28   **12.** 8)48   **13.** 2)18   **14.** 72 ÷ 8   **15.** 24 ÷ 6

Write the fact family for the set of numbers. *(page 182)*

16. 2, 2, 4   **17.** 3, 6, 18   **18.** 7, 7, 49   **19.** 9, 1, 9   **20.** 5, 4, 20

Find the missing factor. *(pages 180, 192)*

21. 3 × ■ = 15   **22.** ■ × 4 = 0   **23.** 9 × ■ = 27   **24.** 4 × (■ × 2) = 40

Name the factors of each pair of numbers. What are the
common factors? *(page 196)*

25. 5, 25   **26.** 9, 12   **27.** 4, 13   **28.** 2, 6

29. Which of the numbers in Exercises 25–28 are even?
    Which are odd? Which are prime? Which are
    composite?

Name a common multiple of the pair of numbers. *(page 196)*

30. 3 and 5   **31.** 7 and 2   **32.** 4 and 6   **33.** 5 and 9

Find the area. *(page 200)*

34. rectangle with length: 4 cm, width: 2 cm   **35.** square with 3 m sides

36. Complete the pictograph. In Green's Orchard there are 30 peach trees, 55 apple trees, 20 pear trees, and 45 orange trees. *(page 202)*

TREES IN GREEN'S ORCHARD

Kind of Tree	Number
Peach	🌳🌳🌳🌳🌳🌳
Apple	
Pear	
Orange	
Each 🌳 = 5 trees	

CRITICAL THINKING

Write *always, sometimes,* or *never* to complete the sentence.

37. The product of a number times 0 is ■ 0. *(page 180)*

38. You can ■ divide a number by zero. *(page 180)*

39. An odd number is ■ a multiple of 3. *(page 197)*

40. The area of a square is ■ the square of a side. *(page 200)*

MIXED APPLICATIONS

41. A trivia contest has 3 competing teams. There are 9 people on each team. What is the total number of people in the contest? What operation did you use to find your answer? *(page 188)*

42. Ada is painting dots on a box. There are 12 dots in the first row, 16 dots in the second, and 20 dots in the third row. If the pattern continues, how many dots will be in the fifth row? *(page 198)*

CHAPTER TEST

Find the missing number.

1. $6 \div 1 = $ ■

2. $7 \times$ ■ $= 7$

3. $2 \times 8 = $ ■ $\times 2$

Multiply or divide.

4. $\begin{array}{r} 4 \\ \times\, 5 \\ \hline \end{array}$

5. $\begin{array}{r} 5 \\ \times\, 8 \\ \hline \end{array}$

6. $\begin{array}{r} 3 \\ \times\, 7 \\ \hline \end{array}$

7. $3\overline{)18}$

8. $6\overline{)42}$

9. $4\overline{)28}$

10. $7 \times 7 = $ ■

11. $9\overline{)81}$

12. $8 \times 9 = $ ■

13. $6\overline{)54}$

Name the factors of the number. Is the number *prime* or *composite?*

Tell if the number is *even* or *odd.*

14. 15

15. 19

16. 23

17. 8

Give 5 multiples for the number.

18. 6

19. 4

Find the area.

20.
3 m
8 m

21.
9 cm
9 cm

Use the pictograph for Questions 22 and 23.

22. Which type of fish are there the most of?

23. How many more goldfish are there than catfish?

Solve.

24. Marta has 56 fish. She puts 8 fish in each tank. How many tanks does she use?

25. Wood is stacked in piles. There are 68 logs in the first pile, 63 in the second pile, and 58 in the third pile. If this pattern continues, how many logs will be in the fifth pile?

FISH AT THE AQUARIUM

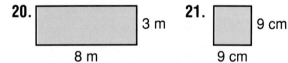

Kind of Fish	Number
Goldfish	🐟🐟🐟🐟🐟
Angelfish	🐟🐟🐟
Swordfish	🐟🐟🐟🐟
Catfish	🐟
Each 🐟 = 5 fish	

ENRICHMENT FOR ALL

SIEVE OF ERATOSTHENES

Eratosthenes was a Greek
mathematician who lived about
2,000 years ago. He used these
steps to separate prime numbers
from composite numbers.

Think: A prime number has exactly
two factors. A composite
number has more than two
factors.

Follow Eratosthenes' steps to find
the prime numbers less than 100.

1. Copy the list of numbers from 1 to 100.

2. Cross out the number 1. It is neither
   prime nor composite. Why?

3. Circle the number 2. It is the first
   prime number.

4. Cross out all the other multiples of 2.
   Why are these not prime numbers?

5. Circle the prime number 3. Cross out
   all the other multiples of 3.

6. What is the next prime number?

7. Circle it and cross out all its multiples.

1	2	3	4	5	6	7	8	9	10
11	12	13	14	15	16	17	18	19	20
21	22	23	24	25	26	27	28	29	30
31	32	33	34	35	36	37	38	39	40
41	42	43	44	45	46	47	48	49	50
51	52	53	54	55	56	57	58	59	60
61	62	63	64	65	66	67	68	69	70
71	72	73	74	75	76	77	78	79	80
81	82	83	84	85	86	87	88	89	90
91	92	93	94	95	96	97	98	99	100

8. Continue to circle the prime numbers and cross out
   their multiples until only prime numbers are left.

9. What is the last prime number that had multiples you
   had to cross out?

10. List all the prime numbers between 1 and 100.

11. How many prime numbers are there less than 100?

12. Why do you think this method is called the Sieve of Eratosthenes?

Multiplication and Division Facts **217**

CUMULATIVE REVIEW

Choose the letter of the correct answer.

1. Which is a factor of 21?

   a. 2   **c.** 7
   b. 4   **d.** not given

2. How much time passes between 2:39 and 3:01?

   a. 21 minutes   **c.** 40 minutes
   b. 22 minutes   **d.** not given

3. ■ × 8 = 40

   a. 4   **c.** 6
   b. 5   **d.** not given

4. The mass of a watermelon is:

   a. 5 g.   **c.** 5 kg.
   b. 50 g.   **d.** 50 kg.

5. Find the area of a square with 7-m sides.

   a. 14 sq m   **c.** 49 sq m
   b. 28 sq m   **d.** not given

6. Find the quotient of 36 ÷ 9.

   a. 4   **c.** 6
   b. 5   **d.** not given

7. What multiplied by 3 is 27?

   a. 6   **c.** 8
   b. 7   **d.** not given

8. Find the perimeter of a square with 3-cm sides.

   a. 6 cm   **c.** 12 cm
   b. 9 cm   **d.** not given

9. 15 is how many more than 9?

   a. 5   **c.** 7
   b. 6   **d.** not given

10. What is 0 divided by 4?

   a. 0   **c.** 4
   b. 1   **d.** not given

11. $24.38 − $.59

   a. $24.89   **c.** $23.79
   b. $23.89   **d.** not given

12. 6 + 2 + 3 + 5

   a. 8   **c.** 15
   b. 11   **d.** not given

13. Which number is prime?

   a. 2   **c.** 9
   b. 4   **d.** not given

14. What is 8 times 4?

   a. 2   **c.** 24
   b. 32   **d.** not given

Multiplying by 1-Digit Factors

MATH CONNECTION: PROBLEM SOLVING

THE SPACE SHUTTLE

★ First Flight: April 12, 1981.

★ Can carry crew of seven astronauts and 65,000 pounds of equipment.

★ Heat protection system consists of 25,000 ceramic tiles.

★ Enters Earth's atmosphere at 16,000 miles per hour.

★ Lands on runway at a speed of 200 miles per hour.

★

★

1. **What is happening in this picture?**

2. **What information do you see?**

3. **Write a story about a space shuttle launch.**

Mental Math: Multiplying 10s and 100s

Bella, Carl, and Dora played a space game. Bella had to pay the fare to Mars 4 times. Carl had to pay the fare to Venus 5 times. Dora had to pay the fare to Mercury 6 times. The players used mental math to find the products.

Bella	Carl	Dora
$4 \times 1 = 4$	$5 \times 3 = 15$	$6 \times 5 = 30$
$4 \times 10 = 40$	$5 \times 30 = 150$	$6 \times 50 = 300$
$4 \times 100 = 400$	$5 \times 300 = 1,500$	$6 \times 500 = 3,000$
$4 \times 1,000 = 4,000$	$5 \times 3,000 = 15,000$	$6 \times 5,000 = 30,000$

The total fares were $4,000; $15,000; and $30,000.

1. Look at Bella's examples. Compare the number of zeros in each product with the number of zeros in each factor. What pattern do you see?

2. Does the pattern work in Carl's examples?

3. Does the pattern work in Dora's examples? Why or why not?

TRY OUT Write the letter of the correct answer. Multiply mentally.

4. 7×20	**a.** 14	**b.** 140	**c.** 1,400	**d.** 14,000
5. 5×800	**a.** 40	**b.** 400	**c.** 4,000	**d.** 40,000
6. 9×700	**a.** 63	**b.** 630	**c.** 6,300	**d.** 63,000
7. $2 \times 5,000$	**a.** 100	**b.** 1,000	**c.** 10,000	**d.** 100,000

PRACTICE

Multiply mentally.

8. $\begin{array}{r} 10 \\ \times\ 3 \\ \hline \end{array}$	**9.** $\begin{array}{r} 10 \\ \times\ 7 \\ \hline \end{array}$	**10.** $\begin{array}{r} 20 \\ \times\ 3 \\ \hline \end{array}$	**11.** $\begin{array}{r} 30 \\ \times\ 4 \\ \hline \end{array}$	**12.** $\begin{array}{r} 40 \\ \times\ 5 \\ \hline \end{array}$
13. $\begin{array}{r} 100 \\ \times\ 2 \\ \hline \end{array}$	**14.** $\begin{array}{r} 200 \\ \times\ 5 \\ \hline \end{array}$	**15.** $\begin{array}{r} 1,000 \\ \times\ 6 \\ \hline \end{array}$	**16.** $\begin{array}{r} 8,000 \\ \times\ 3 \\ \hline \end{array}$	**17.** $\begin{array}{r} 7,000 \\ \times\ 7 \\ \hline \end{array}$

18. 6 × 10   **19.** 9 × 10   **20.** 4 × 100   **21.** 5 × 100

22. 8 × 100   **23.** 7 × 1,000   **24.** 5 × 1,000   **25.** 3 × 1,000

26. 6 × 70   **27.** 9 × 50   **28.** 7 × 400   **29.** 8 × 600

30. 2 × 900   **31.** 6 × 9,000   **32.** 5 × 6,000   **33.** 8 × 8,000

Critical Thinking

34. How would you find the product of 8 × 50,000? What is the product?

Mixed Applications

35. The fare to the Moon was $800. Bella had to pay it 4 times. What was her total fare?

36. Dora bought a book about astronauts for $9.75 and one about astronomy for $11.59. How much did she spend for the two books?

CHALLENGE

You can write multiples of 10 using **exponents**.

There are 3 zeros after the 1.

$10 \times 10 \times 10 = 1,000 = 10^3$ Say: 10 to the 3rd power.

There are 3 factors.   The exponent is 3.

Write the power of 10 using an exponent.

1. 10 × 10   **2.** 10 × 10 × 10 × 10   **3.** 10 × 10 × 10 × 10 × 10

4. 100   **5.** 100,000   **6.** 1,000,000

UNDERSTANDING A CONCEPT

Estimating Products

The State Astronomy Club has 476 members. The secretary sends about 6 books a year to each member. About how many books does the secretary send each year?

Estimate: 6 × 476

Here are two ways to estimate the product.

Rounding

Round the greater number to its greatest place.

$$\begin{array}{r} 476 \\ \times\ \ 6 \\ \hline \end{array}$$

Think:
$$\begin{array}{r} 500 \\ \times\ \ 6 \\ \hline 3,000 \end{array}$$

The estimate is 3,000 books.

Front-End Estimation

Use the front digits.

$$\begin{array}{r} 476 \\ \times\ \ 6 \\ \hline \end{array}$$

Think:
$$\begin{array}{r} 400 \\ \times\ \ 6 \\ \hline 2,400 \end{array}$$

The estimate is 2,400 books.

1. Which method results in an estimate that is greater than the exact answer? Why?

2. Which method results in an estimate that is less than the exact answer? Why?

3. Estimate 9 × 3,174 using both methods. How do the estimates compare? Why?

TRY OUT Write the letter of the correct answer.

Estimate by rounding.

4. 5 × 84   **a.** 40   **b.** 400   **c.** 4,000

5. 8 × $327   **a.** $2,400   **b.** $240   **c.** $24,000

6. 3 × 4,617   **a.** 150,000   **b.** 15,000   **c.** 1,500

Estimate by using the front digits.

7. 7 × $68   **a.** $4,200   **b.** $42   **c.** $420

8. 2 × 843   **a.** 160   **b.** 1,600   **c.** 16,000

9. 4 × 6,809   **a.** 24,000   **b.** 2,400   **c.** 240

PRACTICE

Estimate by rounding.

10. 29
    $\times\ 6$

11. 83
    $\times\ 3$

12. 18
    $\times\ 7$

13. $47
    $\times\ \ 9$

14. 621
    $\times\ \ 5$

15. 869
    $\times\ \ 8$

16. $535
    $\times\ \ 2$

17. 3,095
    $\times\ \ \ 4$

18. 7,730
    $\times\ \ \ 5$

19. $1,927
    $\times\ \ \ \ 6$

20. 4 × 53

21. 6 × 32

22. 3 × $95

23. 7 × 429

24. 5 × 417

25. 8 × $447

26. 9 × 3,350

27. 2 × 5,711

Estimate by using the front digits.

28. 14
    $\times\ 8$

29. 39
    $\times\ 5$

30. 72
    $\times\ 3$

31. $68
    $\times\ \ 4$

32. 226
    $\times\ \ 7$

33. 555
    $\times\ \ 6$

34. $829
    $\times\ \ \ 5$

35. 4,706
    $\times\ \ \ 9$

36. 2,275
    $\times\ \ \ 8$

37. 3,579
    $\times\ \ \ 4$

38. 6 × 16

39. 9 × 87

40. 4 × $93

41. 8 × 177

42. 7 × 342

43. 5 × $716

44. 3 × 4,882

45. 6 × 5,279

Write >, <, or =. Use estimation.

46. 4 × 284 ● 645

47. 6 × 329 ● 2,400

48. 4,511 ● 5 × 838

Mixed Applications

49. The Astronomy Club has $5,000. It wants to buy 3 telescopes that cost $1,397 each. Does the club have enough money?

50. The Planetarium has 786 new members this year. There are 1,427 members. How many people were members last year?

Mixed Review

Find the missing number.

51. 8 × ■ = 56

52. 632 − ■ = 114

53. ■ ÷ 9 = 4

54. 82 + ■ + 71 = 202

55. ■ × 5 = 5

56. 9 − ■ = 9

PROBLEM SOLVING

Strategy: Guess, Test, and Revise

There are 54 candidates at astronaut school. They are divided into 14 teams. Some teams have 3 members each, and the rest have 5 members each. How many teams have only 3 members? 5 members?

One way to solve the problem is to start with a guess; test your guess; and if it is wrong, revise it until you find the correct answer.

Barb first guesses that there are 7 teams with 3 members and 7 teams with 5 members. She tests her guess.

Guess

**7 teams of 3
and
7 teams of 5**

1. Is her guess correct? Why or why not?

2. Why should Barb lower her guess?

Test
$7 \times 3 = 21$
$7 \times 5 = \underline{35}$
56

Barb revises her guess lower. She tries 8 teams with 3 members and 6 teams with 5 members.

Revised Guess

**8 teams of 3
and
6 teams of 5**

3. Is her guess correct? Why or why not?

Test
$8 \times 3 = 24$
$6 \times 5 = \underline{30}$
$\blacksquare$

4. **What if** some teams had 3 members each and the rest had 6 members each? How many of each would there be?

5. How did you make your guess? Did you have to revise it? Why or why not?

PRACTICE

Use the Guess, Test, and Revise strategy to solve the problem.
Use mental math, a calculator, or paper and pencil.

6. There are twice as many men as women among the 54 candidates. How many men are there? How many women?

7. The weightless chamber is used 48 times a day. Only one astronaut may be in the chamber at a time. Senior astronauts use the chamber 6 times a day, and junior astronauts use it 4 times a day. How many senior astronauts are there? junior astronauts?

8. During training only 1 member of a team of 7 astronauts can be asleep at any time during a 24-hour period. Each astronaut may sleep for either 3 hours or 4 hours. How many may sleep for 3 hours? How many for 4 hours?

9. In a 24-hour day astronauts study 3 times as many hours as they exercise. They eat and sleep twice as many hours as they exercise. How many hours do they study? exercise? eat and sleep?

Strategies and Skills Review

Solve.

10. Donna bought a model of a spaceship and a book on space travel for $22. She spent $4 more on the book than on the model. How much did she spend on each?

11. The space center sells booklets of 8 postcards of spaceships. Last month it sold 700 booklets. How many postcards of spaceships did it sell?

12. The Soviet spaceship *Soyuz 7* orbited earth 79 times. *Soyuz 11* orbited 360 times. How many fewer times did *Soyuz 7* orbit earth than *Soyuz 11*?

13. *Write a problem* that can be solved by the Guess, Test, and Revise strategy. Solve the problem. Ask others to solve your problem.

EXPLORING A CONCEPT

Multiplying by 1-Digit Factors

A spaceship can carry 48 people. How many people can it carry in 3 trips?

WORKING TOGETHER

Solve the problem by making a model.

1. How does your model show the number of people the ship can carry in 1 trip?

2. How does your model show the number of trips?

3. How many people can the ship carry in 3 trips?

Here is a way to solve the problem using place-value models.

Step 1 Make 3 groups of models with 4 tens and 8 ones in each group.

Step 2 Combine the ones. Regroup the ones.

4. How many ones are left? How many new tens are there?

Step 3 Combine the tens, including the new tens. Regroup the tens.

5. How many tens are left?

6. How many hundreds are there?

7. How does the model show the total number of people?

Use place-value models to find how many there are in all.

8. 3 groups of 32

9. 4 groups of 23

10. 5 groups of 34

11. 6 groups of 41

SHARING IDEAS

12. In which exercises above did you need to regroup ones? tens? Why?

13. Why is it better to combine the ones before the tens and the tens before the hundreds?

ON YOUR OWN

Solve using place-value models.
Tell how you solved the problem.

14. Each of 5 astronauts on a space flight has 21 experiments to work on. How many experiments will the astronauts perform?

15. On one mission 7 astronauts will spend 13 days in space. They each must write one report a day. How many reports will they write?

16. A space bus has 16 seats in each module. If there are 3 modules, how many seats are there?

17. *Write a problem* that you can solve by finding the total number of things in equal groups.

Multiplying 2-Digit Numbers— Regrouping Once

A. Erik and Kim are planning a school trip to the Space Museum. They know that 3 classes are going. Each class has 26 students. How many students are going on the trip?

Erik estimates: $3 \times 30 = 90$

He uses place-value models to find the exact answer.

Multiply: 3×26

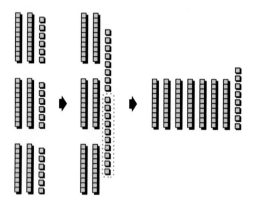

1. Did Erik regroup? Why or why not?

2. How many students are going on the trip?

B. Kim records the way that Erik multiplied. They both see that they can multiply without models.

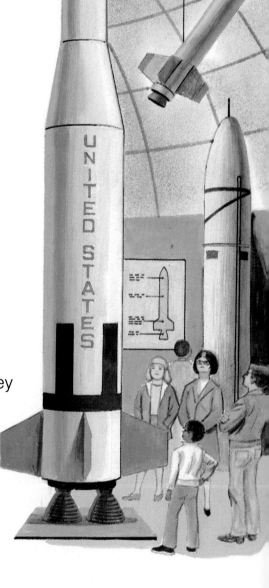

Step 1	Step 2
Multiply the ones. **Regroup if necessary.**	**Multiply the tens.** **Add any new tens.**
$$\begin{array}{r} \overset{1}{2}6 \\ \times\ \ 3 \\ \hline 8 \end{array}$$	$$\begin{array}{r} \overset{1}{2}6 \\ \times\ \ 3 \\ \hline 78 \end{array}$$
Think: 18 ones = 1 ten 8 ones	***Think:*** 3 × 2 tens = 6 tens; 6 tens + 1 ten = 7 tens

SHARING IDEAS

3. Compare Kim's method with Erik's. How are they the same? How are they different?

4. What products do you see when using the models that you do not see when using the numbers?

5. Look at Step 2 of Kim's method. Why do you multiply the tens before you add the new tens?

PRACTICE

Multiply. Use place-value models if needed.

| 6. | 49
× 2 | 7. | 61
× 5 | 8. | 25
× 3 | 9. | 42
× 4 | 10. | 14
× 6 | 11. | 21
× 8 |

| 12. | 25
× 2 | 13. | 81
× 5 | 14. | 36
× 2 | 15. | 42
× 7 | 16. | 29
× 2 | 17. | 52
× 4 |

18. 8×31 19. 5×16 20. 2×35 21. 4×91 22. 6×15

23. 2×39 24. 3×53 25. 2×95 26. 5×19 27. 2×74

28. What is 4 times 16?   29. What is the product of 6 and 51?

Find the missing number. Tell how you found it.

30. $25 \times \blacksquare = 25$ 31. $\blacksquare \times 37 = 0$ 32. $17 \times 4 = \blacksquare \times 17$

Mixed Applications

Solve. Which method did you use?

ESTIMATION
MENTAL MATH
PAPER/PENCIL
CALCULATOR

33. The students spent 4 weeks on a space travel project. They worked on it for 13 hours per week. How many hours did they spend on the project?

34. A rocket is traveling at a speed of 6,139 miles per hour. About how many miles does it travel in 5 hours?

35. Erik read a 179-page book on the history of space travel and a 101-page book on the training of astronauts. How many pages did he read in all?

UNDERSTANDING A CONCEPT

Multiplying 2-Digit Numbers— Regrouping Twice

The students in 2 art classes are making space-station models. There are 57 students in each art class. How many students are making space-station models?

Estimate first. Then multiply: 2 × 57.

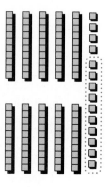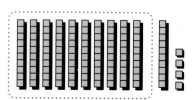

Step 1	Step 2
Multiply the ones. **Regroup if necessary.**	**Multiply the tens.** **Add any new tens.** **Regroup if necessary.**
$\begin{array}{r} \overset{1}{5}\,7 \\ \times\ \ 2 \\ \hline 4 \end{array}$	$\begin{array}{r} \overset{1}{5}\,7 \\ \times\ \ 2 \\ \hline 1\,1\,4 \end{array}$
Think: 14 ones = 1 ten 4 ones	***Think:*** 10 tens + 1 ten = 11 tens

There are 114 students making models.

1. Is the answer reasonable? How do you know?

2. What is the product of 57 × 2? How do you know?

TRY OUT   Write the letter of the correct answer.

3. 4 × 39   **a.** 126   **b.** 156   **c.** 246   **d.** 1,236

4. 5 × 48   **a.** 200   **b.** 240   **c.** 400   **d.** 2,040

PRACTICE

Multiply.

5. 37 × 3	**6.** 65 × 2	**7.** 28 × 4	**8.** 22 × 5	**9.** 34 × 6	**10.** 39 × 4
11. 73 × 5	**12.** 18 × 7	**13.** 16 × 8	**14.** 24 × 9	**15.** 26 × 7	**16.** 32 × 8
17. 62 × 7	**18.** 49 × 3	**19.** 13 × 8	**20.** 54 × 5	**21.** 28 × 7	**22.** 47 × 6

23. 9 × 44   **24.** 4 × 54   **25.** 8 × 76   **26.** 6 × 65   **27.** 7 × 89

28. 6 × 58   **29.** 4 × 44   **30.** 7 × 34   **31.** 3 × 96   **32.** 6 × 47

33. 8 × 35   **34.** 9 × 79   **35.** 3 × 87   **36.** 8 × 46   **37.** 5 × 65

Critical Thinking

38. If you multiply a 2-digit number by a 1-digit number (other than 0), what is the smallest product you can get? What is the largest product you can get?

Mixed Applications

39. Mia designs a space lab with 5 rooms. Each room holds 24 people. How many people could fit in Mia's space lab?

40. Maria designed a shuttle that has 18 seats. Teri's shuttle has 24 seats. How many more seats does Teri's shuttle have?

41. *Write a problem* that you solve by multiplying. Solve your problem. Ask others to solve your problem.

MENTAL MATH

Here is a way to multiply mentally. Find 6 × 23.

Think: 6 × 20 = 120 and 6 × 3 = 18;
so 6 × 23 = 120 + 18 = 138.

Multiply mentally.

1. 4 × 36   **2.** 2 × 75   **3.** 3 × 84   **4.** 7 × 56   **5.** 8 × 67

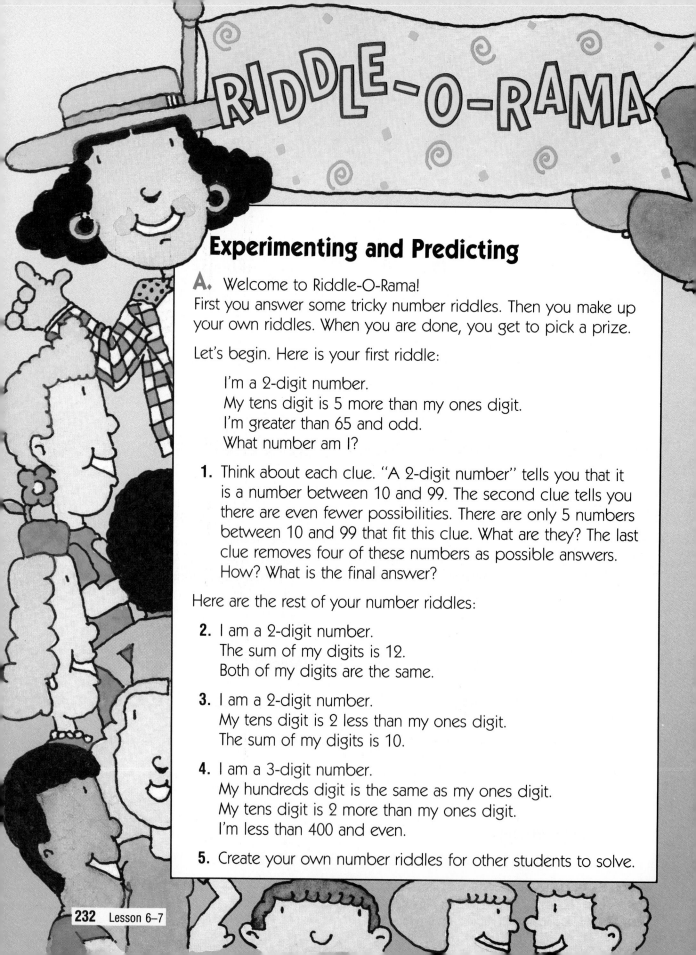

RIDDLE-O-RAMA

Experimenting and Predicting

A. Welcome to Riddle-O-Rama!
First you answer some tricky number riddles. Then you make up your own riddles. When you are done, you get to pick a prize.

Let's begin. Here is your first riddle:

> I'm a 2-digit number.
> My tens digit is 5 more than my ones digit.
> I'm greater than 65 and odd.
> What number am I?

1. Think about each clue. "A 2-digit number" tells you that it is a number between 10 and 99. The second clue tells you there are even fewer possibilities. There are only 5 numbers between 10 and 99 that fit this clue. What are they? The last clue removes four of these numbers as possible answers. How? What is the final answer?

Here are the rest of your number riddles:

2. I am a 2-digit number.
The sum of my digits is 12.
Both of my digits are the same.

3. I am a 2-digit number.
My tens digit is 2 less than my ones digit.
The sum of my digits is 10.

4. I am a 3-digit number.
My hundreds digit is the same as my ones digit.
My tens digit is 2 more than my ones digit.
I'm less than 400 and even.

5. Create your own number riddles for other students to solve.

B. Congratulations! You have solved your number riddles. Now you get to pick your prize. Which one

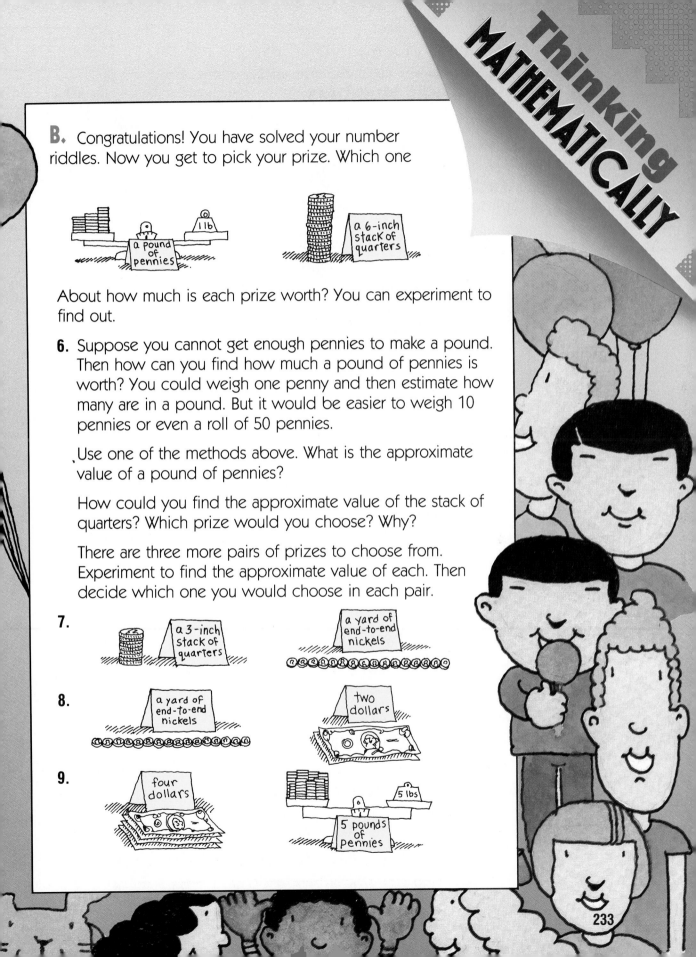

About how much is each prize worth? You can experiment to find out.

6. Suppose you cannot get enough pennies to make a pound. Then how can you find how much a pound of pennies is worth? You could weigh one penny and then estimate how many are in a pound. But it would be easier to weigh 10 pennies or even a roll of 50 pennies.

Use one of the methods above. What is the approximate value of a pound of pennies?

How could you find the approximate value of the stack of quarters? Which prize would you choose? Why?

There are three more pairs of prizes to choose from. Experiment to find the approximate value of each. Then decide which one you would choose in each pair.

7. a 3-inch stack of quarters   a yard of end-to-end nickels

8. a yard of end-to-end nickels   two dollars

9. four dollars   5 pounds of pennies

Multiplying 3-Digit Numbers

The Franklin School students are going to the State Science Fair. It is 156 km from their town. How many km is the round-trip?

Multiply: 2 × 156

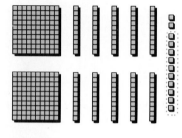 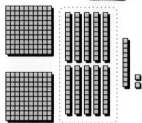 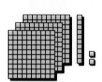

Step 1	**Step 2**	**Step 3**
Multiply the ones. Regroup if necessary.	Multiply the tens. Add any new tens. Regroup if necessary.	Multiply the hundreds. Add any new hundreds. Regroup if necessary.

Step 1

$$\begin{array}{r} \overset{1}{1}5\,6 \\ \times\quad 2 \\ \hline 2 \end{array}$$

Think:
12 ones =
1 ten 2 ones

Step 2

$$\begin{array}{r} \overset{1\;1}{1}5\,6 \\ \times\quad 2 \\ \hline 1\,2 \end{array}$$

Think:
10 tens + 1 ten =
11 tens =
1 hundred, 1 ten

Step 3

$$\begin{array}{r} \overset{1\;1}{1}5\,6 \\ \times\quad 2 \\ \hline 3\,1\,2 \end{array}$$

Think:
2 hundreds +
1 hundred =
3 hundreds

The round-trip is 312 km.

1. How do you know the answer is reasonable?

2. **What if** you multiplied 4 × 207? How many times do you need to regroup? What is the product?

3. **What if** you multiplied 3 × 479? How many times do you need to regroup? What is the product?

TRY OUT   Write the letter of the correct answer.

4. 4 × 413   **a.** 1,652   **b.** 1,661   **c.** 1,682   **d.** 16,412

5. 5 × 706   **a.** 3,503   **b.** 3,530   **c.** 4,050   **d.** 35,030

6. 3 × 589   **a.** 1,632   **b.** 1,767   **c.** 1,547   **d.** 4,007

PRACTICE

Multiply. Solve only for products that are greater than 900.

7. 216 × 4	**8.** 323 × 3	**9.** 611 × 5	**10.** 413 × 7	**11.** 609 × 8	**12.** 534 × 2
13. 409 × 3	**14.** 442 × 2	**15.** 624 × 6	**16.** 509 × 9	**17.** 480 × 4	**18.** 825 × 5
19. 763 × 3	**20.** 196 × 5	**21.** 794 × 7	**22.** 645 × 9	**23.** 842 × 8	**24.** 939 × 9

25. 6 × 340   **26.** 7 × 109   **27.** 6 × 725   **28.** 4 × 813

29. 8 × 423   **30.** 9 × 603   **31.** 3 × 908   **32.** 5 × 467

Mixed Applications

33. There are 5 rooms for showing the students' projects. Each room has 306 projects. How many projects are there?

34. Models of the solar system cost $126. The science teacher bought 2 of them. How much did she spend all together?

35. There are 205 projects on space science and 146 on biology. How many more projects on space science are there than projects on biology?

36. At the fair 355 students saw a show on planets, and 416 students saw a show on fossils. What was the total attendance at the two shows?

Mixed Review

Find the answer. Which method did you use?

37. 3,400 + 200   **38.** $405.37 − $287.89

39. 7 × 6,000   **40.** 72 ÷ 8

41. $3.75 + $5.98   **42.** 709 − 165

MENTAL MATH
CALCULATOR
PAPER/PENCIL

UNDERSTANDING A CONCEPT

Multiplying Money

Anita bought 4 rocket models at the Space Museum souvenir shop. Each model cost $8.59. How much did the 4 models cost?

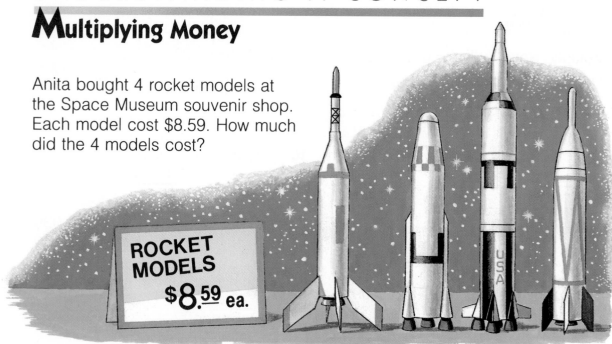

ROCKET MODELS $8.⁵⁹ ea.

Estimate. Round to the nearest dollar. 4 × $9 = $36

Multiply: 4 × $8.59

Step 1

Multiply as you would multiply whole numbers.

```
  2 3
$8.5 9
×    4
3 4 3 6
```

Step 2

Write the dollar sign and decimal point in the product.

```
  2 3
$8.5 9
×    4
$3 4.3 6
```

Think: Money amounts always have 2 decimal places.

The 4 models cost $34.36.

1. **What if** you multiplied 5 × $.87? How could you estimate the product? What would be the product?

2. **What if** you multiplied 3 × $2.08? What would be the product?

TRY OUT Write the letter of the correct answer.

3. 7 × $.65   **a.** $4.25   **b.** $4.55   **c.** $4.73   **d.** $455

4. 3 × $5.26   **a.** $15.68   **b.** $15.78   **c.** $15.98   **d.** $1,578

5. 5 × $3.98   **a.** $15.50   **b.** $19.90   **c.** $45.50   **d.** $199

PRACTICE

Multiply.

6. $.35
× 2

7. $.18
× 7

8. $.20
× 5

9. $.58
× 8

10. $.89
× 6

11. $1.32
× 4

12. $2.03
× 3

13. $6.12
× 8

14. $5.03
× 6

15. $7.28
× 3

16. $4.68
× 5

17. $7.32
× 9

18. $5.75
× 6

19. $4.80
× 7

20. $6.78
× 4

21. 8 × $.37

22. 4 × $.70

23. 9 × $.35

24. 2 × $.98

25. 6 × $5.00

26. 3 × $8.75

27. 7 × $8.69

28. 9 × $6.08

Mixed Applications

Solve. You may need to use the Databank on page 519.

29. Lenny bought 3 packets of astronaut food. How much did they cost all together?

30. How much more do the kites cost than the Frisbees?

31. **What if** you had $10 to spend in the souvenir shop. What would you buy? How much money would you have left?

32. **Write a problem** about buying things in the souvenir shop. Ask others to solve your problem.

Mixed Review

Find the answer. Which method did you use?

33. $3.75
+ 2.98

34. $6.80
− 2.00

35. 80
× 5

36. 506
− 178

37. 200
× 7

MENTAL MATH
CALCULATOR
PAPER/PENCIL

Multiplying Larger Numbers

The 4 new Discovery space satellites will be launched together. They will explore every planet in the solar system. Each satellite weighs 1,958 lb. How much do they weigh all together?

WORKING TOGETHER

Work with a group. Have each one in the group use a different method to find the answer. You can use mental math, paper and pencil, or a calculator.

1. Is it easy to find the product mentally? Why or why not?

2. Show how to use paper and pencil to find the product.

3. Check to see if your answer is reasonable. How can you do this?

4. Which method was fastest?

5. **What if** the satellites weighed 2,000 lb? Which method would be fastest?

Use all three methods to find 3 × 4,203.

6. Can you find the product mentally? Why or why not?

7. Use paper and pencil to find the product.

8. Which method was fastest: mental math, a calculator, or paper and pencil?

Find the product. Which method did you use?

9.	10.	11.	12.	13.
2,415	6,001	$40.20	7,369	$56.89
× 7	× 5	× 3	× 8	× 2

SHARING IDEAS

14. How is multiplying 4-digit numbers like multiplying 3-digit or 2-digit numbers?

15. Tell how you decided which method to use in Exercises 9–13.

16. When do you think it is easier to use mental math? paper and pencil? a calculator? Tell why.

17. Is the calculator always easier to use with larger numbers? Why or why not?

ON YOUR OWN

MENTAL MATH
CALCULATOR
PAPER/PENCIL

Solve. Which method did you use?

18. Attendance at the Space Museum is about 4,500 on Saturdays in the summer. About what is the total attendance for 5 Saturdays in the summer?

19. One day 1,020 people each paid $8 to see a show about travel to Venus. What was the total amount paid to see the show?

20. A research satellite circles Earth 5,419 times a year. How many times will it circle Earth if it stays up 6 years?

21. *Write a problem* involving multiplication with a large factor. Choose which method you think should be used to solve it. Trade problems with other students and compare methods.

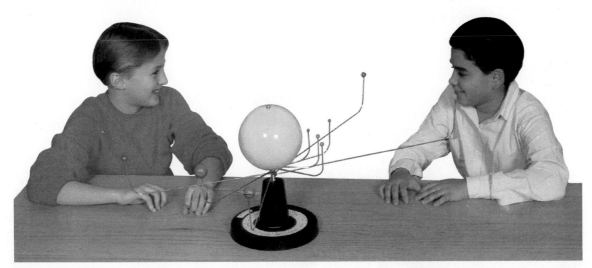

PROBLEM SOLVING

Strategy: Using Estimation

Sally manages the Space Museum gift shop. She wants to have at least 500 models of spaceships in stock. She has 243 models and expects a shipment of 125 next week. She plans to order another 230 models. Will she have enough models?

To make sure there will be enough models, Sally plans to **underestimate.**

Sally tries her plan. She rounds *down* to get an underestimate.

$$
\begin{array}{rcr}
243 & \to & 200 \\
125 & \to & 100 \\
230 & \to & +\,200 \\
\hline
& & 500
\end{array}
$$

1. Will she have enough models?

2. Why can Sally be sure she has enough if she underestimates?

Robert is in charge of selling tickets to the sky show. The auditorium seats 400 people. The museum sells 197 tickets through the mail, 72 over the phone, and 83 at the box office. Will there be enough seats for all ticket holders?

To make sure there are enough seats, Robert plans to **overestimate** the number of tickets sold. So he rounds *up.*

$$
\begin{array}{rcr}
197 & \to & 200 \\
72 & \to & 100 \\
83 & \to & +\,100 \\
\hline
& & 400
\end{array}
$$

3. Will there be enough seats?

4. Why can Robert be sure he has enough if he overestimates?

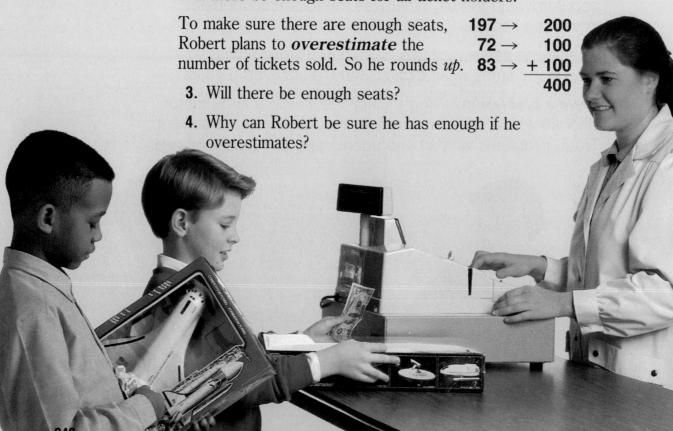

240

PRACTICE

Use estimation to solve the problem. Did you overestimate or underestimate?

5. Three astronauts who will travel together must not weigh more than 500 pounds. If two of the astronauts weigh 167 pounds and 153 pounds, can the third one weigh 205 pounds?

6. The record one-day attendance at the museum was 732. This morning 512 people came. If 245 people come this afternoon, will this break the attendance record?

7. The museum cafeteria plans to give away at least 500 samples of space food. It has 135 cookies, 163 pieces of fruit, and 348 nuts. Does it have enough samples?

8. Guido wants to buy a space set for $10.95, a planet book for $7.75, and a telescope for $109.50. Is $130.00 enough to pay for these items?

Strategies and Skills Review

9. Sally calculates that 180 people came into the store today. She knows that 22 people came in each hour and that the store is open for 6 hours. Is her answer reasonable? Why or why not?

10. Sally sells 45 posters of Jupiter. She also sells 69 buttons and 3 books on Jupiter. How many posters and buttons does she sell?

11. Robert is looking at the attendance figures for the sky show. Last month 357 people saw the show. This month 402 people saw the show. How many more people saw the show this month?

12. *Write a problem* that can be solved by estimation. Solve your problem. Ask others to solve your problem.

DECISION MAKING

Problem Solving: Planning a Trip to Mars

SITUATION

The Sanger School Rocket Club wrote to NASA for data to plan a trip to Mars. They found that Earth and Mars pass near each other about every 780 days. For a long trip, the club could fly to Mars as the planets pass near each other. They could explore a long time, flying home as the planets get close again. For a short trip, the club could fly to Mars just before the planets pass near each other. They could stay a short time and fly back before the planets get too far apart.

PROBLEM

Should they take the long trip or the short trip?

DATA

DATA FOR MARS TRIP

	LONG TRIP	SHORT TRIP
Travel Time	6 months one way	7 months one way
Exploration Time	18 months	1 month
Total Time	30 months (912 days)	15 months (456 days)
Daily Food Supply	3 pounds per person	3 pounds per person
Daily Water Supply	5 pounds per person	5 pounds per person
Daily Oxygen Supply	2 pounds per person	2 pounds per person

COMPARISON OF TRIPS

LONG TRIP
o more back-up equipment needed
o less room for research equipment
o more time to research and explore
o more strain on equipment and people

SHORT TRIP
o less back-up equipment needed
o more room for research equipment
o less time to research and explore
o less strain on equipment and people

RESEARCH PROJECTS FOR TRIP

1) Collect and study soil samples
2) Look for signs of life, past or present
3) Draw map of entire planet
4) Study the effects of weightlessness and lower Martian gravity on the human body

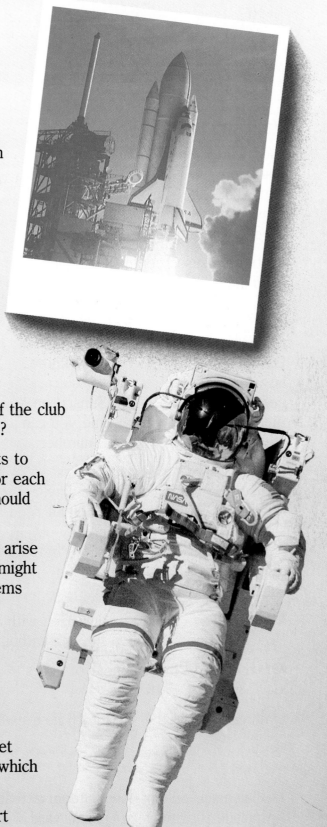

USING THE DATA

1. How many pounds of food will each person need for the long trip? the short trip?

2. How many pounds of water will each person need for the long trip? the short trip?

3. How many pounds of oxygen will each person need for the long trip? the short trip?

4. How many pounds of supplies will each person need in all for the long trip? the short trip?

MAKING DECISIONS

5. On which trip would the members of the club be able to find out more about Mars?

6. Consider the research the club wants to complete and the supplies needed for each crew member. How many people should the club send to Mars?

7. What are some problems that might arise in 30 months away from Earth that might not arise in 15 months? What problems might not become as serious in a shorter time?

8. A lot of time, money, and effort are invested in a trip to another planet. Which trip will offer more in return for all that time, money, and effort?

9. **Write a list** of things that the Rocket Club should think about in deciding which trip to take.

10. Would you take a long trip or a short trip if you were going to Mars? How many people would you take? Why?

CURRICULUM CONNECTION

Math and Health

Have you ever had a sunburn? Then you know that the sun's rays can hurt your skin. The sun's rays can also cause wrinkles or more serious skin problems in the years to come.

Putting on sunscreen protects skin from sunburn. Sunscreens are labeled with **SPF** numbers. SPF stands for **Sun Protection Factor**. The higher the SPF number, the more protection the sunscreen gives. You may know how long you can stay in the sun before you start to burn. Sunscreen multiplies the amount of time you can stay in the sun by the SPF number.

Lynn can stay in the sun 2 hours without burning. She plans to play baseball for 4 hours. Using a sunscreen with an SPF of 2, Lynn can stay out 2 times as long, or 4 hours. To be safe, she should use a sunscreen with an even higher SPF.

What if Lynn chooses a sunscreen with an SPF of 8? How long can she stay in the sun without burning?

Think: $2 \times \text{SPF} = \text{number of hours}$
$2 \times 8 = 16$

Lynn can stay in the sun for 16 hours without burning.

ACTIVITY

1. Plan a schedule of your outdoor activities for a weekend day. Include activities like sports, yard work, or watching a sporting event. After you have planned your schedule, decide what sunscreen you will need. How did you decide?

Computer Spreadsheets

Six space missions need to be equipped with food packs for the astronauts. You want to know how many food packs each mission needs. You can organize your data on a spreadsheet. If you input your data into a computer spreadsheet program, the computer will arrange the data in rows and columns and compute the totals.

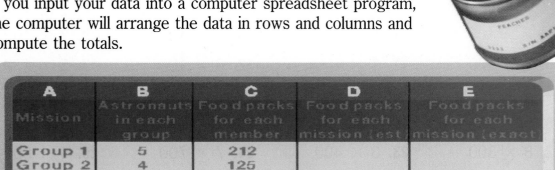

A Mission	B Astronauts in each group	C Food packs for each member	D Food packs for each mission (est	E Food packs for each mission (exact
Group 1	5	212		
Group 2	4	125		
Group 3	8	87		
Group 4	3	129		
Group 5	4	175		
Group 6	6	99		

DATA

On a sheet of paper, copy the computer spreadsheet above. Fill in columns A, B, and C with the data given. Compute the amounts for columns D and E using mental math, paper and pencil, or a calculator.

THINKING ABOUT COMPUTERS

1. What parts of the spreadsheet would change if mission 3 had one more member? Which mission would then need the most food packs?

2. What parts of the spreadsheet would change if each member needed twice as many food packs?

3. Why is it useful to have column D? Would column D be necessary if you were using a computer? Why?

4. NASA wants to use the spreadsheet for missions that will involve hundreds of astronauts. Why would a computer be the most efficient way to make the computations?

EXTRA PRACTICE

Mental Math: Multiplying 10s and 100s, page 221 ....

Multiply mentally.

1.　10　　**2.**　100　　**3.**　1,000　　**4.**　30　　**5.**　600
　　× 3　　　　× 9　　　　× 5　　　　× 2　　　　× 5

6. 2 × 10　　　　**7.** 8 × 10　　　　**8.** 4 × 10　　　　**9.** 7 × 10

10. 3 × 100　　　**11.** 1 × 100　　　**12.** 6 × 100　　　**13.** 5 × 100

14. 4 × 1,000　　**15.** 2 × 1,000　　**16.** 8 × 1,000　　**17.** 7 × 1,000

18. 5 × 20　　　**19.** 8 × 50　　　**20.** 5 × 30　　　**21.** 8 × 60

22. 8 × 300　　　**23.** 3 × 400　　　**24.** 7 × 700　　　**25.** 4 × 800

Estimating Products, page 223 ....

Estimate. Use rounding.

1.　38　　**2.**　72　　**3.**　19　　**4.**　$.38　　**5.**　318
　　× 7　　　　× 4　　　　× 5　　　　× 8　　　　× 3

6.　987　　**7.**　$612　　**8.**　4,012　　**9.**　8,842　　**10.**　$2,841
　　× 4　　　　× 3　　　　× 5　　　　× 7　　　　× 6

11. 3 × 47　　　**12.** 6 × 53　　　**13.** 4 × $33　　　**14.** 8 × 317

15. 6 × 82　　　**16.** 5 × 466　　　**17.** 8 × $419　　　**18.** 3 × 8,728

Estimate. Use the front digits.

19.　23　　**20.**　49　　**21.**　62　　**22.**　$57　　**23.**　385
　　× 9　　　　× 6　　　　× 5　　　　× 8　　　　× 7

24.　333　　**25.**　$689　　**26.**　4,307　　**27.**　4,189　　**28.**　6,487
　　× 7　　　　× 5　　　　× 8　　　　× 4　　　　× 9

29. 5 × 14　　　**30.** 8 × 76　　　**31.** 3 × $85　　　**32.** 2 × 899

33. 7 × 91　　　**34.** 6 × 577　　　**35.** 9 × $529　　　**36.** 4 × 9,228

Problem Solving Strategy: Guess, Test, and Revise, page 225 ....

Use the guess, test and revise strategy to solve the problem.

1. Chris, Bret and Greg make 14 fliers to advertise the garage sale. Each boy makes a different number of fliers. Each boy makes fewer than 7 fliers. How many does each boy make?

2. Harry was sailing and fishing at the pond for 6 hours. He sailed for twice as long as he fished. How long did he sail?

3. Think of two numbers whose sum is 25 and whose difference is 3. What are the numbers?

4. Liza is two times as old as her brother. The sum of their ages is 21. How old is Liza?

Multiplying 2-Digit Numbers—Regrouping Once, page 229 ....

Multiply.

1. 12 × 2

2. 19 × 2

3. 11 × 3

4. 14 × 3

5. 11 × 7

6. 25 × 2

7. 13 × 3

8. 49 × 2

9. 31 × 3

10. 27 × 3

11. 33 × 3

12. 17 × 5

13. 32 × 4

14. 46 × 2

15. 17 × 2

16. 62 × 3

17. 13 × 7

18. 23 × 4

19. 17 × 3

20. 16 × 5

21. 13 × 7

22. 15 × 4

23. 36 × 2

24. 21 × 5

25. 7 × 13   **26.** 6 × 18   **27.** 9 × 21   **28.** 4 × 19

29. 5 × 18   **30.** 9 × 31   **31.** 2 × 22   **32.** 4 × 32

33. 3 × 14   **34.** 4 × 14   **35.** 2 × 49   **36.** 5 × 15

EXTRA PRACTICE

Multiplying 2-Digit Numbers—Regrouping Twice, page 231 ....

Multiply.

1. 22 $\times$ 7	**2.** 32 $\times$ 9	**3.** 43 $\times$ 5	**4.** 28 $\times$ 7	**5.** 42 $\times$ 8	**6.** 22 $\times$ 9
7. 43 $\times$ 7	**8.** 37 $\times$ 4	**9.** 58 $\times$ 5	**10.** 46 $\times$ 3	**11.** 55 $\times$ 7	**12.** 36 $\times$ 6
13. 32 $\times$ 5	**14.** 62 $\times$ 8	**15.** 26 $\times$ 7	**16.** 34 $\times$ 9	**17.** 44 $\times$ 3	**18.** 43 $\times$ 6
19. 73 $\times$ 9	**20.** 44 $\times$ 8	**21.** 46 $\times$ 3	**22.** 93 $\times$ 5	**23.** 97 $\times$ 8	**24.** 99 $\times$ 9

25. 3 $\times$ 66   **26.** 7 $\times$ 42   **27.** 9 $\times$ 53   **28.** 5 $\times$ 42

29. 6 $\times$ 69   **30.** 2 $\times$ 57   **31.** 4 $\times$ 63   **32.** 9 $\times$ 66

Multiplying 3-Digit Numbers, page 235 ....

Multiply.

1. 113 $\times$ 3	**2.** 206 $\times$ 8	**3.** 417 $\times$ 5	**4.** 522 $\times$ 7	**5.** 370 $\times$ 5
6. 302 $\times$ 5	**7.** 819 $\times$ 6	**8.** 638 $\times$ 3	**9.** 409 $\times$ 8	**10.** 537 $\times$ 4
11. 738 $\times$ 3	**12.** 603 $\times$ 9	**13.** 431 $\times$ 7	**14.** 240 $\times$ 5	**15.** 652 $\times$ 8

16. 4 $\times$ 313   **17.** 6 $\times$ 385   **18.** 7 $\times$ 517   **19.** 8 $\times$ 652

20. 2 $\times$ 783   **21.** 6 $\times$ 329   **22.** 5 $\times$ 438   **23.** 9 $\times$ 849

24. 7 $\times$ 413   **25.** 4 $\times$ 865   **26.** 9 $\times$ 989   **27.** 5 $\times$ 565

Multiplying Money, page 237 ....

Multiply.

1. $.42
   × 3

2. $.16
   × 8

3. $.22
   × 7

4. 38¢
   × 5

5. 73¢
   × 4

6. $1.23
   × 6

7. $2.12
   × 8

8. $4.37
   × 5

9. $5.18
   × 3

10. $8.62
    × 9

11. $6.54
    × 2

12. $3.89
    × 4

13. $4.75
    × 7

14. $7.31
    × 9

15. $8.95
    × 5

16. 3 × $.47

17. 8 × $.60

18. 7 × 55¢

19. 2 × 89¢

20. 5 × $6.00

21. 4 × $5.85

22. 3 × $7.09

23. 6 × $8.73

Problem Solving Strategy: Using Estimation, page 241 ....

Use estimation to solve. Did you overestimate or underestimate?

1. Michael has $45 to spend on books. He wants to buy a computer book for $14.95, a history book for $28.85, and a science fiction novel for $10.90. Does he have enough money to buy all three books?

2. The sponsors of the Book Fair projected that 187 students would attend the three-day fair. There were 67 students on Monday, 55 on Tuesday, and 43 on Wednesday. Did the Book Fair attract as many students as projected?

3. The author of a new fitness book hoped to speak to at least 70 students at the Book Fair. She talked to 26 students on Monday, 24 on Tuesday, and 39 on Wednesday. Did she meet her goal?

4. Kayla bought a book cover for $2.19 and a calendar for $7.69. Is $10 enough money to pay for these items?

Practice PLUS

KEY SKILL: Multiplying 2-Digit Numbers: Regrouping Twice
(Use after page 231.)

Level A

Multiply.

1. 23 × 5	**2.** 19 × 7	**3.** 39 × 3	**4.** 55 × 2	**5.** 66 × 3
6. 54 × 5	**7.** 22 × 7	**8.** 38 × 4	**9.** 25 × 4	**10.** 76 × 3

11. Robin has 7 bags of apples. Each bag has 16 apples. How many apples are there in all?

Level B

Multiply.

12. 83 × 8	**13.** 26 × 5	**14.** 27 × 4	**15.** 23 × 9	**16.** 86 × 2

17. 8 × 42   **18.** 5 × 45   **19.** 3 × 37   **20.** 7 × 29

21. Terry has 6 packages of baseball cards. There are 24 cards in each package. How many baseball cards does Terry have?

Level C

Multiply.

22. 7 × 65   **23.** 8 × 47   **24.** 4 × 75   **25.** 6 × 87

26. 8 × 28   **27.** 7 × 98   **28.** 5 × 57   **29.** 9 × 74

30. 9 × 94   **31.** 8 × 35   **32.** 6 × 52   **33.** 7 × 49

34. Jake filled 69 pages in his photo album. Each page holds 9 photos. How many photos are there in all?

PRACTICE *PLUS*

KEY SKILL: Multiplying Money (Use after page 237.)

Level A
Multiply.

1. $.20
    × 2

2. $.16
    × 3

3. $.24
    × 4

4. $.62
    × 3

5. $.35
    × 6

6. $.39
    × 7

7. $.42
    × 3

8. $.55
    × 6

9. $.80
    × 3

10. $1.04
    × 2

11. Cleo bought 4 softballs. Each ball cost $4.89. How much money did she spend in all?

Level B
Multiply.

12. $.85
    × 4

13. $.93
    × 2

14. $.59
    × 7

15. $.45
    × 8

16. $.63
    × 7

17. $.89
    × 5

18. $8.25
    × 6

19. $4.12
    × 7

20. $8.04
    × 6

21. $9.33
    × 4

22. 2 × $4.92   23. 4 × $5.19   24. 5 × $3.88   25. 3 × $9.75

26. Jeremy bought 6 tickets to the school play. Each ticket cost $5.95. How much money did he spend?

Level C
Multiply.

27. 3 × $7.98   28. 9 × $2.73   29. 7 × $6.08   30. 6 × $6.35

31. 5 × $8.13   32. 8 × $5.25   33. 9 × $4.24   34. 7 × $9.36

35. 6 × $3.15   36. 9 × $7.28   37. 7 × $4.52   38. 9 × $8.96

39. Linda bought 9 compact discs. Each disc cost $9.98. How much money did she spend in all?

CHAPTER REVIEW

LANGUAGE AND MATHEMATICS

Complete the sentences. Use the words in the chart.

1. The ■ is the answer in a multiplication problem. *(page 220)*

2. Converting 20 ones to 2 tens is an example of ■. *(page 228)*

3. ■ estimation results in an estimate that is less than the exact answer. *(page 222)*

4. In the amount $10.75 the 7 is a ■ place representing tenths of a dollar. *(page 236)*

5. *Write a definition* or give an example of the words you did not use from the chart.

VOCABULARY
regrouping
front-end
product
factor
decimal
rounding

CONCEPTS AND SKILLS

Multiply mentally. *(page 220)*

6.	**7.**	**8.**	**9.**	**10.**
10	20	30	10	70
× 4	× 2	× 6	× 9	× 8

11. 5 × 400   **12.** 7 × 600   **13.** 3 × 500   **14.** 7 × 2,000

15. 9 × 1,000   **16.** 1 × 9,000   **17.** 3 × 8,000   **18.** 8 × 5,000

Estimate. Use rounding. *(page 222)*

19.	**20.**	**21.**	**22.**	**23.**
28	87	37	62	$74
× 5	× 4	× 8	× 7	× 6

24. 3 × 92   **25.** 5 × $78   **26.** 9 × 324   **27.** 6 × 4,812

Estimate. Use the front digits. *(page 222)*

28.	**29.**	**30.**	**31.**	**32.**
18	49	82	$58	367
× 4	× 6	× 3	× 7	× 5

33. 2 × 137   **34.** 8 × 423   **35.** 9 × $811   **36.** 4 × 5,976

Multiply. *(pages 228–231, 234–239)*

37. 16
  × 4

38. 34
  × 5

39. 84
  × 2

40. 23
  × 8

41. $.17
  × 7

42. 39
  × 6

43. $.45
  × 4

44. 77
  × 2

45. $13
  × 9

46. 48
  × 6

47. 612
  × 4

48. 423
  × 3

49. $517
  × 5

50. 736
  × 6

51. $7.68
  × 3

52. 399
  × 9

53. $8.52
  × 7

54. $961
  × 6

55. 279
  × 8

56. 744
  × 8

57. 9 × 327

58. 2 × $.39

59. 7 × $5.02

60. 4 × 286

61. 8 × $.46

62. 3 × $4.25

63. 9 × $.73

64. 5 × 139

CRITICAL THINKING

Find the missing number. Tell how you found it. *(page 228)*

65. 35 × ■ = 35

66. 19 × 6 = ■ × 19

67. If you multiply a three-digit number by a one-digit number other than 0, what is the smallest product you can get? *(page 234)*

MIXED APPLICATIONS

68. There are twice as many dogs as cats at the kennel. If there are 48 animals at the kennel, how many are dogs and how many are cats? *(page 224)*

69. Julio wants to buy a bicycle pump for $59.95, a tire for $12.25, and a tube for $3.75. Is $80 enough to pay for these items? *(page 240)*

70. Bill planted trees for the Forest Service for 5 days. Each day, he planted 455 trees. How many trees did he plant? *(page 234)*

71. Alva bought 3 shirts for $7.99 each. How much did she spend? *(page 236)*

CHAPTER TEST

Multiply mentally.

1. 50
  × 3

2. 20
  × 6

3. 5 × 600

4. 8 × 500

Estimate by rounding.

5. 592
  × 7

6. $82
  × 3

7. $211 × 9

8. 3,690 × 4

Estimate by using front digits.

9. 13
  × 7

10. 57
  × 5

11. 432 × 6

12. 225 × 8

Multiply.

13. 51
  × 3

14. 223
  × 2

15. 19
  × 5

16. $.84
  × 2

17. 42
  × 8

18. 292
  × 9

19. $5.86
  × 2

20. 105
  × 7

21. 6 × 10

22. 5 × 466

23. 8 × $4.19

Solve.

24. The theater seats 450 people. There are twice as many children as adults in it. How many children are there? How many adults?

25. Karla collects $95 for charity. Sam and Dave collect $30 each. If Sid collects $60, will they reach their goal of collecting $200 dollars?

NAPIER'S RODS

About 400 years ago a Scottish mathematician named John Napier invented this method of multiplication.

Here is how Napier would have used the rods to find the product 3 × 698.

Step 1 Place the rods for 6, 9, and 8 next to the "times" rod.

Step 2 Use the numbers in row 3.

Step 3 Start at the right and add along the diagonals. Regroup when necessary.

3 × 698 = 2,094

Make your own set of Napier's Rods using strips of paper. Then use the rods to find the product.

1. 5 × 274   **2.** 8 × 175   **3.** 4 × 549   **4.** 3 × 637

5. Can you think of a way to use Napier's Rods to find the product of 40 × 387? Experiment with the rods to see if your idea works.

CUMULATIVE REVIEW

Choose the letter of the correct answer.

1. Estimate by rounding: 6 × $324
   - **a.** $18
   - **b.** $180
   - **c.** $1,800
   - **d.** $8,000

2. How many are in 3 groups of 5?
   - **a.** 2
   - **b.** 2 R2
   - **c.** 15
   - **d.** not given

3. How long does a movie last?
   - **a.** seconds
   - **b.** minutes
   - **c.** days
   - **d.** not given

4. $55.95 + $3.08 + $6.11
   - **a.** $15.14
   - **b.** $65.86
   - **c.** $147.85
   - **d.** not given

5. 4 × $6.28
   - **a.** $25.82
   - **b.** $25.12
   - **c.** $24.82
   - **d.** not given

6. Find the product of 8 and 72.
   - **a.** 9
   - **b.** 566
   - **c.** 567
   - **d.** not given

7. Subtract 312 from 5,000.
   - **a.** 4,688
   - **b.** 4,798
   - **c.** 4,968
   - **d.** not given

8. What is 3 times 34?
   - **a.** 102
   - **b.** 92
   - **c.** 72
   - **d.** not given

9. 6 × 900
   - **a.** 5,600
   - **b.** 5,400
   - **c.** 560
   - **d.** not given

10. Multiply 9 × 452.
    - **a.** 3,658
    - **b.** 3,668
    - **c.** 4,068
    - **d.** not given

11. What is 24 ÷ 6?
    - **a.** 3
    - **b.** 4
    - **c.** 144
    - **d.** not given

12. 5 × 308
    - **a.** 1,540
    - **b.** 1,530
    - **c.** 15,040
    - **d.** not given

13. What is 28 minutes after 4:53 P.M.?
    - **a.** 4:25 P.M.
    - **b.** 5:20 P.M.
    - **c.** 5:21 P.M.
    - **d.** not given

14. 7 × $.61
    - **a.** $.42
    - **b.** $4.27
    - **c.** $42.70
    - **d.** not given

Dividing by 1-Digit Divisors

MATH CONNECTIONS: STATISTICS • PROBLEM SOLVING

Grapefruit
$1.00
each

Pineapple
$3.00
each

Apples
$ .75
per pound

Kiwi Fruit
2 for
$1.00

Oranges
$1.00
per pound

Pears
$1.25
per pound

1. What information do you see in this picture of a vegetable stand?

2. How can you use this information?

3. Write a problem about this picture.

DEVELOPING A CONCEPT
Dividing with Remainders

A. Earl has 13 pumpkins. He packs 4 pumpkins in each carton. He sees that he has 3 full cartons and 1 pumpkin left over.

Sometimes when you try to divide a number of objects into groups of equal size, you have some objects left over. The number of objects left over is called the **remainder**.

WORKING TOGETHER

You can do this activity to learn more about remainders.

Step 1 Copy the table at the right. Notice that the box for 13 shows a remainder of 1.

Step 2 Place 14 counters on your desk. Divide them into equal groups of 4. Record the remainder in the table.

Step 3 Make equal groups of 4 using 15, 16, and 17 counters. Record the remainders in the table.

4 EQUAL GROUPS

Total Number of Counters	13	14	15	16	17
Remainder	1				

1. What is the largest remainder you found? Is this the largest remainder you could get when making equal groups of 4? Why?

B. You can find a quotient and remainder without using counters.

Divide: 13 ÷ 4

Step 1

Think: How many fours are in 13?

$$2 \times 4 = 8 \qquad 8 < 13$$
$$3 \times 4 = 12 \qquad 12 < 13$$
$$4 \times 4 = 16 \qquad 16 > 13$$

There are 3 fours in 13.

Step 2

Divide.

Write 3 in the quotient.
Multiply. $3 \times 4 = 12$
Subtract. $13 - 12 = 1$
Compare. $1 < 4$
Write the remainder.

$$\begin{array}{r} 3 \text{ R1} \\ 4\overline{)13} \\ -12 \\ \hline 1 \end{array}$$

SHARING IDEAS

2. Can the remainder be greater than the divisor? Why or why not?

3. How can you use multiplication and subtraction to help you find the remainder in a division problem?

PRACTICE

Find the quotient and remainder.

4. $4\overline{)6}$ **5.** $8\overline{)9}$ **6.** $5\overline{)9}$ **7.** $2\overline{)9}$ **8.** $3\overline{)8}$ **9.** $6\overline{)8}$

10. $6\overline{)15}$ **11.** $9\overline{)20}$ **12.** $7\overline{)16}$ **13.** $2\overline{)17}$ **14.** $8\overline{)12}$ **15.** $5\overline{)11}$

16. $3\overline{)10}$ **17.** $5\overline{)27}$ **18.** $6\overline{)14}$ **19.** $4\overline{)19}$ **20.** $2\overline{)19}$ **21.** $4\overline{)18}$

22. $8\overline{)33}$ **23.** $6\overline{)35}$ **24.** $7\overline{)52}$ **25.** $9\overline{)80}$ **26.** $8\overline{)50}$ **27.** $5\overline{)47}$

28. $65 \div 8$ **29.** $41 \div 7$ **30.** $39 \div 5$ **31.** $16 \div 6$ **32.** $30 \div 4$

33. $26 \div 9$ **34.** $60 \div 8$ **35.** $42 \div 8$ **36.** $32 \div 5$ **37.** $52 \div 6$

38. What is 9 divided by 2? **39.** What is 30 divided by 8?

Write the missing number.

40. $31 \div 4 = 7\ R\blacksquare$ **41.** $23 \div 3 = \blacksquare\ R2$ **42.** $35 \div \blacksquare = 5\ R5$

43. $25 \div \blacksquare = 6\ R1$ **44.** $36 \div 7 = 5\ R\blacksquare$ **45.** $19 \div 2 = \blacksquare\ R1$

Mixed Applications

46. There are 35 oranges in a box. If 8 children each get the same number of oranges, how many oranges will be left over?

47. Sam is putting 34 tomatoes in each carton. He fills 6 cartons. How many tomatoes does he pack all together?

48. There is room to park 5 cars in each row of the parking lot. How many rows are full if there are 32 cars?

49. Millie's Market opens at 12:30 P.M. each day. It closes at 6:00 P.M. For how much time is the market open?

UNDERSTANDING A CONCEPT

Mental Math: Dividing 10s and 100s

Curtis, Fran, and Gary just finished playing a game. Each player scored an equal number of points in each move he or she made. Curtis scored 600 points in 3 moves, Fran scored 2,800 points in 4 moves, and Gary scored 3,000 points in 5 moves. Who scored the most points in each move?

Curtis	Fran	Gary
6 ÷ 3 = 2	28 ÷ 4 = 7	30 ÷ 5 = 6
60 ÷ 3 = 20	280 ÷ 4 = 70	300 ÷ 5 = 60
600 ÷ 3 = 200	2,800 ÷ 4 = 700	3,000 ÷ 5 = 600

Fran scored 700 points, the most points in each move.

1. Look at Curtis's examples. Compare the number of zeros in each dividend with the number of zeros in each quotient. What pattern do you see?

2. Does the pattern work in Fran's examples? in Gary's examples? Why or why not?

TRY OUT Write the letter of the correct answer.

3. 2)‾80̅   **a.** 4   **b.** 40   **c.** 400   **d.** 4,000

4. 5)‾1,000̅   **a.** 2   **b.** 20   **c.** 200   **d.** 2,000

5. 420 ÷ 7   **a.** 6   **b.** 60   **c.** 600   **d.** 6,000

6. 3,000 ÷ 6   **a.** 5   **b.** 50   **c.** 500   **d.** 5,000

PRACTICE

Divide. Use mental math.

7. 4)40   **8.** 2)60   **9.** 9)90   **10.** 3)90   **11.** 7)70

12. 4)80   **13.** 2)20   **14.** 5)50   **15.** 2)40   **16.** 3)30

17. 8)320   **18.** 4)160   **19.** 3)210   **20.** 7)560   **21.** 6)300

22. 5)200   **23.** 2)400   **24.** 4)800   **25.** 3)600   **26.** 8)400

27. 8)720   **28.** 6)480   **29.** 9)630   **30.** 7)490   **31.** 8)640

32. 4)2,000   **33.** 3)9,000   **34.** 6)4,200   **35.** 9)8,100   **36.** 7)4,200

37. 60 ÷ 6   **38.** 80 ÷ 2   **39.** 300 ÷ 5   **40.** 600 ÷ 2

41. 3,500 ÷ 7   **42.** 2,000 ÷ 5   **43.** 5,400 ÷ 9   **44.** 4,000 ÷ 8

CRITICAL THINKING

45. How would you find the quotient 30,000 ÷ 5? What is the quotient?

Mixed Applications

46. Fran earned 800 points in 2 moves. She made the same number of points in each move. what is that number?

47. The friends started playing the game at 3:15 P.M. They played for 3 hours and 15 minutes. When did they stop playing?

48. Gary made 1,400 points in each of the first 2 games. What was his total score for the 2 games?

49. There are 60 game cards. Each of the 3 players got the same number of cards. How many cards did each player get?

Mixed Review

Find the answer. Which method did you use?

MENTAL MATH
CALCULATOR
PAPER/PENCIL

50. 3 × $6.95   **51.** 8,400 − 200   **52.** $32.65 + $15.98

53. 64 ÷ 8   **54.** 420 + 600   **55.** $5.05 − $3.78

Estimating Quotients

Ed has 6 trucks and 255 crates of potatoes. He puts the same number of crates on each truck. About how many crates does he put on each truck?

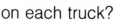

Estimate: 255 ÷ 6
Here are two methods you can use.

Rounding	**Compatible Numbers**
Round the dividend to its greatest place.	Use a basic fact.
Think: 255 rounds to 300.	***Think:*** 255 is close to 240.
300 ÷ 6 = 50	**240 ÷ 6 = 40**
He puts about 50 crates on each truck.	He puts about 40 crates on each truck.

1. Which estimate is greater than the exact answer? Which estimate is less? How do you know?

2. Estimate 2,315 ÷ 3 by using compatible numbers. Then estimate by rounding. Which method is more helpful in finding an estimate? Why? What is your estimate?

T**RY OUT** Write the letter of the correct answer.

Estimate by rounding.

3. 4)78   **a.** 2   **b.** 20   **c.** 80   **d.** 200

4. 6,318 ÷ 2   **a.** 3   **b.** 300   **c.** 3,000   **d.** 6,000

Estimate by using compatible numbers.

5. 3)716   **a.** 2   **b.** 20   **c.** 200   **d.** 700

6. 3,059 ÷ 7   **a.** 40   **b.** 400   **c.** 3,000   **d.** 4,000

PRACTICE

Estimate by rounding.

7. 2)38 **8.** 4)84 **9.** 3)87 **10.** 5)376 **11.** 2)635

12. 8)763 **13.** 6)295 **14.** 3)5,718 **15.** 4)2,206 **16.** 2)7,703

Estimate by using compatible numbers.

17. 4)71 **18.** 3)72 **19.** 5)61 **20.** 2)79 **21.** 6)75

22. 6)538 **23.** 8)495 **24.** 3)595 **25.** 7)320 **26.** 2)750

27. 4)1,575 **28.** 9)7,138 **29.** 6)2,068 **30.** 5)6,310 **31.** 8)3,844

Estimate the quotient.

32. 62 ÷ 3   **33.** 82 ÷ 3   **34.** 88 ÷ 4   **35.** 73 ÷ 4

36. 514 ÷ 5   **37.** 758 ÷ 8   **38.** 434 ÷ 6   **39.** 392 ÷ 8

40. 346 ÷ 5   **41.** 906 ÷ 4   **42.** 1,235 ÷ 2   **43.** 4,106 ÷ 3

44. 6,218 ÷ 3   **45.** 4,475 ÷ 5   **46.** 2,602 ÷ 5   **47.** 4,376 ÷ 9

Mixed Applications

Find the answer. Did you estimate or find an exact answer?

48. Carl had 70 crates of potatoes on his truck. He sold them for $9 per crate. How much money did he get?

49. One of Ed's drivers drove 596 km in 3 days. About how many kilometers did the driver travel each day?

50. Ed has $1,150 to give to the 6 drivers. He gives each driver the same amount. About how much money does he give each driver?

51. The workers picked 1,768 potatoes the first week. They picked 2,158 potatoes the second week. How many potatoes did they pick in the two weeks?

PROBLEM SOLVING

UNDERSTAND
PLAN
TRY
CHECK
EXTEND

Strategy: Making a Table

Nancy, Art, Mica, and Tyrone are picking fruit on Mr. Lyman's farm. Each picks a different kind of fruit. Nancy picks apples. Art does not pick plums or grapes. Tyrone picks plums. Who picks grapes?

You can solve the problem by making a table.

You know Nancy picks apples. Write *Yes* in the box for Nancy.

1. Why can you write *No* in the other boxes next to Nancy?

2. Why can you write *No* in the other boxes under Apples?

	Apples	Plums	Pears	Grapes
Nancy	Yes	No	No	No
Art	No			
Mica	No			
Tyrone	No			

You know Art does not pick plums or grapes.

3. Why can you write *Yes* in the box under Pears for Art?

You know Tyrone picks plums.

4. Complete the rest of the table. Who picks grapes?

5. How does making a table help you solve the problem?

	Apples	Plums	Pears	Grapes
Nancy	Yes	No	No	No
Art	No	No	Yes	No
Mica	No		No	
Tyrone	No		No	

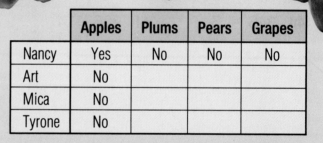

264 Lesson 7–4

PRACTICE

Make a table to solve.

6. Mr. Lundgren did a different chore each day. He did not clean his tools on Tuesday. He went to the market on Monday. He fixed the fence on Thursday. He did not paint the barn door on Wednesday. He took the rest of the week off. On what day did he paint the barn door?

7. Anna, Jo, and Sue each planted one kind of flower. Anna planted daisies. Walter will not touch roses. Who planted mums?

8. Bob, Lucy, Carl, and Rose each feed one kind of animal. Rose feeds the pigs. Bob does not feed the chickens. Lucy feeds the rabbits. Who feeds the chickens?

9. The farmer's children know what they want to be when they grow up. Bob wants to be a farmer. Lucy does not want to teach. Carl and Rose do not want to be movie stars. Carl does not want to be a truck driver. Who wants to be a teacher?

Strategies and Skills Review

Solve. Use mental math, a calculator, or paper and pencil.

10. Nancy, Art, and Tyrone each worked one day last week. They could work only on Monday, Tuesday, or Wednesday. Art did not work on Tuesday, and Tyrone did not work on Wednesday. Art worked before Tyrone. On what day did Nancy work?

11. Mrs. Gray plants 240 corn plants on her 16-acre farm. She plants them in rows of 8. How many rows does she plant?

12. Mrs. Gray buys 6 cows for $325 each and 1 horse for $275. How much does she pay for the animals?

13. **Write a problem** that can be solved by making a table. Solve your problem. Ask others to solve it.

EXPLORING A CONCEPT

Dividing by 1-Digit Divisors

Liane picked 56 apples. She packed them in 3 baskets. She put the same number in each basket. How many apples did she put in each basket? How many extra apples does she have?

You can divide to find the number of apples in each basket.

WORKING TOGETHER

Estimate. Then solve the problem by using place-value models.

1. How did you model the 56 apples?

2. Which models did you separate first, tens or ones?

3. How many tens did you put in each group? How many were left?

4. What did you do with the tens that were left? Why?

5. How many ones did you put in each group? How many were left?

6. How many apples did she put in each basket? How many extra apples does she have?

7. **What if** Liane picked 57 apples? How many apples would she put in each basket? How many extra apples would she have?

8. How many extra apples would Liane have if she picked 54 apples? 55 apples?

Use models to find each quotient and remainder.

9. $84 \div 4$ 10. $98 \div 8$ 11. $100 \div 6$ 12. $319 \div 2$

SHARING IDEAS

13. How do your models compare with those of others?

14. Why should you start with the largest place in the dividend when dividing?

ON YOUR OWN

Solve. Use counters if needed.

15. Treat has 46 tomatoes to pack. He packs the same number of tomatoes in each of 4 bags. How many tomatoes does he pack in each bag? How many tomatoes are extra?

16. Kay has 60 melons. She puts them into 2 cartons. She puts the same number in each carton. How many melons does she put in each carton?

17. Last harvest, 3 helpers picked a total of 114 cucumbers. Each helper picked the same number of cucumbers. How many cucumbers did each helper pick?

18. *Write a problem* that can be solved by using division. Solve your problem. Then ask others to solve it.

DEVELOPING A CONCEPT

Dividing 2-Digit Numbers

A. Juan and Holly have 41 plants to put in 3 boxes. They put the same number of plants in each box. To find how many plants will be in each box, Juan used models.

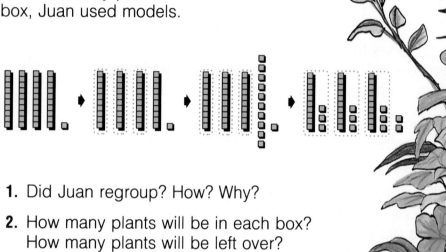

1. Did Juan regroup? How? Why?

2. How many plants will be in each box? How many plants will be left over?

B. Holly divided without using models. Here is what she did.

Divide: 41 ÷ 3

Step 1	Step 2	Step 3
Decide where to place the first digit of the quotient.	**Divide the tens.**	**Bring down the ones. Divide the ones.**

Step 1

■
3)41

Think: 3 < 4
There are enough tens to divide.

Step 2

$$\begin{array}{r} 1 \\ 3\overline{)41} \\ -3 \\ \hline 1 \end{array}$$

Think: $3\overline{)\overset{1}{4}}$
Multiply. $1 \times 3 = 3$
Subtract. $4 - 3 = 1$
Compare. $1 < 3$

Step 3

$$\begin{array}{r} 13 \text{ R2} \\ 3\overline{)41} \\ -3\downarrow \\ \hline 11 \\ -9 \\ \hline 2 \end{array}$$

Think: $3\overline{)11}$
Multiply. $3 \times 3 = 9$
Subtract. $11 - 9 = 2$
Compare. $2 < 3$
Write the remainder.

Check. Multiply, then add: $3 \times 13 = 39$; $39 + 2 = 41$

divisor ↓   dividend ↓
         ↑ quotient   ↑ remainder

SHARING IDEAS

3. Is the answer reasonable? How do you know?

4. Why can you multiply to check division?

5. How are Juan's and Holly's methods alike? How are they different?

6. How does each method show regrouping?

PRACTICE

Divide.

7. $4\overline{)23}$ **8.** $2\overline{)17}$ **9.** $6\overline{)49}$ **10.** $3\overline{)22}$ **11.** $7\overline{)30}$

12. $3\overline{)51}$ **13.** $6\overline{)79}$ **14.** $2\overline{)65}$ **15.** $3\overline{)36}$ **16.** $8\overline{)89}$

17. $5\overline{)75}$ **18.** $6\overline{)84}$ **19.** $4\overline{)29}$ **20.** $8\overline{)93}$ **21.** $9\overline{)50}$

22. $4\overline{)84}$ **23.** $5\overline{)71}$ **24.** $6\overline{)36}$ **25.** $2\overline{)76}$ **26.** $8\overline{)70}$

27. $58 \div 7$ **28.** $48 \div 4$ **29.** $78 \div 6$ **30.** $25 \div 3$ **31.** $47 \div 5$

32. $49 \div 5$ **33.** $63 \div 4$ **34.** $94 \div 6$ **35.** $25 \div 4$ **36.** $96 \div 8$

Mixed Applications

37. Holly has 70 new flowerpots to sell. She puts them in stacks of 5. How many stacks does she make?

38. Juan is making a new sign. He will use 6 different colors. Each can of paint costs $1.39. How much will Juan pay for paint?

39. Holly started the day with 65 rosebushes. By the end of the day, she had sold all but 7 of them. How many had she sold?

Mixed Review

Compare. Write $>$, $<$, or $=$.

40. $3 + 8 + 7 \bullet 20$ **41.** $210 \div 7 \bullet 90 \div 3$ **42.** $1 \times 8 \bullet 8 - 0$

43. $650 - 310 \bullet 400$ **44.** $6 \times 175 \bullet 4 \times 230$ **45.** $13 - 4 \bullet 72 \div 9$

UNDERSTANDING A CONCEPT

Dividing 3-Digit Numbers

The owners of Fancy Plants are growing 458 marigold plants. They want to make equal-size groups to sell in 4 different markets. How many plants will they put in each group?

You can divide to find the answer.

First estimate. $400 \div 4 = 100$

Divide: $458 \div 4$

Step 1

Decide where to place the first digit of the quotient.

$4\overline{)458}$

Think: $4 = 4$
There are enough hundreds to divide.

Step 2

Divide the hundreds.

$$\begin{array}{r} 1 \\ 4\overline{)458} \\ -4 \\ \hline 0 \end{array}$$

Think: $4\overline{)4}^{\,1}$
Multiply. $1 \times 4 = 4$
Subtract. $4 - 4 = 0$
Compare. $0 < 4$

Step 3

Bring down the tens.
Divide the tens.

$$\begin{array}{r} 11 \\ 4\overline{)458} \\ -4\downarrow \\ \hline 05 \\ -4 \\ \hline 1 \end{array}$$

Think: $4\overline{)5}^{\,1}$
Multiply. $1 \times 4 = 4$
Subtract. $5 - 4 = 1$
Compare. $1 < 4$

Step 4

Bring down the ones.
Divide the ones.

$$\begin{array}{r} 114 \text{ R2} \\ 4\overline{)458} \\ -4 \quad \\ \hline 05 \quad \\ -4\downarrow \\ \hline 18 \\ -16 \\ \hline 2 \end{array}$$

Think: $4\overline{)18}^{\,4}$
Multiply. $4 \times 4 = 16$
Subtract. $18 - 16 = 2$
Compare. $2 < 4$

Check: $4 \times 114 = 456; \quad 456 + 2 = 458$

They will put 114 plants in each group.

1. Is the answer reasonable? How do you know?

2. ***What if*** they had 2 more plants? What would the quotient and remainder be?

TRY OUT Divide.

3. 3)532 **4.** 2)785 **5.** 906 ÷ 7 **6.** 891 ÷ 6

PRACTICE

Divide.

7. 5)626 **8.** 2)835 **9.** 3)663 **10.** 6)991 **11.** 2)376

12. 7)928 **13.** 5)575 **14.** 8)976 **15.** 4)910 **16.** 2)933

17. 3)956 **18.** 7)879 **19.** 4)898 **20.** 3)688 **21.** 2)709

22. 5)664 **23.** 6)953 **24.** 7)791 **25.** 8)929 **26.** 4)954

27. 689 ÷ 6 **28.** 896 ÷ 8 **29.** 919 ÷ 5 **30.** 999 ÷ 9

31. 870 ÷ 7 **32.** 956 ÷ 2 **33.** 907 ÷ 8 **34.** 888 ÷ 3

Mixed Applications Solve. Which method did you use?

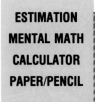

ESTIMATION
MENTAL MATH
CALCULATOR
PAPER/PENCIL

35. Jon has 435 tulip bulbs to put in packages. He puts 3 bulbs in each package. How many packages does he fill?

36. There are 8 iris plants in each box. How many iris plants are in 40 boxes?

CALCULATOR

Rena used a calculator to divide 17 by 6 and got this: $\boxed{2.8333333}$
To find the remainder, Rena followed these steps.

Step 1 Multiply the divisor by the whole-number part of the answer. $2 \times 6 = 12$

Step 2 Subtract the product from the original dividend. The answer is the remainder. $17 - 12 = 5$

Step 3 Write the quotient and remainder. $17 \div 6 = 2 \text{ R}5$

Divide.

1. 4)30 **2.** 8)196 **3.** 6)201 **4.** 5)718

2, 4, 6, 8, IF YOU'RE PUZZLED CALCULATE!

Investigating Patterns

A. You can use a calculator to multiply and divide. Think about this problem: $8 \times 12 = \blacksquare$. How can you find this answer with a calculator?

You can use repeated addition to find an answer.

Press ⟨1⟩⟨2⟩⟨+⟩⟨1⟩⟨2⟩⟨=⟩⟨=⟩⟨=⟩⟨=⟩⟨=⟩⟨=⟩⟨=⟩.

But it is easier to use the multiplication key.

⟨8⟩⟨×⟩⟨1⟩⟨2⟩⟨=⟩

1. What does the display show?

2. How do you know the calculator is right?

3. Find the product of 465 and 899 with a calculator.

Now think about this problem: $72 \div 9 = \blacksquare$.

You can use constant subtraction to find an answer. Keep subtracting until the display shows 0 or a number less than 9. Count how many times you press the ⟨=⟩ key.

⟨7⟩⟨2⟩⟨−⟩⟨9⟩⟨=⟩⟨=⟩⟨=⟩⟨=⟩⟨=⟩⟨=⟩⟨=⟩⟨=⟩

But it is easier to use the division key.

Press ⟨7⟩⟨2⟩⟨÷⟩⟨9⟩⟨=⟩.

4. What does the display show?

5. Use a calculator to find the number of times that 18 goes into 558.

B. A calculator can help you solve problems and puzzles. You can work quickly with a calculator. Try to solve each of the puzzles on this page. Why is it helpful to use a calculator?

6. Use the numbers 6, 7, 8, and 9 to complete each problem. What is the greatest possible answer for each? The least possible answer?

8. The product of these page numbers is 420. Which two page numbers can you multiply to get each of these products?

a. 1,806

b. 3,422

c. 12,656

7. Find the greatest whole number that can be multiplied by 7 to get a product less than 25,000.

9. What is the pattern? Use a calculator to check your answer.

$(1,089 \times 9) \div 1 = 9,801$
$(2,178 \times 8) \div 2 = 8,712$
$(\underline{\quad} \times \underline{\quad}) \div \underline{\quad} = 7,623$
$(\underline{\quad} \times \underline{\quad}) \div \underline{\quad} = 6,534$
$(\underline{\quad} \times \underline{\quad}) \div \underline{\quad} = 5,445$
$(\underline{\quad} \times \underline{\quad}) \div \underline{\quad} = 4,356$
$(\underline{\quad} \times \underline{\quad}) \div \underline{\quad} = 3,267$
$(8,712 \times \underline{\quad}) \div \underline{\quad} = \underline{\mathbf{2,178}}$
$(\underline{\quad} \times \underline{\quad}) \div \underline{\quad} = \underline{\mathbf{1,089}}$

UNDERSTANDING A CONCEPT

More Dividing 3-Digit Numbers

Ben and Tess are decorating the gym for the Future Farmers party. They have 215 gourds to put in bowls. They put 4 gourds in each bowl. How many bowls can they fill? How many gourds will they have left over?

Divide: 215 ÷ 4

Step 1	Step 2	Step 3
Decide where to place the first digit of the quotient.	Divide the tens.	Bring down the ones. Divide the ones. Write the remainder.

Step 1

$$4\overline{)215}$$

Think: 4 > 2 There are not enough hundreds.

4 < 21 There are enough tens to divide.

Step 2

$$\begin{array}{r} 5 \\ 4\overline{)215} \\ -20 \\ \hline 1 \end{array}$$

Think: $4\overline{)21}^{\,5}$
Multiply. 5 × 4 = 20
Subtract. 21 − 20 = 1
Compare. 1 < 4

Step 3

$$\begin{array}{r} 53 \text{ R3} \\ 4\overline{)215} \\ -20\downarrow \\ \hline 15 \\ -12 \\ \hline 3 \end{array}$$

Think: $4\overline{)15}^{\,3}$
Multiply. 3 × 4 = 12
Subtract. 15 − 12 = 3
Compare. 3 < 4

Check: 4 × 53 = 212; 212 + 3 = 215

They can fill 53 bowls. They will have 3 gourds left over.

1. How can you tell when there will be only 2 digits in the quotient?

TRY OUT Write the letter of the correct answer.

2. 5)347   **a.** 69   **b.** 69 R2   **c.** 72   **d.** 161 R2

3. 8)913   **a.** 11 R4   **b.** 114   **c.** 114 R1   **d.** 182 R2

4. 6)282   **a.** 47   **b.** 47 R5   **c.** 52   **d.** 313 R3

5. 827 ÷ 3   **a.** 27 R2   **b.** 212 R1   **c.** 275 R2   **d.** 311 R1

PRACTICE

Divide.

6. $3\overline{)149}$ **7.** $2\overline{)129}$ **8.** $7\overline{)397}$ **9.** $6\overline{)209}$ **10.** $4\overline{)284}$

11. $4\overline{)886}$ **12.** $6\overline{)933}$ **13.** $7\overline{)476}$ **14.** $3\overline{)937}$ **15.** $5\overline{)709}$

16. $6\overline{)496}$ **17.** $7\overline{)801}$ **18.** $4\overline{)593}$ **19.** $8\overline{)745}$ **20.** $4\overline{)372}$

21. $3\overline{)524}$ **22.** $6\overline{)385}$ **23.** $5\overline{)446}$ **24.** $8\overline{)995}$ **25.** $3\overline{)740}$

26. $655 \div 4$ **27.** $299 \div 9$ **28.** $818 \div 3$ **29.** $325 \div 5$

30. $758 \div 6$ **31.** $200 \div 3$ **32.** $896 \div 8$ **33.** $894 \div 9$

34. Divide 807 by 5. **35.** Divide 472 by 8.

36. What is the quotient of 197 and 3? What is the remainder?

37. What is the quotient of 777 and 6? What is the remainder?

Critical Thinking

38. When you divide a 3-digit number by a 1-digit number, what is the greatest number of digits you can have in the quotient? What is the least number of digits?

Mixed Applications

39. The party committee collected $956 for tickets. The tickets cost $4 each. How many tickets did the committee sell?

40. Tess bought a new skirt for the party. It cost $23.89. She gave the clerk $30.00. What was her change?

Mixed Review

Find the answer. Which method did you use?

41. $7\overline{)52}$ **42.** $\begin{array}{r} 146 \\ \times \quad 8 \\ \hline \end{array}$ **43.** $9\overline{)3,600}$ **44.** $9\overline{)4,727}$

| MENTAL MATH |
| CALCULATOR |
| PAPER/PENCIL |

45. $59.77 + $86.95 **46.** $7,900 - 500$ **47.** $1,000 - 721$

UNDERSTANDING A CONCEPT

Zeros in the Quotient

Greta is giving away apples at the Farmers Fair. She brought 325 apples. How many people could each get 3 free apples? How many apples would Greta have left?

Divide: 325 ÷ 3

Step 1

Decide where to place the first digit of the quotient.

3)325   *Think:* 3 = 3
There are enough hundreds to divide.

Step 2

Divide the hundreds.

$$\begin{array}{r} 1 \\ 3\overline{)325} \\ -3 \\ \hline 0 \end{array}$$

Think: 3)3
Multiply. 1 × 3 = 3
Subtract. 3 − 3 = 0
Compare. 0 < 3

Step 3

Bring down the tens.
Divide the tens.

$$\begin{array}{r} 10 \\ 3\overline{)325} \\ -3\downarrow \\ \hline 02 \end{array}$$

Think: 2 < 3
There are not enough tens to divide.
Write 0 in the quotient.

Step 4

Bring down the ones.
Divide the ones.
Write the remainder.

$$\begin{array}{r} 108 \text{ R1} \\ 3\overline{)325} \\ -3\downarrow\downarrow \\ \hline 025 \\ -24 \\ \hline 1 \end{array}$$

Think: 3)25
Multiply. 8 × 3 = 24
Subtract. 25 − 24 = 1
Compare. 1 < 3

So 108 people could each get 3 free apples.
Greta would have 1 apple left.

1. Divide 62 by 3. Where do you put a 0 in the quotient? What does the 0 mean?

TRY OUT Write the letter of the correct answer.

2. 4)363   **a.** 18 R1   **b.** 90   **c.** 90 R3   **d.** 900 R1

3. 541 ÷ 3   **a.** 10 R5   **b.** 113   **c.** 180   **d.** 180 R1

4. 861 ÷ 8   **a.** 107   **b.** 107 R5   **c.** 112   **d.** 118

PRACTICE

Divide.

5. $4\overline{)82}$ **6.** $9\overline{)90}$ **7.** $3\overline{)61}$ **8.** $5\overline{)53}$ **9.** $2\overline{)61}$

10. $2\overline{)816}$ **11.** $4\overline{)522}$ **12.** $8\overline{)324}$ **13.** $5\overline{)453}$ **14.** $7\overline{)910}$

15. $3\overline{)92}$ **16.** $5\overline{)540}$ **17.** $8\overline{)86}$ **18.** $3\overline{)662}$ **19.** $6\overline{)628}$

20. $2\overline{)60}$ **21.** $3\overline{)901}$ **22.** $4\overline{)83}$ **23.** $6\overline{)843}$ **24.** $2\overline{)619}$

25. $75 \div 7$ **26.** $802 \div 4$ **27.** $31 \div 3$ **28.** $813 \div 9$

29. $483 \div 4$ **30.** $92 \div 3$ **31.** $624 \div 6$ **32.** $83 \div 4$

33. What is 83 divided by 8? **34.** What is 304 divided by 6?

35. Divide 76 by 7. **36.** Divide 187 by 9.

Mixed Applications

37. There were 455 people sitting on benches at the hog show. If 9 people sat on each bench, how many benches did they fill?

38. Prizes were given for 5 different kinds of pie. There were 52 pies of each kind. How many pies were there?

MENTAL MATH

You can often divide mentally.

$$3\overline{)7^{1}1\,5}\quad{}^{2}$$

Think: $2 \times 3 = 6$
$7 - 6 = 1$

$$3\overline{)7^{1}1^{2}5}\quad{}^{2\ 3}$$

Think: $3 \times 3 = 9$
$11 - 9 = 2$

$$3\overline{)7^{1}1^{2}5}\quad{}^{2\ 3\ 8\ R1}$$

Think: $8 \times 3 = 24$
$25 - 24 = 1$

Use this method to divide.

1. $2\overline{)428}$ **2.** $3\overline{)693}$ **3.** $2\overline{)351}$ **4.** $4\overline{)257}$

Dividing Money

A. Tina spent $7.98 for 6 pairs of woolen socks. Each pair cost the same amount. How much did each pair cost?

Divide: $7.98 ÷ 6

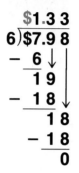

Step 1

Divide as you would divide whole numbers.

```
      1 3 3
6)$7.9 8
 −  6 ↓ |
    1 9 |
 −  1 8 ↓
      1 8
    − 1 8
        0
```

Step 2

Write the dollar sign and the decimal point.

```
    $1.3 3
6)$7.9 8
 −  6 ↓ |
    1 9 |
 −  1 8 ↓
      1 8
    − 1 8
        0
```

Each pair of socks cost $1.33.

1. Is the answer reasonable? How do you know?

B. Money amounts must always have 2 places to the right of the decimal point. Sometimes you have to write a 0.

Divide: $.49 ÷ 7

Step 1

Divide as you would divide whole numbers.

```
       7
7)$.49
```

Step 2

Write the dollar sign and the decimal point.

```
     $.07
7)$.49
```

Think: Write 0 in the place after the decimal point.

TRY OUT Write the letter of the correct answer.

2. 4)$.64   **a.** $.11   **b.** $.16   **c.** 16   **d.** $16

3. 6)$.42   **a.** $.07   **b.** $.70   **c.** $7   **d.** $70

4. $3.76 ÷ 8   **a.** $.42   **b.** $.47   **c.** $4.70   **d.** $47

5. $6.18 ÷ 3   **a.** $.29   **b.** $2.06   **c.** $290   **d.** 209

PRACTICE

Divide.

6. $5\overline{)\$.40}$ **7.** $4\overline{)\$.36}$ **8.** $7\overline{)\$.84}$ **9.** $6\overline{)\$.78}$ **10.** $3\overline{)\$.27}$

11. $8\overline{)\$.96}$ **12.** $9\overline{)\$.90}$ **13.** $5\overline{)\$.65}$ **14.** $7\overline{)\$.98}$ **15.** $8\overline{)\$.64}$

16. $3\overline{)\$2.79}$ **17.** $5\overline{)\$8.50}$ **18.** $4\overline{)\$8.56}$ **19.** $2\overline{)\$2.08}$ **20.** $6\overline{)\$5.22}$

21. $5\overline{)\$6.75}$ **22.** $8\overline{)\$6.32}$ **23.** $7\overline{)\$8.05}$ **24.** $5\overline{)\$6.00}$ **25.** $3\overline{)\$9.27}$

26. $.84 \div 6$ **27.** $.63 \div 9$ **28.** $.36 \div 3$ **29.** $.75 \div 5$

30. $1.80 \div 3$ **31.** $8.14 \div 2$ **32.** $8.16 \div 4$ **33.** $5.44 \div 8$

Mixed Applications Solve. Which method did you use?

34. The price of hickory chips is 2 bags for $7.98. What is the cost of each bag?

35. Mark spent $16.89 for maple syrup and $21.65 for a snow shovel. How much did he spend all together?

36. Shirts cost $13.95. Zena wants to buy two of them. Is $30.00 enough money to buy the shirts?

> ESTIMATION
> MENTAL MATH
> CALCULATOR
> PAPER/PENCIL

CHALLENGE

Which is the **better buy**: 3 shirts for $9.78 or $3.35 for each shirt?

Divide to find the cost of each shirt: $9.78 \div 3 = \$3.26$
Compare: $\$3.26 < \$3.35$

So 3 for $9.78 is the better buy.

Find the better buy.

1. 7 patches for $3.01 or $.42 for each patch

2. 5 towels for $8.65 or $1.75 for each towel

EXPLORING A CONCEPT

Dividing Larger Numbers

Prime Press prints 6,475 copies of a farm journal every week. The company's workers load the same number of journals on each of 5 trucks. How many journals do they load on each truck?

WORKING TOGETHER

Work with a group. Have each person in the group use a different method to find the answer. You can use mental math, a calculator, or paper and pencil.

1. Is it easy to find the quotient mentally? Why or why not?

2. Show how to use paper and pencil to find the quotient.

3. Check to see that your answer is reasonable. How can you do this?

4. Which method was fastest?

5. **What if** Prime Press printed 5,000 journals weekly? Which method would be fastest?

Use all the methods to divide 16,365 by 4.

6. Is it easy to divide mentally?

7. Show how to use paper and pencil to divide.

8. Which method was fastest—mental math, paper and pencil, or calculator?

Find the quotient. Which method did you use?

9. $3\overline{)6,025}$ 10. $7\overline{)1,894}$ 11. $2\overline{)84,066}$ 12. $8\overline{)39,507}$ 13. $9\overline{)\$650.88}$

SHARING IDEAS

14. How is dividing 4-digit and 5-digit numbers like dividing smaller numbers?

15. Tell how you decided which method to use in Exercises 9–13.

16. When do you think it is easier to use mental math? paper and pencil? a calculator? Tell why.

17. Is the calculator always easier to use with larger numbers? Why or why not?

ON YOUR OWN

Solve. Which method did you use?

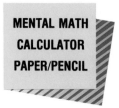

MENTAL MATH
CALCULATOR
PAPER/PENCIL

18. Nat ordered 3 belts from an ad in the journal. Each belt cost the same amount. He paid $27.15 in all. How much did each belt cost?

19. Marty paid $51.75 to put an ad in the journal. The ad was 9 lines long. What was the cost per line?

20. Teri writes about 3,000 words each week for the journal. She writes 5 days a week and about the same number of words each day. About how many words does she write each day?

21. *Write a division problem* using a large dividend. Choose a method and solve it. Ask others to solve it and compare methods.

PROBLEM SOLVING

Interpreting the Quotient and Remainder

Farmer Jones has 43 chickens. He wants to put 8 chickens in each of his chicken coops. Into how many coops will he put 8 chickens?

You know that Farmer Jones has 43 chickens and that he will put 8 chickens in each coop. You can use division to find how many coops he needs.

1. Into how many coops will he put 8 chickens?

$$\begin{array}{r} 5 \\ 8\overline{)43} \\ -40 \\ \hline 3 \end{array}$$

Farmer Jones decides to sell the chickens he did not put in a coop.

2. How many chickens will he sell?

3. *What if* he changed his mind and put the leftover chickens in a coop of their own? How many coops would have chickens then?

There are three ways to interpret the quotient and the remainder to answer division problems:

- The quotient answers questions like Problem 1.

- The remainder answers questions like Problem 2.

- The next greater number than the quotient answers questions like Problem 3.

4. How can you tell which type of answer is needed?

PRACTICE

Solve the problem. Tell how you interpreted the remainder. Use mental math, a calculator, or paper and pencil.

5. Juanita is planting tomato plants. She wants to put exactly 8 plants in each row. She has 52 tomato plants. How many rows will she plant?

6. Whitney is laying down a white gravel garden path. She needs to cover 55 square feet. Each bag will cover 3 square feet. How many bags does she need?

7. Tomas has 36 pounds of apples to pack in boxes. Each box must have 5 pounds of apples. Any leftover apples are packed in bags. How many pounds of apples does Tomas pack in bags?

8. A dairy cow needs 3 acres of grazing land. Mr. Jones wants to buy as many cows as he can. If his farm has 160 acres of grazing land, how many cows should he buy?

Strategies and Skills Review

9. Horace wants to buy a bike that costs $119. He earns $9 a day working after school at Mrs. Blair's farm. How many days does he have to work to earn enough to pay for the bike?

10. Mr. McCall picked 273 pounds of peaches. He picked 65 more pounds of apples than peaches. He also picked 491 pounds of pears. How many more pounds of pears than peaches did he pick?

11. Mrs. Blair bought 24 ears of corn at the farmer's market. She paid $.75 for each bag of 6 ears. How much money did she pay?

12. **Write a problem** that can be solved by interpreting the quotient and remainder. Solve it. Ask others to solve your problem.

Median, Range, and Average

Maria sold handmade dolls during the two weeks of the Green County Fair. She recorded the sales in a table.

You can describe a group of numbers in several ways.

NUMBER OF DOLLS SOLD AT GREEN COUNTY FAIR

Day	Week 1	Week 2
1	6	12
2	10	12
3	8	9
4	6	12
5	7	15
6	15	14
7	25	17

WORKING TOGETHER

Use graph paper to make a strip for the number of dolls she sold each day during Week 1.

Arrange the strips in order from least to greatest.

The middle number is the **median** of the group of numbers.

1. How many squares does the middle strip have? What is the median?

The difference between the least number and the greatest number is the **range**.

2. How many more squares are in the longest strip than in the shortest? What is the range?

3. Tape all of the strips together to form one long strip. Cut this strip into 7 equal strips. How many squares are in each strip?

You have just found the **average** of the group of numbers.

4. Use graph paper to find the median, range, and average for the number of dolls Maria sold during the second week.

5. Compare the medians, ranges, and averages of the two weeks. What do you notice?

6. Pick 5 numbers from 1 to 20. Use graph paper to find the median, range, and average of the numbers picked.

SHARING IDEAS

7. Tell how you find the median, range, and average of a group of numbers.

8. How does the median compare with the other numbers in a group if the range is small? if the range is large?

9. How does the average compare with the numbers in a group if the range is small? if the range is large?

ON YOUR OWN

10. Find the median, range, and average of the data in the graph below.

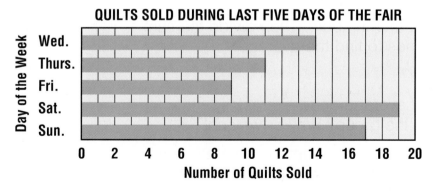

QUILTS SOLD DURING LAST FIVE DAYS OF THE FAIR

11. Measure the handspan in centimeters of 5 people in your class. Find the median, range, and average of the measures. Compare your results with those of others.

More Median, Range, and Average

Steve keeps track of the number of vegetables entered in the competition at the fair. What is the median, range, and average of this group of numbers?

Recall that the median is the number in the middle when you order the numbers from greatest to least.

1. What is the median?

You can subtract the least number from the greatest number to find the range.

2. What is the range?

You can find the sum of the numbers. Then divide the sum by the number of addends to find the average.

3. What is the average?

4. Are the median and average good examples of the numbers that are in the group? Why or why not?

5. *What if* 77 vegetables had been entered in 1986 and 78 vegetables in 1987? What would the median, range, and average be?

6. *What if* the group of numbers were 67, 75, 92, 88, and 103? What would their median, range, and average be?

VEGETABLES IN COMPETITION

Year	Number
1986	68
1987	72
1988	93
1989	88
1990	104

Which statements below do you agree with? Which do you disagree with? Give examples to support your answers.

7. The average must equal one of the numbers in a group.

8. The median and the average can be the same number.

9. Different groups of numbers must have different medians, ranges, and averages.

PRACTICE

Find the median, range, and average.

10. 4, 11, 6, 9, 5

11. $5, $10, $6, $9, $10

12. 3, 5, 2, 8, 5, 3, 2

13. 30¢, 27¢, 40¢, 28¢, 35¢

14. 35, 26, 18, 25, 31

15. 25, 28, 20, 25, 21, 25, 24

16. 105, 100, 104

17. 112, 64, 67, 103, 79

18. $101, $100, $118, $95, $101

19. 96, 120, 102, 124, 95, 101, 125

Critical Thinking

20. To estimate the sum of 65 + 62 + 58 + 49 + 59, Ann wrote: 5 × 60 = 300. So the sum is about 300. Why does Ann's method work?

Mixed Applications

Solve. You may need to use the Databank on page 520.

21. What is the average number of goats entered in the Green County Fair competition for the 5 years?

22. How many goats, hogs, and cows were entered in all in 1988?

23. What is the median number of hogs entered in the 5 years?

24. ***Write a problem*** about the number of animals entered in the fair competition. Have another student solve it.

Dividing by 1-Digit Divisors **287**

Problem Solving: Giving a Fruit Basket

SITUATION

Max's family wants to give their new neighbors a fruit basket as a welcome gift. They can buy a ready-made gift basket at the market. For an extra charge the market will deliver the basket. The market also sells all the things they need to make a fruit basket.

PROBLEM

Should Max's family buy a ready-made gift basket, or should they make a fruit basket themselves?

DATA

GIFT BASKETS
Skillfully arranged and decorated ■ ■ ■ ■ Delivery $5 extra

SMALL $12

1 pound each:
apples, oranges, pears
½ pound of nuts

MEDIUM $20

2 pounds each:
apples, oranges, pears
1 pineapple
½ pound of nuts

LARGE $30

3 pounds each:
apples, oranges, pears
1 pineapple
1-pound box of dates
1 pound of nuts

Make Your Own Basket

Baskets:
Large $6
Medium $4
Small $2

Decorating Kits: $2
(Bows, cellophane, etc.)

Premium Fruits / Fresh Produce

Apples $.75 per pound

Oranges $1 per pound

Grapefruit $1 each

Pineapple $3 each

Pears $1.25 per pound

Kiwi Fruit 2 for $1

Mixed Nuts $4 per pound

Dates 1-pound box $3

USING THE DATA

What will it cost to buy these baskets and decorate them?

1. large   **2.** medium   **3.** small

Find the cost of matching the contents of the ready-made baskets.

4. large   **5.** medium   **6.** small

What will it cost to buy each ready-made basket and have it delivered?

7. large   **8.** medium   **9.** small

MAKING DECISIONS

10. Is it cheaper for Max's family to make a medium fruit basket or to buy it ready-made?

11. *What if* the family knows one of their neighbors cannot eat pineapple? What should they do? Why?

12. *What if* they can spend only $20? What choices do they have?

13. *What if* they cannot deliver the basket? What are their choices?

14. What advantages are there in buying a fruit basket? in making a fruit basket?

15. *Write a list* of things the family should think about before buying or making a fruit basket.

16. What would you do if you had $27.00 to spend on a fruit basket?

Math and Science

A **solar system** is a group of planets and other objects that move around a star, or sun. Our solar system is circular in shape and has nine planets, including the Earth.

A planet's path around the Sun is called its **orbit.** The time a planet takes to orbit the Sun is the length of its year. The Earth orbits the Sun in 365 days.

The chart to the right shows the planets arranged in order from the closest to the Sun to the farthest away. You can see that the farther a planet is from the Sun, the longer its year is.

Planets	Length of Year (in Earth days)
Mercury	88
Venus	225
Earth	365
Mars	687
Jupiter	4,333
Saturn	10,759
Uranus	30,685
Neptune	60,188
Pluto	90,700

What if a satellite sends pictures of Mars back to Earth every 7 Earth days for a full Mars year? About how many times will pictures be sent during the year?

Think: Divide. 687 ÷ 7 = 98.14285714

Pictures will be sent about 98 times.

ACTIVITIES

1. How old would you be on Mercury, Venus, and Mars? First find your age in days by multiplying your age by 365. Then divide the number of days you have lived by the number of days in each planet's year.

2. How many Earth years equal one year on planets that are farther from the Sun? Divide the number of days in each planet's year by 365, the number of days in Earth's year.

Computer Applications: Line Graphs

For the past 8 weeks, Anne has been walking for exercise. She has kept a log of the distance she has walked each week. She wants to make a line graph of the data. A computer would be helpful for drawing the graph. You can input the data and the computer will draw the graph. You can also use the computer to extend the graph so you can predict what will happen in the future.

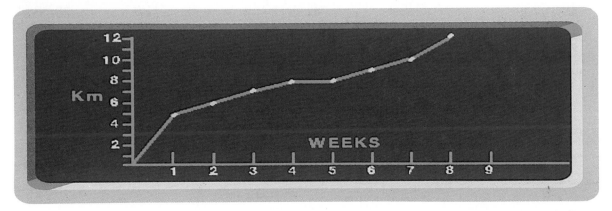

DATA

Here are data showing how many kilometers Anne walked. Use this information to draw a line graph.

THINKING ABOUT COMPUTERS

Refer to the line graph you drew to answer these questions.

1. During which week did Anne walk the most kilometers? the fewest kilometers?

2. How would you describe Anne's trend in walking?

3. How could Anne use the computer to estimate how far she will walk the following week?

4. **What if** Anne wants to graph the number of kilometers each student in her class walked for the 8 weeks? How could a computer be useful?

DISTANCE WALKED

Week	Km
1	5
2	6
3	7
4	8
5	8
6	9
7	10
8	12

EXTRA PRACTICE

Dividing with Remainders, page 259

Find the quotient and remainder.

1. $4\overline{)5}$

2. $7\overline{)8}$

3. $5\overline{)8}$

4. $2\overline{)7}$

5. $3\overline{)10}$

6. $8\overline{)17}$

7. $9\overline{)19}$

8. $6\overline{)19}$

9. $2\overline{)15}$

10. $8\overline{)11}$

11. $5\overline{)23}$

12. $3\overline{)7}$

13. $6\overline{)13}$

14. $8\overline{)34}$

15. $4\overline{)18}$

16. $6\overline{)34}$

17. $9\overline{)52}$

18. $3\overline{)19}$

19. $7\overline{)11}$

20. $8\overline{)42}$

21. $26 \div 3$

22. $33 \div 4$

23. $65 \div 9$

24. $48 \div 5$

25. $32 \div 5$

26. $31 \div 6$

27. $43 \div 7$

28. $75 \div 9$

Mental Math: Dividing 10s and 100s, page 261

Divide. Use mental math.

1. $2\overline{)10}$

2. $5\overline{)100}$

3. $7\overline{)700}$

4. $9\overline{)3,600}$

5. $30 \div 3$

6. $80 \div 2$

7. $10 \div 5$

8. $60 \div 3$

9. $60 \div 6$

10. $40 \div 2$

11. $800 \div 4$

12. $240 \div 8$

13. $900 \div 3$

14. $640 \div 8$

15. $480 \div 6$

16. $2,500 \div 5$

17. $8,100 \div 9$

18. $6,000 \div 2$

19. $3,200 \div 8$

20. $1,800 \div 6$

Estimating Quotients, page 263

Estimate by rounding.

1. $2\overline{)41}$

2. $4\overline{)83}$

3. $5\overline{)276}$

4. $2\overline{)763}$

5. $59 \div 3$

6. $61 \div 6$

7. $896 \div 3$

8. $8,322 \div 4$

Estimate by using compatible numbers.

9. $6\overline{)537}$

10. $3\overline{)585}$

11. $5\overline{)5,212}$

12. $2\overline{)650}$

13. $523 \div 5$

14. $637 \div 9$

15. $4,073 \div 3$

16. $4,417 \div 9$

Problem Solving Strategy: Making a Table, page 265..........................

Make a table to solve the problem.

1. Bill, Ron, and Cindy have different jobs after school. They deliver newspapers, babysit, or walk dogs. Ron is allergic to dogs. Cindy delivers newspapers. What after-school job does Bill have?

2. Paul, Carlos, Monica and her sister, Anna, are each buying different color sweatshirts. Monica's sister prefers red. Carlos does not like yellow. Paul buys a blue sweatshirt. Who buys the yellow one? the green one?

Dividing 2-Digit Numbers, page 269..

Divide.

1. 4)13 **2.** 2)15 **3.** 6)77 **4.** 3)53

5. 7)89 **6.** 3)54 **7.** 6)75 **8.** 2)87

9. 3)39 **10.** 7)79 **11.** 5)56 **12.** 6)72

13. 4)81 **14.** 8)92 **15.** 9)97 **16.** 4)86

17. 57 ÷ 8 **18.** 72 ÷ 5 **19.** 58 ÷ 5 **20.** 85 ÷ 4

Dividing 3-Digit Numbers, page 271..

Divide.

1. 2)435 **2.** 5)606 **3.** 6)990 **4.** 3)669

5. 7)876 **6.** 5)778 **7.** 4)952 **8.** 4)928

9. 4)573 **10.** 6)685 **11.** 2)673 **12.** 7)877

13. 6)795 **14.** 7)884 **15.** 2)825 **16.** 5)899

17. 500 ÷ 4 **18.** 986 ÷ 3 **19.** 935 ÷ 2 **20.** 674 ÷ 5

EXTRA PRACTICE

More Dividing 3-Digit Numbers, page 275............................

Divide.

1. 2)131 **2.** 7)297 **3.** 6)111 **4.** 3)148

5. 8)624 **6.** 6)560 **7.** 9)623 **8.** 8)249

9. 5)437 **10.** 3)197 **11.** 5)906 **12.** 7)787

13. 347 ÷ 5 **14.** 469 ÷ 3 **15.** 379 ÷ 3 **16.** 246 ÷ 9

Zeros in the Quotient, page 277............................

Divide.

1. 3)62 **2.** 2)41 **3.** 5)52 **4.** 3)62

5. 2)410 **6.** 5)603 **7.** 6)484 **8.** 9)993

9. 5)103 **10.** 4)883 **11.** 8)484 **12.** 2)81

13. 8)83 **14.** 6)602 **15.** 3)605 **16.** 4)202

17. 123 ÷ 6 **18.** 516 ÷ 5 **19.** 421 ÷ 7 **20.** 313 ÷ 3

Dividing Money, page 279............................

Divide.

1. 6)$3.60 **2.** 6)$.78 **3.** 3)$.39 **4.** 2)$1.68

5. 3)$6.18 **6.** 6)$8.10 **7.** 3)$6.18 **8.** 4)$1.28

9. 9)$7.02 **10.** 6)$4.08 **11.** 6)$.90 **12.** 7)$9.80

13. 2)$7.92 **14.** 4)$6.24 **15.** 5)$5.45 **16.** 7)$7.98

17. $6.24 ÷ 6 **18.** $9.12 ÷ 8 **19.** $5.15 ÷ 5 **20.** $8.32 ÷ 4

EXTRA PRACTICE

Problem Solving: Interpreting the Quotient and Remainder, page 283........

Solve the problem. Tell how you interpreted the quotient and remainder.

1. Tania's dad is putting up bookshelves. He can fit only 8 books on each shelf. Tania has 36 books. How many shelves will be filled?

2. Harry has 327 stamps in his collection. He puts 4 rows of stamps on each page in his album. How many of the pages in Harry's album are full?

3. Parent volunteers took 118 fourth graders on a field trip to the science museum. Each parent was responsible for 8 students. How many parents went on the trip?

4. Roy and Tomas have saved $31 to buy videotapes. The tapes they like are on sale at $9 each. How much money will they have left after they buy the videotapes?

Median, Range, and Average, page 287...

Find the median, range, and average.

1. 3; 7; 12; 8; 5

2. $8; $12; $10; $4; $6

3. 2; 4; 1; 3; 8; 7; 3

4. 20¢; 30¢; 26¢; 14¢; 5¢

5. 42; 16; 35; 32; 25

6. 16; 18; 25; 37; 22; 23; 34

7. 108; 107; 103

8. 114; 63; 81; 107; 85

9. $118; $100; $107; $85; $75

10. 83; 118; 107; 139; 85; 101; 74

11. 33; 87; 52; 65; 83

12. 89; 107; 125; 47; 63; 49; 80

13. 80; 40; 60; 110; 75

14. 62; 72; 22; 35; 44

15. $214; $75; $138; $99; $159

16. 28; 42; 36; 48; 33; 41; 38

17. 109; 99; 108; 110; 89

18. 42; 37; 58; 29; 48; 32; 55

19. 111; 102; 125; 136; 121

20. $63; $78; $92; $104; $63

Practice PLUS

KEY SKILL: Zeros in the Quotient (Use after page 277.)

Level A

Divide.

1. 3)31 **2.** 6)60 **3.** 4)80 **4.** 3)62 **5.** 4)42

6. 5)53 **7.** 2)41 **8.** 3)32 **9.** 8)85 **10.** 7)73

11. 4)804 **12.** 5)554 **13.** 3)962 **14.** 3)182 **15.** 7)213

16. Emma has 240 beads. She can fit 8 beads on a string to make a necklace. How many necklaces can she make?

Level B

Divide.

17. 9)96 **18.** 6)63 **19.** 2)815 **20.** 5)545 **21.** 6)424

22. 2)100 **23.** 6)305 **24.** 6)484 **25.** 5)540 **26.** 3)272

27. Divide 86 by 8. **28.** What is 507 divided by 5?

29. Paul has 604 stickers. He fits 6 stickers on each page of a book. How many pages does he fill? How many stickers are left over?

Level C

Divide.

30. 52 ÷ 5 **31.** 82 ÷ 4 **32.** 122 ÷ 3 **33.** 154 ÷ 5

34. 723 ÷ 4 **35.** 653 ÷ 5 **36.** 646 ÷ 8 **37.** 544 ÷ 6

38. 304 ÷ 6 **39.** 960 ÷ 9 **40.** 740 ÷ 7 **41.** 843 ÷ 6

42. Divide 546 by 9. **43.** What is 701 divided by 5?

44. Shane has 818 stamps in his collection. He puts 9 stamps in each envelope. How many envelopes does he fill? How many stamps are left over?

KEY SKILL: Dividing Money (Use after page 279.)

Level A ....

Divide.

1. 6)$.30   **2.** 8)$.40   **3.** 5)$.45   **4.** 6)$.72   **5.** 3)$.21

6. 8)$.80   **7.** 4)$.56   **8.** 2)$.34   **9.** 3)$.51   **10.** 7)$.84

11. 7)$.70   **12.** 3)$.24   **13.** 9)$.54   **14.** 7)$.42   **15.** 5)$.65

16. The cost of two grapefruits is $.88. How much does each grapefruit cost?

Level B ....

Divide.

17. 7)$.98   **18.** 6)$.78   **19.** 7)$.63   **20.** 3)$1.41   **21.** 2)$8.16

22. 4)$6.80   **23.** 3)$9.18   **24.** 5)$7.15   **25.** 4)$4.16   **26.** 7)$4.55

27. 5)$6.95   **28.** 7)$1.05   **29.** 5)$2.65   **30.** 6)$6.54   **31.** 3)$7.41

32. Charlie bought 5 pineapples for $7.25. What is the cost of each pineapple?

Level C ....

Divide.

33. $8.36 ÷ 4   **34.** $4.55 ÷ 7   **35.** $5.10 ÷ 6   **36.** $3.28 ÷ 8

37. $3.85 ÷ 7   **38.** $9.16 ÷ 4   **39.** $1.71 ÷ 9   **40.** $5.31 ÷ 9

41. $9.84 ÷ 8   **42.** $5.85 ÷ 5   **43.** $8.37 ÷ 9   **44.** $7.07 ÷ 7

45. $6.48 ÷ 9   **46.** $7.12 ÷ 8   **47.** $9.90 ÷ 6   **48.** $9.31 ÷ 7

49. Tina lives in Quebec, Canada. She makes a 9-minute phone call to Tampa Bay, Florida, that costs $8.91. What is the cost per minute for the phone call?

CHAPTER REVIEW

LANGUAGE AND MATHEMATICS

Complete the sentences. Use the words in the chart.

1. Use ■ to check division. *(page 268)*

2. The answer in a division problem is the ■.
   (page 260)

3. The middle number of a group of numbers is
   the ■. *(page 286)*

4. The ■ of a group of numbers is the sum of the
   numbers divided by the number of addends.
   (page 286)

5. ***Write a definition*** or give an example of the words
   you did not use from the chart.

CONCEPTS AND SKILLS

Find the quotient and remainder. *(page 258)*

6. $8\overline{)34}$ 7. $9\overline{)30}$ 8. $7\overline{)40}$ 9. $8\overline{)49}$ 10. $4\overline{)18}$ 11. $6\overline{)25}$

12. $87 \div 9$ 13. $75 \div 8$ 14. $57 \div 7$ 15. $48 \div 9$

Divide. Use mental math. *(page 260)*

16. $9\overline{)720}$ 17. $8\overline{)480}$ 18. $7\overline{)4,900}$ 19. $6\overline{)5,400}$

20. $400 \div 5$ 21. $600 \div 3$ 22. $4,000 \div 2$ 23. $3,600 \div 6$

Estimate the quotients by rounding. *(page 264)*

24. $3\overline{)92}$ 25. $5\overline{)47}$ 26. $6\overline{)3,122}$ 27. $4\overline{)2,450}$

28. $515 \div 5$ 29. $432 \div 6$ 30. $3,210 \div 8$ 31. $7,190 \div 7$

Estimate the quotients by using compatible numbers. *(page 264)*

32. $4\overline{)374}$ 33. $7\overline{)215}$ 34. $9\overline{)555}$ 35. $8\overline{)734}$

36. $437 \div 7$ 37. $378 \div 9$ 38. $641 \div 7$ 39. $282 \div 3$

Divide. *(pages 268–271, 274–279)*

40. $8\overline{)75}$ **41.** $2\overline{)49}$ **42.** $7\overline{)88}$ **43.** $6\overline{)\$.90}$ **44.** $6\overline{)735}$

45. $3\overline{)724}$ **46.** $7\overline{)\$6.86}$ **47.** $5\overline{)725}$ **48.** $8\overline{)756}$ **49.** $4\overline{)277}$

50. $2\overline{)\$6.42}$ **51.** $6\overline{)836}$ **52.** $3\overline{)92}$ **53.** $9\overline{)364}$ **54.** $4\overline{)961}$

55. $759 \div 7$ **56.** $\$819 \div 9$ **57.** $\$368 \div 4$ **58.** $723 \div 8$

59. $300 \div 4$ **60.** $499 \div 6$ **61.** $541 \div 5$ **62.** $637 \div 7$

Find the median, range, and average. *(page 286)*

63. 14, 27, 12, 33, 9   **64.** $104, $99, $101, $106, $100

65. 32¢, 45¢, 23¢, 30¢, 25¢, 40¢, 36¢

CRITICAL THINKING

66. Amalfi says it is faster to divide 2,893 by 2,893 mentally than by using a calculator. Do you agree? Why or why not? *(page 280)*

67. Bridget says that the average of a group of numbers can be greater than the greatest number in the group? Do you agree? Why or why not? *(page 286)*

MIXED APPLICATIONS

68. Kara, Bill, and Marc are each in a different school club. Bill is in the Track Club. Kara does not play chess. Who is in the Chess Club? *(page 264)*

69. There are 148 jars of paint on the shelves in the closet. Each shelf holds 16 jars. How many full shelves of paint are there? *(page 282)*

70. In five days, students ate 57, 63, 79, 81, and 105 granola bars. What is the average number of granola bars eaten each day? *(page 286)*

71. The price of blank cassette tapes is 5 for $6.25. What is the cost of a single tape? *(page 278)*

CHAPTER TEST

Divide mentally.

1. $8\overline{)80}$ **2.** $4\overline{)200}$ **3.** $2\overline{)1,000}$

4. $560 \div 8$ **5.** $2,400 \div 4$

Estimate by rounding.

6. $3\overline{)94}$ **7.** $5\overline{)5,104}$

8. $408 \div 8$ **9.** $8,491 \div 2$

Estimate by using compatible numbers.

10. $3\overline{)220}$ **11.** $6\overline{)490}$

12. $1,957 \div 4$ **13.** $1,107 \div 5$

Divide.

14. $5\overline{)60}$ **15.** $7\overline{)61}$ **16.** $7\overline{)672}$ **17.** $8\overline{)914}$

18. $82 \div 3$ **19.** $627 \div 6$

Find the median, range, and average.

20. $110; $120; $103   **21.** 9; 5; 11; 3; 6; 10; 12

Solve.

22. Roz has collected $216 selling concert tickets. The tickets cost $8. How many tickets has she sold?

23. There are 200 people coming to a party. If 9 people can sit at one table, how many tables are needed?

24. Gail, Karla, and Gene each have a pet. Karla has a snake. Gene is allergic to cats. Gail has a dog. Who has a pet turtle?

25. The river guide allows 6 people on each raft. There are 81 people going down the river. How many rafts are needed?

DIVISIBILITY RULES FOR 2, 5, AND 10

You can tell when a number is divisible by 2, 5, or 10 without having to divide the number.

Think: A number is divisible by another number when the remainder is zero.

Use the numbers in the box with the questions below to write some divisibility rules for 2, 5, and 10.

Divide each of the numbers in the box by 2.

21
22
23
24
25
26
27
28
29
30

1. Which numbers are divisible by 2? How do you know?

2. Look at the ones place of the numbers divisible by 2. What digits do you find?

3. Try another number where the digit in the ones place is one of the digits you found in Problem 2. Is it divisible by 2?

4. Write a rule that can help you decide, without dividing, if a number is divisible by 2. Test your rule on other numbers.

Divide each of the numbers in the box by 5.

5. Which numbers are divisible by 5?

6. Write a rule that can help you decide, without dividing, if a number is divisible by 5. Test your rule on other numbers.

Divide each of the numbers in the box by 10.

7. Write a rule that can help you decide, without dividing, if a number is divisible by 10.

Copy the table. Complete by writing *yes* or *no*.

Number	59	84	175	210	414	3,102
Divisible by 2?	8. ■	9. ■	10. ■	11. ■	12. ■	13. ■
Divisible by 5?	14. ■	15. ■	16. ■	17. ■	18. ■	19. ■
Divisible by 10?	20. ■	21. ■	22. ■	23. ■	24. ■	25. ■

26. Look at the table. What do you notice about the numbers that are divisible by 10?

CUMULATIVE REVIEW

Choose the letter of the correct answer.

1. Find the range: 25, 11, 9

   a. 11   **c.** 16
   b. 15   **d.** not given

2. Estimate the capacity of a spoonful of oil.

   a. 30 L   **c.** 30 mL
   b. 3 L   **d.** 3 mL

3. Find the remainder: 45 ÷ 7

   a. 3   **c.** 5
   b. 4   **d.** not given

4. $2.64 ÷ 8

   a. $.14   **c.** $.44
   b. $.33   **d.** not given

5. Estimate by rounding: 734 ÷ 7

   a. 80   **c.** 100
   b. 90   **d.** 150

6. 4 × ■ × 3 = 36

   a. 1   **c.** 3
   b. 2   **d.** not given

7. What is 2 minutes before 3 A.M.?

   a. 2:52 A.M.   **c.** 3:02 A.M.
   b. 2:58 A.M.   **d.** not given

8. 478 ÷ 9

   a. 43   **c.** 53 R1
   b. 43 R1   **d.** not given

9. 532 × 7

   a. 3,514   **c.** 3,714
   b. 3,524   **d.** not given

10. Find the common factor of 6 and 27.

   a. 2   **c.** 6
   b. 3   **d.** not given

11. 4,900 ÷ 7

   a. 7   **c.** 700
   b. 70   **d.** not given

12. What is 8 times 26?

   a. 168   **c.** 228
   b. 208   **d.** not given

13. Find the median: 3, 9, 1, 2, 7

   a. 1   **c.** 3
   b. 2   **d.** not given

14. What is 625 divided by 3?

   a. 208 R1   **c.** 280 R1
   b. 208 R2   **d.** not given

CHAPTER 8

Geometry

MATH CONNECTIONS: VOLUME • PROBLEM SOLVING

1. What do you see in this picture?

2. What different geometric shapes do you see?

3. Why do you think these shapes have been used?

4. Describe what a modern home might look like in the year 2025.

EXPLORING A CONCEPT
Geometry Around Us

Look around. Everywhere there are lines and shapes like circles, triangles, squares, and rectangles. Geometry is in the world around us, in the objects we use every day, in the rooms where we live, work, and play.

WORKING TOGETHER

The picture above shows what a classroom might look like in the future.

1. List the objects in the picture that have geometric shapes you know. Sort these objects according to their shapes.

2. Which objects have only parts of geometric figures? Which combine two or more geometric figures?

3. Which objects are made of straight lines? of curves?

4. Where can you find lines that cross or meet?

5. What other objects do you think might be in a classroom of the future? What shapes could they have?

SHARING IDEAS

6. Which shapes did you see most often? Why do you think these shapes are used so much?

7. Look at your list of circular and curved objects. Why are they shaped that way?

8. Compare the objects that have straight parts with those that have curves. How are they used? Why do you think they are shaped the way they are?

ON YOUR OWN

9. Think about objects in the world around you. What shapes can you add to the list you made? How would you sort them? Compare your list with those of others.

10. Draw a picture of something you see every day. Identify the shapes in your picture. Compare your picture with those of others.

UNDERSTANDING A CONCEPT

Lines, Line Segments, and Rays

To design houses, architects need to know about lines, line segments, and rays.

The floor of the house is shown by a **line segment**. It is a straight figure with two endpoints.

A   B

Read: line segment *AB* or line segment *BA*
Write: $\overline{AB}$ or $\overline{BA}$

1. Name other line segments shown in the design.

A **line** is also a straight figure, but it has no endpoints. It goes on forever in both directions.

Read: line *DC* or line *CD*
Write: $\overleftrightarrow{DC}$ or $\overleftrightarrow{CD}$

2. Is the ceiling drawn in the design a good example of a line? Why or why not?

A **ray** is another straight figure. It has one endpoint. It goes on forever in just one direction.

3. Name something you know that suggests a ray.

Read: ray *ZY* (Name the endpoint first.)
Write: $\overrightarrow{ZY}$ (Direct the arrow to the right.)

Sometimes lines, line segments, and rays meet, or **intersect**.

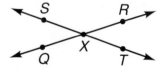

$\overleftrightarrow{QR}$ intersects $\overleftrightarrow{ST}$ at point *X*.

If they do not meet and if they remain the same distance apart, they are **parallel**.

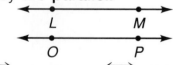

$\overleftrightarrow{LM}$ is parallel to $\overleftrightarrow{OP}$. **Write:** $\overleftrightarrow{LM} \parallel \overleftrightarrow{OP}$.

4. Name something in your classroom that suggests intersecting lines or parallel lines.

Write the name of the figure.

5.

6.

7.

8.

PRACTICE

Write the name of the figure.

9.

10.

11.

12.

13.

14.

Tell if the figures are *intersecting* or *parallel*.

15.

16.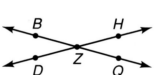

17.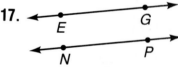

Draw the figure.

18. ray *RT*

19. line *DF*

20. line segment *KL*

21. line *AB* intersecting line *MR* at point *D*

22. ray *HN* parallel to ray *DE*

23. line segment *ET* intersecting line *SD* at point *A*.

24. a capital letter made from only two intersecting line segments

Mixed Review

Find the answer. Which method did you use?

25. 3,005 − 978

26. 6 × 20

27. 9 + 8

28. 458 ÷ 6

29. 5,789 + 6,857

30. 32 ÷ 4

MENTAL MATH
CALCULATOR
PAPER/PENCIL

EXTRA Practice, page 332

UNDERSTANDING A CONCEPT

Angles

A. Schools in the future may be covered by a dome like this. The dome would keep out bad weather. This type of dome is made of many angles.

An **angle** is formed by two rays with the same endpoint. The rays are the **sides** of the angle. The endpoint is its **vertex**.

Read: angle *ABC* or angle *CBA* or angle *B*
Write: ∠*ABC* or ∠*CBA* or ∠*B*

1. What do you notice about the vertex and the name of the angle?

B. An angle that forms a square corner is a **right angle.** Intersecting lines that form right angles are called **perpendicular** lines.

∠*UXY* and ∠*WXZ* are right angles.
$\overleftrightarrow{UZ}$ is perpendicular to $\overleftrightarrow{WY}$.
Write: $\overleftrightarrow{UZ} \perp \overleftrightarrow{WY}$

2. Can you think of something that suggests perpendicular lines?

3. Why would you not name any of these angles ∠*X*?

4. Name another right angle in this figure.

C. The size of an angle depends on the space between the sides. It does not matter how much of the rays are drawn. You can sort angles by their size.

less than a right angle   right angle   greater than a right angle

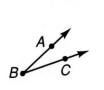

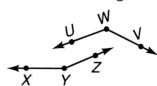

TRY OUT

5. Name the right angle.

6. Name the angle that is less than a right angle.

7. Name the angle that is greater than a right angle.

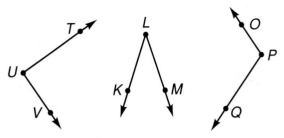

PRACTICE

Write the name of the angle. Tell if it is a right angle. If not, write *more* or *less*.

8.
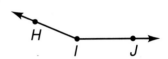

9.

10.
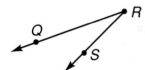

Use the figure to answer the problem.

11. Name a pair of perpendicular lines.

12. Name a pair of parallel lines.

13. How many right angles have point *D* as a vertex? Name one.

14. Name an angle with vertex *D* that is less than a right angle.

15. Name an angle with vertex *D* that is greater than a right angle.

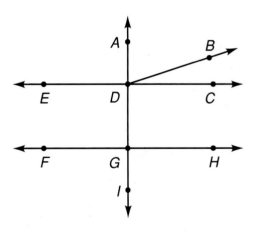

Draw the figure.

16. right angle *RST*

17. $\overleftrightarrow{UZ} \perp \overleftrightarrow{WY}$

18. $\overleftrightarrow{UZ} \parallel \overleftrightarrow{WY}$

Critical Thinking

Write *true* or *false.* Tell how you made your decision.

19. Longer rays make greater angles.

20. Two angles cannot share a vertex.

21. The corner of a book suggests a right angle.

22. The hands of a clock form angles.

UNDERSTANDING A CONCEPT

Plane Figures

Peggy is learning to figure skate. The blades of her skates carve shapes into the flat surface of the ice.

A flat surface that goes on in all directions is a **plane**.

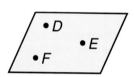

Read: plane *DEF*

Shapes drawn on a plane are called **plane figures.**

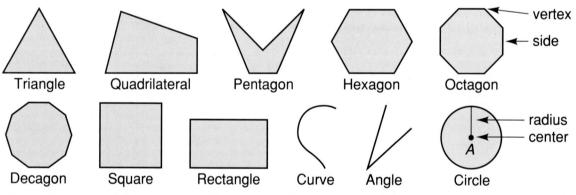

Triangle   Quadrilateral   Pentagon   Hexagon   Octagon — vertex — side

Decagon   Square   Rectangle   Curve   Angle   Circle — radius — center

Plane figures can be **open** like an angle, or **closed** like a circle. **Polygons** are closed plane figures with sides that are line segments.

1. Which plane figures are not polygons? Why?

2. Count the sides and vertices for each polygon. What do you notice?

3. How are rectangles and squares the same? How are they different?

T**RY OUT**

Write *yes* or *no* to tell if the figure is a polygon. Name the polygon.

4.

5.

6.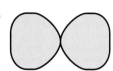

7.

P**RACTICE**

Name the figure.

8.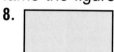

9.

10.

11.

12. a polygon with 4 sides

13. a polygon with 10 sides

14. a quadrilateral with 4 right angles and 4 equal sides

15. a polygon with 8 vertices

Draw the figure.

16. a pentagon with 1 right angle

17. a triangle with 3 equal sides

18. a quadrilateral with at least 2 right angles

19. a triangle with 1 angle greater than a right angle

20. *Describe a figure.* Have others draw and name your figure.

C**HALLENGE**

You can use a **compass** to draw a circle.

Step 1 Put the compass where you want the center of the circle to be.

Step 2 Open the compass to the length you want the radius to be.

Step 3 Turn the pencil around the center point to draw the circle.

Use a compass to draw the circle.

1. radius = 3 cm

2. radius = 5 cm

3. radius = 7 cm

PROBLEM SOLVING

Solving a Multistep Problem

Ryan is building a house that uses solar-heating window panels. He has 35 panels. He needs 4 panels for each of the 8 rooms in the house and 2 panels for the garage. How many panels will he have left?

1. What information do you know?

2. What do you need to find out?

Sometimes it takes several steps to solve a problem. When there is more than one step, it helps to make a plan.

3. What do you need to find out *before* you can decide how many panels will be left?

Try your plan.

4. What can you do first?

5. What do you do next?

6. What do you do last?

7. How many panels are left?

8. Could you have made a different plan to solve the problem? Try it.

PLAN

Step 1 $8 \times 4 = 32$

Step 2 $32 + 2 = 34$

Step 3 $35 - 34 = \blacksquare$

PRACTICE

Make a plan. Then solve the problem. Use mental math, a calculator, or paper and pencil.

9. There are 87 homes in Spruceville and 16 homes in Greenleaf. Tandyville has 4 streets with 15 homes on each. How many more homes are there in Spruceville than in Tandyville and Greenleaf together?

10. In Carol's model castle there are 3 floors with 12 doors on each floor, 2 floors with 14 doors on each floor, and 1 floor with 11 doors. How many doors are there in all?

11. Jason is going to paint his model house. He buys 3 cans of paint for $6.00 each and 1 brush for $.95. How much change should he get from a $20 bill?

12. Mr. Roland is building a 6-room house with 2 large windows and 3 small windows in each room. He has already made 14 windows. How many windows does he have left to make?

Strategies and Skills Review

13. All 137 light bulbs in the halls of an office building are being replaced. Light bulbs come in packages of 8. How many packages are needed?

14. Sir Christopher Wren was a famous English architect. He was born in 1632 and died in 1723. How long did he live?

15. In the Architectural Museum one room has 9 models of seaside homes and 14 blueprints. Another room has 3 times as many models and half as many blueprints. How many models and blueprints are there in the two rooms?

16. **Write a problem** that takes several steps to solve. Solve your problem. Ask others to solve it. Talk about whether there is more than one way to solve the problem.

EXTRA Practice, page 333

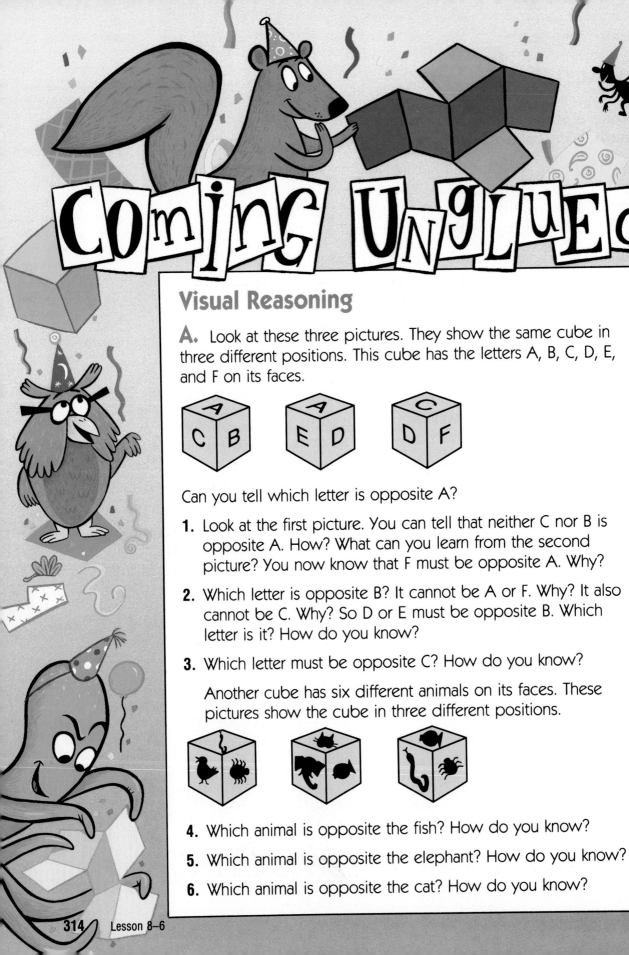

Coming Unglued

Visual Reasoning

A. Look at these three pictures. They show the same cube in three different positions. This cube has the letters A, B, C, D, E, and F on its faces.

Can you tell which letter is opposite A?

1. Look at the first picture. You can tell that neither C nor B is opposite A. How? What can you learn from the second picture? You now know that F must be opposite A. Why?

2. Which letter is opposite B? It cannot be A or F. Why? It also cannot be C. Why? So D or E must be opposite B. Which letter is it? How do you know?

3. Which letter must be opposite C? How do you know?

   Another cube has six different animals on its faces. These pictures show the cube in three different positions.

4. Which animal is opposite the fish? How do you know?

5. Which animal is opposite the elephant? How do you know?

6. Which animal is opposite the cat? How do you know?

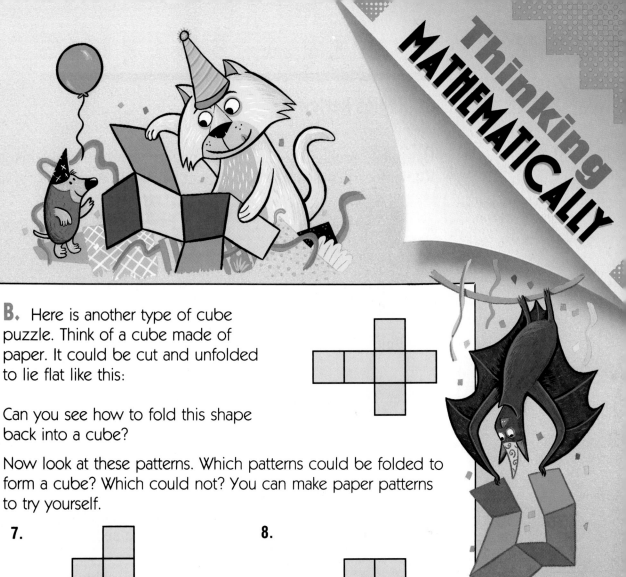

B. Here is another type of cube puzzle. Think of a cube made of paper. It could be cut and unfolded to lie flat like this:

Can you see how to fold this shape back into a cube?

Now look at these patterns. Which patterns could be folded to form a cube? Which could not? You can make paper patterns to try yourself.

7.

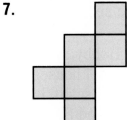

8.

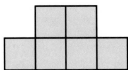

9.

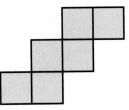

10.

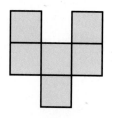

11. Now design 3 new patterns that could be folded into cubes. How many squares will you use for each pattern? Use paper models to help you make sure that your patterns work.

PROBLEM SOLVING

Strategies Review

Use these problem-solving strategies to solve the problems. Remember that some problems can be solved using more than one strategy.

- Using Number Sense
- Drawing a Diagram
- Choosing the Operation
- Solving a Two-Step Problem
- Finding a Pattern

- Guess, Test, and Revise
- Using Estimation
- Making a Table
- Solving a Multistep Problem

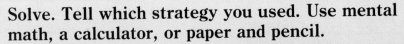

Solve. Tell which strategy you used. Use mental math, a calculator, or paper and pencil.

1. Chester is assembling his model house. He uses 3 hinges to attach 2 boards. How many hinges does he need to attach 4 boards in a row?

2. The stairs of the library have 18 steps between floors. How many steps must you walk to get to the sixth floor?

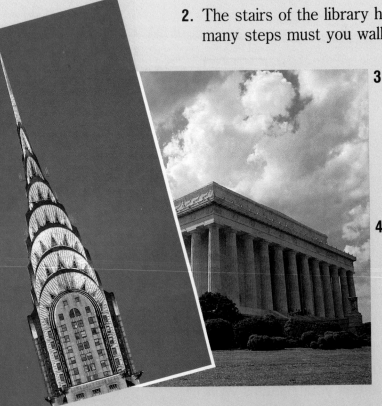

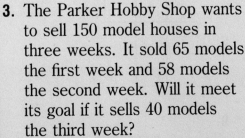

3. The Parker Hobby Shop wants to sell 150 model houses in three weeks. It sold 65 models the first week and 58 models the second week. Will it meet its goal if it sells 40 models the third week?

4. Brietta's apartment has 3 French windows and 5 regular windows. The French windows have 18 panes each, and the regular windows have 2 panes each. How many windowpanes does her apartment have in all?

5. A designer bought sheets of blueprint and transparent paper. The blueprint paper cost $2 a sheet, and the transparent paper cost $.50 a sheet. He paid $33 for 30 sheets. How many sheets of each kind of paper did he buy?

6. Cheri is designing a tile wall. She puts 2 red tiles in the first row, 4 red tiles in the second row, and 8 red tiles in the third row. How many tiles are in each of the next two rows?

7. Mike and Roberto are bricklayers. Mike finished 21 rows of bricks in one day, and Roberto finished 24 rows. Each row has 9 bricks. How many more bricks did Roberto lay than Mike?

7. The Scout troop needs to raise $300 for new equipment. They raised $148 from a card sale and $210 from a bake sale. Have they raised enough money?

9. The architect Henry Bacon, who built the Lincoln Memorial, lived from 1866 to 1924. The inventor of the geodesic dome, R. Buckminster Fuller, lived from 1895 to 1983. Who lived longer?

10. The Eiffel Tower in Paris weighs about 9,000 tons. The CN Tower in Toronto weighs about 160,000 tons. Does the CN Tower weigh more than 150,000 tons more than the Eiffel Tower?

11. Hal needs 30 round pegs to assemble his model house. If the pegs are packed 8 to a box, how many boxes of pegs should Hal buy?

12. *Write a problem* that can be solved by using one of the strategies. Ask others to solve your problem.

DEVELOPING A CONCEPT
Slides, Flips, and Turns

Many artists make patterns by moving a geometric figure.

You can move a geometric figure in different ways.

You can **slide** a figure across a line.

You can **flip** a figure over a line.

You can **turn** a figure around a point on a line.

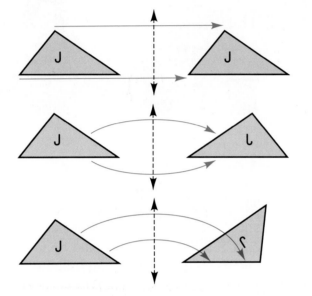

WORKING TOGETHER

Step 1 Draw a geometric figure. Trace it and cut it out.

Step 2 Place the cutout over your drawing. Slide the cutout and trace its new position. Label this move on the back.

Step 3 Show a flip on another piece of paper. Show a turn on a third piece of paper.

Step 4 Share your figures with others. Tell how their figures have been moved.

SHARING IDEAS

1. Tell how each move changes the position of a figure.

2. Can you always tell how a figure was moved? Why or why not?

3. Does the shape of a figure change when you move it? Why or why not?

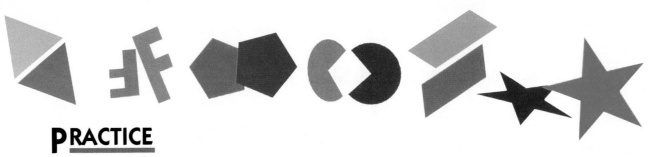

PRACTICE

Write *slide, flip,* or *turn* to tell how the figure was moved.

4.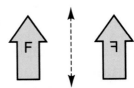

5.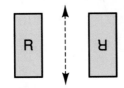

6.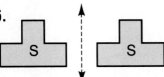

7.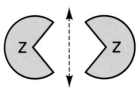

8.

9.

Critical Thinking

10. Tell how you can move the triangle to make the complete design.

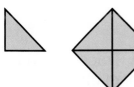

Mixed Applications

11. Lotte uses 479 tiles to cover the wall. She uses 503 tiles to cover the floor. How many more tiles does she use on the floor than on the wall?

12. Ben uses 5 boxes of colored tiles to make a design on a wall. Each box holds 25 tiles. How many tiles does Ben use?

CHALLENGE

Look at the picture. A triangle has been slid, flipped, and turned to create a pattern. The triangles fit together with no space between them. A pattern that repeats one figure of the same size and shape in this way is called a **tessellation.**

Which pattern blocks can be used to make a tessellation?

Congruence and Similarity

At the Lyndon B. Johnson Space Center in Houston, Jorge bought a poster of a planned space colony. He also bought two postcards of the same picture.

Figures that are the same shape and size are **congruent.** Figures that have the same shape but are different sizes are **similar.**

1. Which of Jorge's pictures are congruent?

2. Which of Jorge's pictures are similar?

WORKING TOGETHER

Step 1 Trace the figures on the page. Cut them out.

Step 2 Sort the figures by their shapes.

Step 3 Match the parts of the figures to see if they are similar.

Step 4 Show which figures are congruent by placing them on top of each other.

3. Which figures are similar?

4. Which figures are congruent?

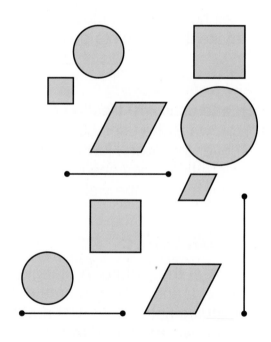

SHARING IDEAS

Tell if the statement is *true* or *false.* Tell how you made your decision.

5. All figures that are the same size are congruent.

6. Figures must be in the same position to be congruent.

7. All line segments are similar.

8. All circles are congruent.

PRACTICE

Tell which figure is congruent and which is similar.

9.
 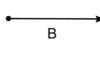 A B C  D

10.
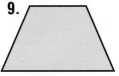 A B C D

11.
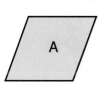 A B C D

Write *congruent* or *similar*.

12. footprints in the sand

13. a photo and its enlargement

14. a newspaper ad and a billboard

15. covers of the same book

Mixed Applications

16. Miko has $15. She wants to buy a poster for $12.99 and a map for $1.98. Does she have enough money? How do you know?

17. **Make a list** of objects that have congruent shapes. Make another list of objects with similar shapes. Share your list with others.

Mixed Review

Find the answer. Which method did you use?

18. 630 ÷ 7   **19.** 6 × 352   **20.** 402 − 298   **21.** 642 + 968

MENTAL MATH
CALCULATOR
PAPER/PENCIL

Line of Symmetry

Brad is making a model of a house. He knows a fast way to cut some of the pieces he needs. They have special shapes.

You can use Brad's method of cutting to explore the shape of figures.

WORKING TOGETHER

Step 1

Fold a piece of paper so that the two parts overlap exactly.

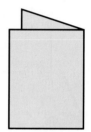

Step 2

From the fold, cut a triangle like the one in the picture.

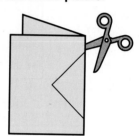

Step 3

Unfold the paper.

1. Look at the two parts of the square. Are they congruent? How do you know?

A figure that can be folded so that its two parts match exactly is a **symmetric figure.** The fold line is a **line of symmetry.**

2. Fold the square in different ways. Can you find other lines of symmetry? How many?

3. Trace and cut out the figures on the right. Fold them to find how many lines of symmetry they have. Draw your own figures. Are they symmetrical?

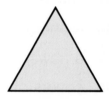

SHARING IDEAS

4. Which figures have 1 line of symmetry? more than 1 line of symmetry?

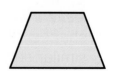

5. Are all figures symmetrical? How can you show this?

6. Which figure has the most lines of symmetry?

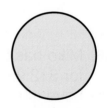

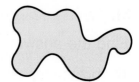

PRACTICE

Tell whether the dotted line is a line of symmetry.

7.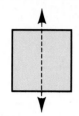

8.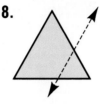

9.

10.

11.

12.

13.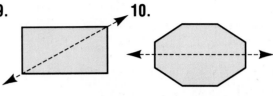

14.

Tell how many lines of symmetry the figure has.

15.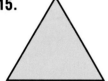

16.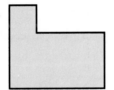

17.

18.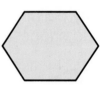

19.

20.

21.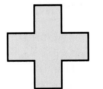

22.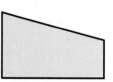

Critical Thinking

23. Would you slide, flip, or turn a copy of this figure to make a complete symmetric figure?

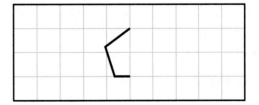

Mixed Applications

24. Which capital letters of the alphabet are symmetrical? Which have more than 1 line of symmetry?

25. Brad cuts out 24 pieces of paper for each house. How many pieces must he cut out for 7 houses?

UNDERSTANDING A CONCEPT

Space Figures

Many buildings and objects in the world suggest **space figures.**

straight edge
flat face
vertex

Pyramid

  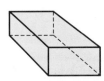 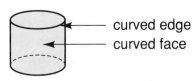

curved edge
curved face

Cube   Sphere   Cone   Rectangular prism   Cylinder

1. Which figures have only flat faces? How many does each have?

2. Which figures have only curved faces? How many does each have?

3. Which figures have both flat and curved faces? How many of each type does each figure have?

4. How many edges does each figure have? How many vertices?

TRY OUT   Name the space figure suggested by each object.

5.

6.

7.

PRACTICE

Name the space figure suggested.

8.

9.

10.

11.

12.

13.

14.

15.

16. It has one vertex and a curved face.

17. Some of its faces are shaped like triangles.

18. Two of its faces are circles.

19. All of its faces are congruent.

Critical Thinking

Write *true* or *false.* Tell how you made your decision.

20. All cubes are congruent figures.

21. All spheres are similar.

22. All rectangular prisms are congruent.

Mixed Applications

Solve. You may need to use the Databank on page 520.

23. Which building is shaped like a pyramid? a rectangular prism? a sphere?

24. A square floor in Kiel's house is 5 m long. What is the area of the floor?

25. Marie's building has 5 floors. There are 75 apartments in it. Each floor has the same number of apartments. How many apartments are on each floor?

26. *Write a description* of a space figure. Have others identify the figure.

Volume

Colin is building a twenty-first century house with geometric blocks. First he builds the base with 3 rows of 4 cubes. How many blocks does Colin use?

Each block is one **cubic unit.** The number of cubic units in a figure is its **volume.**

WORKING TOGETHER

You can use centimeter cubes to find the volume of the blocks Colin uses.

Use centimeter cubes to model Colin's blocks.

1. How many rows of cubes are there? How many cubes are in each row?

2. Count the total number of cubes.

3. How many cubic units does Colin use?

4. *What if* Colin builds a figure 3 cubes long, 2 cubes wide, and 2 cubes high? What would the volume be?

5. See how many other solid figures you can make using this number of cubic units. What is the volume of each figure?

6. How many of the figures you made are shaped like rectangular prisms? Record the length, width, height, and volume of each in a table.

7. Use different numbers of cubes to make other rectangular prisms. Record their measurements in the table.

Length	Width	Height	Volume
3	4	1	■
3	2	2	■

SHARING IDEAS

8. Can different figures have the same volume? Why or why not?

9. Can you use the same number of cubes to make different figures that have different volumes? Why or why not?

10. What pattern do you see in the table? How can you use this pattern to find the volume of rectangular prisms without counting?

PRACTICE

Find the volume.

11.

12.

13.

14.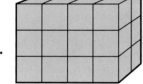

15.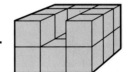

16.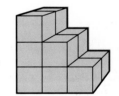

Find the volume of each of the following rectangular prisms. Use centimeter cubes if needed.

17. length: 2 cm; width: 4 cm; height: 1 cm

18. length: 5 cm; width: 3 cm; height: 2 cm

19. length: 3 cm; width: 5 cm; height: 3 cm

Mixed Applications Solve. Which method did you use?

20. Erin builds a large cube out of centimeter cubes. Each side is 3 cubes long. How many centimeter cubes did Erin use?

21. A box of blocks costs $14.37. Colin pays for it with a $20.00 bill. How much change will he get?

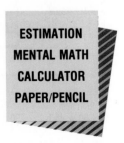

ESTIMATION
MENTAL MATH
CALCULATOR
PAPER/PENCIL

DECISION MAKING

Problem Solving: Arranging Your Room

SITUATION

The Esposito family has just moved to a new home. Mrs. Esposito told Jackie to decide how she wants to arrange the furniture in her room.

PROBLEM

Where should Jackie put each piece of furniture? Will everything fit in her room?

DATA

l = length w = width h = height
The side of each square = 30 cm.
All measurements are in cm.

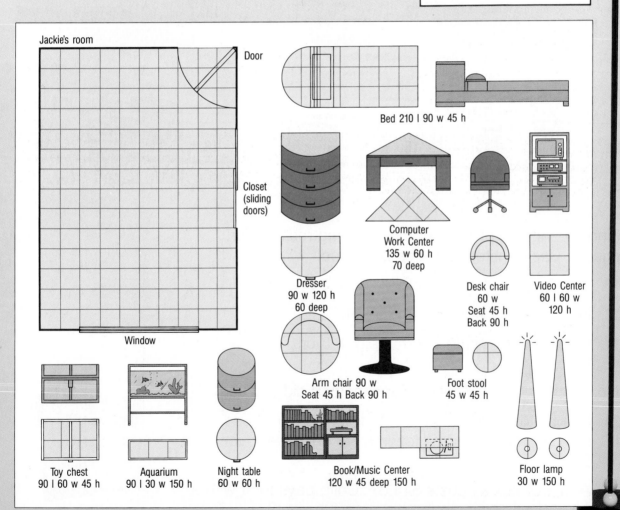

Jackie's room

Door

Closet (sliding doors)

Window

Bed 210 l 90 w 45 h

Computer Work Center 135 w 60 h 70 deep

Dresser 90 w 120 h 60 deep

Desk chair 60 w Seat 45 h Back 90 h

Video Center 60 l 60 w 120 h

Arm chair 90 w Seat 45 h Back 90 h

Foot stool 45 w 45 h

Toy chest 90 l 60 w 45 h

Aquarium 90 l 30 w 150 h

Night table 60 w 60 h

Book/Music Center 120 w 45 deep 150 h

Floor lamp 30 w 150 h

USING THE DATA

1. What size is Jackie's room?

2. The window sill is 1 m from the floor. Which pieces of furniture can fit under the window?

Can all these pieces of furniture fit on the wall opposite the closet?

3. dresser
   aquarium (width)
   book/music center
   bed (width)

4. dresser,
   bed (length)
   book/music center

5. bed (length)
   floor lamp
   dresser
   video center

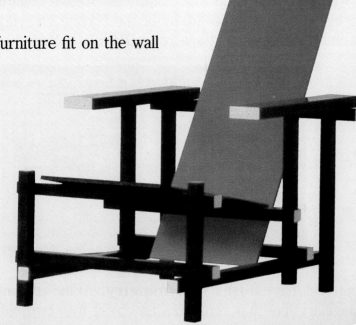

MAKING DECISIONS

6. **Write a list** of the things Jackie should think about when deciding how to arrange the furniture in her room.

7. Can Jackie put all the furniture in the room and still have room to walk around comfortably? What pieces could Jackie leave out?

8. Which pieces do you think Jackie needs in her room?

9. Where should Jackie put her lamps? Why?

10. **What if** Jackie shares a room with her sister? What other furniture will she need? How can she arrange the room?

11. How would you arrange Jackie's room?

12. Suppose you could design a bedroom any way you chose. Use grid paper to draw a picture of how it would look.

Math and Art

Navajo women in the American Southwest weave beautiful clothes, blankets, and rugs. Many of the Navajo weavings have geometric shapes that are repeated in a pattern.

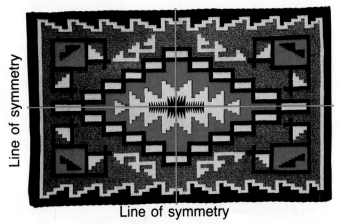

Line of symmetry

Line of symmetry

Many patterns have a **line of symmetry.** A line of symmetry is an imaginary line that separates an object into two halves that are exactly alike. The rug above has two lines of symmetry.

What if you saw this blanket? What line or lines of symmetry does it have?

Think: A vertical line through the middle of the blanket would separate it into two halves that are exactly alike.

The blanket has one line of symmetry.

ACTIVITIES

1. Use graph paper and crayons to draw a geometric pattern that has a line of symmetry. Draw the line of symmetry.

2. Find pictures of weavings or quilts that use geometric shapes. Do they have lines of symmetry? Discuss with your class how they are the same as the Navajo weavings and how they are different.

Computer Graphics

Many designers use geometric shapes in the things they design. They start with some basic shapes, then copy them and change their size and position to create a pattern. A computer drawing program makes it easy to draw, copy, or change shapes.

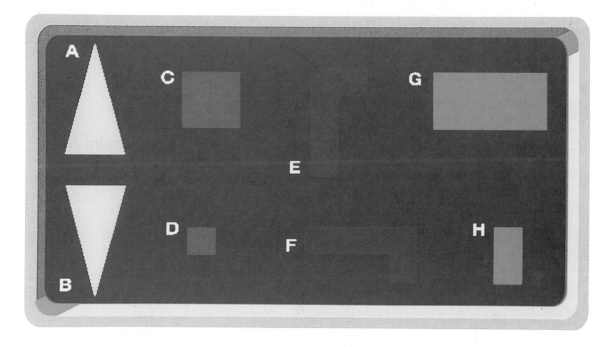

DATA

Use graph paper and pencil to copy the figures above.

THINKING ABOUT COMPUTERS

1. Which figures are congruent? similar?

2. Which figures have been flipped? turned?

3. Make a design using some of the shapes. Compare your design to those of others.

4. Why would it be easier to use a computer to make or change a design?

EXTRA PRACTICE

Lines, Line Segments, and Rays, page 307 ....

Name each figure.

1.

2.

3.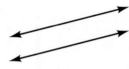

Tell if the figures are *intersecting* or *parallel*.

4.

5.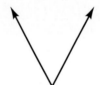

6.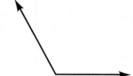

Draw the figure.

7. ray *CB*   **8.** line *MD*   **9.** line segment *RN*

10. line *RS* intersects line *PQ* at point *B*

Angles, page 309 ....

Name the angle. Tell if it is a *right angle*.
If not, write *greater* or *less*.

1.

2.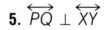

3.

Draw the figure.

4. right angle *BCD*   **5.** $\overleftrightarrow{PQ} \perp \overleftrightarrow{XY}$ **6.** $\overleftrightarrow{RS} \parallel \overleftrightarrow{WY}$

Plane Figures, page 311 ....

Name the figure.

1.

2.

3.

4.

Problem Solving Strategy: Solving a Multistep Problem, page 313............

Make a plan. Then solve the problem.

1. On Jean's block there are 15 houses. There are 14 houses on Tim's block and 17 houses on Ray's block. How many more houses are there on Tim and Ray's blocks together than on Jean's block?

2. On Tim's block there are 3 families with 5 children, 4 families with 3 children, and 7 families with 2 children. How many children live on Tim's block?

3. Ray is washing the windows in his house. He lives in a 7-room house with 2 large windows and 2 small windows in each room. He has already washed 15 windows. How many windows does he have left to wash?

4. Jean is decorating her bedroom. She buys 2 pillows for $12 each and 1 curtain for $19.95. How much change should she get from a $50 bill?

Problem Solving: Strategies Review, page 317.................................

Solve. Tell which strategy you used.

1. Ray takes 3 minutes to wash each window. If there are 18 windows in his house, how many minutes does it take him to wash all of the windows?

2. Jean needs 50 paper cups to serve fruit punch at her garden party. The cups are packaged 12 to a pack. How many packages of cups does Jean need to buy?

3. The keys on a piano are 1 inch in width. Serge's hand can span 6 inches. How many piano keys can Serge reach?

4. Lars, Daisy, and Fred each have a different pet. Lars raises rabbits. Fred does not own a dog. Who owns a dog?

EXTRA PRACTICE

Slides, Flips, and Turns, page 319 ....

Tell whether each is a *slide, flip,* or *turn.*

1.

2.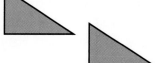

3.

Congruence and Similarity, page 321 ....

Tell which figure is *congruent* and which is *similar.*

1.
  a.
  b.
  c.
  d.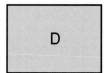

2.
  a.
  b.
  c.
  d.

Line of Symmetry, page 323 ....

Tell whether the dotted line is a line of symmetry.

1.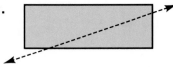

2.

3.

Tell how many lines of symmetry each figure has.

4.
5.
6.
7.

Space Figures, page 325

Name the space figure suggested.

1.

2.

3.

4.

5.

6.

7.

8.

Volume, page 327

Find the volume.

1.

2.

3.

4.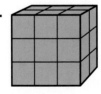

5.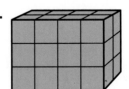

6.

Complete the table. Each figure is a rectangular prism. Use centimeter cubes if needed.

	Length (cm)	Width (cm)	Height (cm)	Volume (cubic cm)
7.	2	3	1	■
8.	3	4	4	■
9.	3	4	3	■
10.	4	6	2	■

Practice *PLUS*

KEY SKILL: Plane Figures (Use after page 311.)

Level A ....

Name the figure.

1.   **2.**   **3.**   **4.**

5. a polygon with 5 sides   **6.** a polygon with 4 sides

Draw the figure.

7. a rectangle   **8.** a triangle with 3 equal sides

Level B ....

Name the figure.

9.   **10.**   **11.**   **12.**

13. a polygon with 10 sides

14. a quadrilateral with 4 right angles and 4 equal sides

Draw the figure.

15. a pentagon

16. a triangle with one angle smaller than a right angle

Level C ....

Name the figure.

17. a polygon with 5 sides and 5 angles

18. a figure with no sides and no angles

Draw the figure.

19. a decagon

20. a polygon with 4 sides and 4 right angles

21. a circle with a radius of 4 cm

PRACTICE PLUS

KEY SKILL: Space Figures (Use after page 325.)

Level A ....

Name the space figure suggested.

1.

2.

3.

4.

5. Jenny uses a pastry bag to decorate a cake. The tip has 1 curved face and 1 edge. What shape is this?

Level B ....

Name the space figure suggested.

6.

7.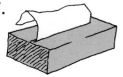

8.

9.

10. David makes a model of a skyscraper. It has 6 faces, 8 vertices, and 12 edges. What is the shape of the model?

Level C ....

Name the space figure suggested.

11.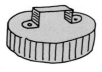

12.

13.

14.

15. Priscilla makes a wind sock to hang outside. It has 2 flat faces and 2 curved edges. What is the shape of the wind sock?

CHAPTER REVIEW

LANGUAGE AND MATHEMATICS

Complete the sentences. Use the words in the chart.

1. If lines do not cross, they are ■. *(page 306)*

2. The rays of an angle meet at the ■. *(page 308)*

3. A flat surface that goes on in all directions is a ■. *(page 310)*

4. ■ figures have the same shape and size. *(page 320)*

5. **Write a definition** or give an example of the words you did not use from the chart.

VOCABULARY
vertex
plane
polygon
congruent
intersect
parallel
flip

CONCEPTS AND SKILLS

Name each figure. Tell whether the figure is a line, line segment, or ray. *(page 306)*

6.

A B

7.

L   M

8.

J   K

Tell if the figures are *intersecting* or *parallel*. *(page 306)*

9.

10.

11.

Use the figure to answer. *(pages 306–309)*

12. Name a pair of perpendicular lines. What is the point of intersection?

13. Name an angle with vertex *O* that is less than a right angle.

14. Name a right angle.

15. Name an angle with vertex *O* that is greater than a right angle.

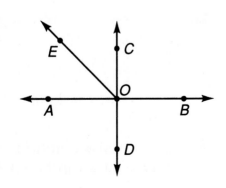

Name the figure. *(pages 310, 324)*

16.

17.

18.

19.

20. Is the figure the result of a *flip, slide,* or *turn*? *(page 318)*

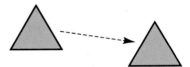

21. Tell how many lines of symmetry the figure has. *(page 322)*

a. **b.**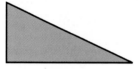

Which figure is *congruent* and which is *similar*? *(page 320)*

22. **a.** **b.** **c.** **d.**

Find the volume. *(page 326)*

23.

24.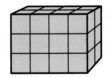

25.

CRITICAL THINKING

26. Do congruent space figures have the same volume? Why or why not? *(pages 320, 326)*

27. Are all triangles similar? Why or why not? *(pages 310, 320)*

MIXED APPLICATIONS

28. Mo's house has 6 rooms with 2 windows each. Mo sews 2 curtains for each window. He finishes 10 curtains. How many more must he sew? *(page 312)*

29. Al uses 4 posts to attach 3 sections of fence. How many posts does he need to attach 15 sections of fence in a row? *(page 316)*

30. A box made of plastic blocks is 9 blocks long, 5 blocks wide, and 5 blocks high. What is its volume? *(page 326)*

CHAPTER TEST

Name the figure.

1.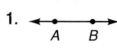
A　B

2.

3.
F　G

4.
L　　　M

5.

6.

7.

8.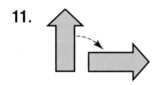

Is the figure the result of a *flip, slide,* or *turn?*

9.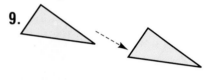

10.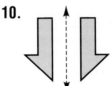

11.

Tell which figure is *congruent* and which is *similar.*

12. 　

13.

14.

15.

Find the volume.

16.

17.

18.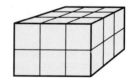

Solve.

19. Hal is planting 90 trees on his land. He has planted 3 acres with 12 trees each and 2 acres with 15 trees each. How many trees does he have left to plant?

20. Dana has $35.00. She gives $11.50 to her brother and buys a book for $18.95. How much money does she have left?

ENLARGING FIGURES ON A GRID

You can use coordinate grids to enlarge geometric figures.

Measure side *AB* of triangle *ABC* and line segment *DE*.

1. Compare their lengths.

2. What are the ordered pairs for points *A* and *B*?

3. What are the ordered pairs for points *D* and *E*?

4. Compare the ordered pairs.

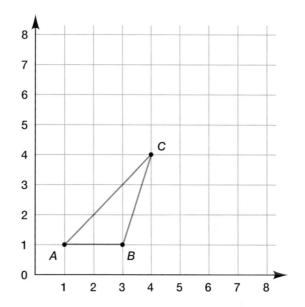

Suppose that you wanted to use line segment *DE* to draw a triangle *DEF* that is twice as large as triangle *ABC*.

5. What would be the ordered pair for point *F*?

6. Copy triangle *ABC*. Draw triangle *DEF* twice as large as triangle *ABC*.

7. How should the length of sides *AC* and *BC* compare with the length of sides *DF* and *EF*? Measure your drawing to check your answer.

8. Are the two triangles similar?

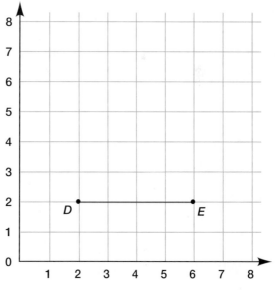

9. Draw and label this rectangle on grid paper: *W* (3, 1), *X* (3, 2), *Y* (5, 2), *Z* (5, 1). Draw a rectangle three times as large as rectangle *WXYZ*. Label the rectangle *RSTU*.

10. **What if** you wanted to reduce a figure? How would you find the ordered pairs of the reduced figure? Check your method by drawing a smaller triangle similar to triangle *FGH*: *F* (3, 3), *G* (6, 3), *H* (6, 6).

CUMULATIVE REVIEW

Choose the letter of the correct answer.

1. What is $\overline{AC}$?

   a. a line   **c.** a line segment

   b. a ray   **d.** not given

2. Name the angle.

   a. $\angle QRS$ **c.** $\angle RSQ$

   b. $\angle RQS$ **d.** not given

3.

   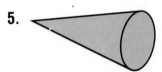

   The lines are:

   a. intersecting   **c.** perpendicular

   b. parallel   **d.** not given

4. 4×52

   a. 88   **c.** 218

   b. 208   **d.** not given

5.

   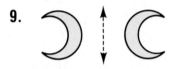

   What figure is shown?

   a. cone   **c.** pyramid

   b. cylinder   **d.** not given

6. Which figure is not a polygon?

   a. decagon   **c.** triangle

   b. circle   **d.** rectangle

7. $6\overline{)896}$

   a. 149   **c.** 149 R2

   b. 149 R1   **d.** not given

8. Divide 18 by 3.

   a. 6   **c.** 15

   b. 54   **d.** not given

9.

   What is shown?

   a. slide   **c.** flip

   b. turn   **d.** circle

10. Find the median: 8, 7, 3, 6, 1.

    a. 5   **c.** 7

    b. 6   **d.** not given

11. 6×73

    a. 428   **c.** 498

    b. 438   **d.** not given

12. Estimate: $369 \div 4$

    a. 8   **c.** 80

    b. 9   **d.** 90

13. $\blacksquare \times 8 = 72$

    a. 8   **c.** 576

    b. 80   **d.** not given

14. $63 \div 7$

    a. 9   **c.** 70

    b. 56   **d.** not given

Understanding Fractions and Mixed Numbers

CHAPTER 9

MATH CONNECTIONS: LENGTH • PROBLEM SOLVING

MUSIC LESSONS FOR CHILDREN

Violin
1 hour
Monday and
Tuesday

Flute
1 hour
Wednesday
only

Drums
½ hour
Tuesday and
Wednesday

1. What is happening in this picture?

2. What information do you see?

3. How can you use this information?

4. Write a problem about the picture.

The Meaning of Fractions

Uri is on the swim team. He uses the pool in the school gym to practice. Each of the lanes in the pool is the same size. Uri practices in one of the 4 lanes. How much of the pool does Uri use for practice?

You can use geoboards to make a model of the pool.

WORKING TOGETHER

Step 1 Use a rubber band to make a rectangle on the geoboard. Use other rubber bands to separate the rectangle into 4 equal parts.

Step 2 Record your geoboard design on dot paper. Shade one of the equal parts.

1. How do you know each part is equal?

Fractions name equal parts of a whole. Each equal part is **one-fourth** of the whole pool.

Uri uses one-fourth of the pool for practice.

2. Show one-fourth of the geoboard another way. Record your design on dot paper. Compare it with those of others.

3. Try making models of other fractions on the geoboard. Record your designs on dot paper.

Fractions can also name parts of a set.

Put out 4 cubes and 1 counter. One of the 5 objects is a counter.

One-fifth of the objects are counters.

4. What fraction of the objects are cubes?

5. Pick other sets of cubes and counters. Tell what fraction of the set are counters and what fraction are cubes.

SHARING IDEAS

6. How are finding parts of a whole and finding parts of a set the same? How are they different?

7. What information do you need to name a fraction?

8. As you separate a whole into more and more parts, what happens to the size of the parts?

PRACTICE

Tell which of these figures show sixths. Tell why or why not.

9.  **10.** 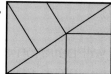 **11.**   **12.**

Match the fraction with a diagram.

13. one-sixth   **14.** two-thirds

15. two-fifths   **16.** one-fourth

Mixed Applications

17. There are 6 swimmers in the five-mile race. Only 2 swimmers finish the race. What fraction of the swimmers finish the race?

18. The bulletin board is separated into 8 equal sections. The results of the swim meet cover 5 of these sections. What fraction of the board do they cover?

19. Uri started to practice at 3:30 P.M. Practice lasted 2 hours. When did Uri finish practicing?

20. *Write a list* of things that are separated into fractions. Compare your list with those of others.

EXTRA Practice, page 374

UNDERSTANDING A CONCEPT

Parts of a Whole

Troy is planning the mural the Art Club will paint on a wall of the school. Troy separates the wall into 6 equal parts. He plans to paint one of the parts himself. What fraction of the wall will Troy paint?

You can draw a diagram to show the part of the wall Troy plans to paint.

He plans to paint one-sixth of the wall.

You can write the fraction.

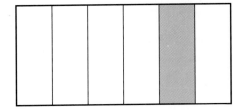

part Troy paints → **1** ← numerator
total number of parts → **6** ← denominator

1. What fraction of the wall will other students paint?

2. What fraction of the wall has been painted so far?

3. What fraction of the wall will be painted when the mural is finished? What whole number is this equal to?

TRY OUT   Write the letter of the correct answer.
What part of the figure is shaded?

4.

a. $\frac{3}{8}$ **b.** $\frac{1}{5}$ **c.** $\frac{5}{8}$ **d.** $\frac{8}{5}$

5.

a. $\frac{2}{3}$ **b.** $\frac{2}{5}$ **c.** $\frac{3}{5}$ **d.** $\frac{5}{2}$

Practice

Write a fraction for the part that is shaded.

6. 　　**7.** 　　**8.** 　　**9.**

Match the picture with a fraction.

10. 　　**11.** 　　**12.** 　　**13.**

a. $\frac{4}{6}$　　　　　　**b.** $\frac{2}{3}$　　　　　**c.** $\frac{1}{2}$　　　　　**d.** $\frac{2}{6}$

Draw a shape. Then color in the fraction.

14. $\frac{3}{4}$　　　**15.** $\frac{4}{10}$　　　**16.** $\frac{7}{8}$　　　**17.** $\frac{1}{2}$　　　**18.** $\frac{6}{9}$

Write the fraction.

19. one-twelfth　　**20.** six-eighths　　**21.** nine-tenths　　**22.** two-sevenths

Write the word name.

23. $\frac{1}{16}$　　　**24.** $\frac{4}{9}$　　　**25.** $\frac{3}{10}$　　　**26.** $\frac{3}{5}$　　　**27.** $\frac{11}{12}$

Critical Thinking

28. Kevin says he can show fourths like this: Is he correct? Why or why not?

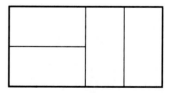

Mixed Applications

29. Troy separates his part of the wall into 8 equal sections. He paints 3 of the sections. What fraction of his part has he painted?

30. There are 20 paintbrushes in a package. Nancy uses 3 of the brushes. How many brushes are left?

31. Li draws a shape that has 4 right angles and 4 sides that are the same length. What shape did Li draw?

32. *Write a problem* about separating something into fractions. Solve your problem. Ask others to solve it.

UNDERSTANDING A CONCEPT

Parts of a Set

Teresa asked 10 friends about their favorite way to exercise. She made this list of their answers. What fraction of Teresa's friends like swimming best?

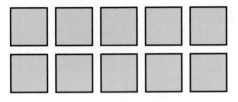

FAVORITE EXERCISE

Swimming 3

Biking 5

Running 2

You can use cubes to make a model of the set.

You can write the fraction.

number who like to swim → **3** ← numerator
number in all → **10** ← denominator

Teresa knows that $\frac{3}{10}$ of her friends like swimming best.

1. What fraction of her friends like biking best?

2. What fraction like running best?

3. What fraction like push-ups best?

4. **What if** all 10 friends said biking was their favorite exercise? What fraction is this?

TRY OUT

Write the letter of the correct answer.
What fraction names the part?

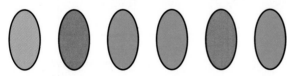

5. red   a. $\frac{1}{6}$ b. $\frac{2}{4}$ c. $\frac{2}{6}$ d. $\frac{3}{6}$

6. blue   a. $\frac{1}{6}$ b. $\frac{1}{5}$ c. $\frac{3}{6}$ d. $\frac{6}{6}$

7. green   a. $\frac{1}{6}$ b. $\frac{2}{6}$ c. $\frac{3}{5}$ d. $\frac{3}{6}$

8. yellow   a. $\frac{0}{6}$ b. $\frac{1}{6}$ c. $\frac{3}{3}$ d. $\frac{6}{6}$

PRACTICE

Write a fraction that shows the part of the set that is shaded.

9.

10.

11.

12.

13.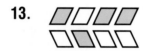

14.

Write the fraction.

15. Seven out of ten answers are correct.

16. Three of the seven balls are soccer balls.

17. Eight of the twelve baskets are full.

18. All of the 4 tennis courts are full.

19. None of the 6 swimming lanes are being used.

Mixed Applications

20. Cindy went to the Bike Club meeting at 4:15 P.M. She stayed until 5:45 P.M. How long did she stay at the meeting?

21. Phil is playing basketball. He takes 8 shots and makes 5 of them. What fraction of his shots does he make?

22. Glen did 10 push-ups before breakfast and 12 more at bedtime. How many push-ups did he do?

23. Jesse is practicing golf. He makes 6 putts. Only 2 of them go in the hole. How many did not go in?

LOGICAL REASONING

Write a fraction for the part described.

1. red or square shapes

2. blue and circular

3. four sides and not blue

4. three sides and green

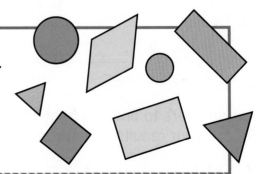

DEVELOPING A CONCEPT

Finding the Fraction of a Number

Eight scouts are going to the country for a hike. After the equipment is loaded, the first van has room for $\frac{1}{4}$ of the group. How many scouts ride in the first van?

WORKING TOGETHER

You can make a model to solve the problem.

Step 1 Use counters to model the total number of scouts.

Step 2 Separate the counters into 4 groups of equal size.

1. How many counters do you need to show the total number of scouts?

2. How many counters are in one of the groups? What is $\frac{1}{4}$ of 8?

3. How many scouts rode in the first van?

4. How many counters are in two of the groups? What is $\frac{2}{4}$ of 8?

5. What is $\frac{3}{4}$ of 8? $\frac{4}{4}$ of 8?

6. Record your answers in a table like the one below.

Part	Total Number in the Set	Number of Groups	Number in Each Group	Number of Groups in Part	Number in Part of Set
$\frac{1}{4}$	of 8	4	2	1	■
$\frac{2}{4}$	of 8	4	2	■	■

Use counters to find the fraction of the number.
Record your results in your table.

7. $\frac{1}{3}$ of 12

8. $\frac{2}{3}$ of 12

9. $\frac{1}{4}$ of 16

10. $\frac{3}{4}$ of 16

SHARING IDEAS

11. Look at the table. How can you use multiplication and division to find the fraction of a number?

12. Can you use counters to find $\frac{1}{3}$ of 8? Why or why not?

PRACTICE

Find the part.

13. $\frac{1}{6}$ of 18   **14.** $\frac{3}{6}$ of 18   **15.** $\frac{1}{3}$ of 18

16. $\frac{2}{3}$ of 18   **17.** $\frac{1}{9}$ of 18   **18.** $\frac{1}{2}$ of 18

Find the fraction of the number. Use counters if needed.

19. $\frac{1}{2}$ of 12   **20.** $\frac{1}{4}$ of 24   **21.** $\frac{1}{6}$ of 24   **22.** $\frac{1}{3}$ of 15

23. $\frac{2}{3}$ of 3   **24.** $\frac{4}{5}$ of 10   **25.** $\frac{3}{8}$ of 32   **26.** $\frac{3}{4}$ of 16

Critical Thinking

27. Ten hikers arrived early. This is $\frac{2}{3}$ of the total number of hikers in a group. How many hikers are in the group? How did you find the answer?

Mixed Applications   Solve. Which method did you use?

28. Mary saw 20 different types of trees in the forest. She knew the names of $\frac{2}{5}$ of them. How many trees could Mary name?

29. One school bus can take 36 hikers on a trip. How many hikers can go on 5 buses?

30. There are 115 hikers in the camp today. They are in 5 equal groups. Are there more than 25 hikers in each group?

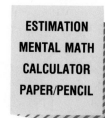

ESTIMATION
MENTAL MATH
CALCULATOR
PAPER/PENCIL

Mixed Review

Compare. Write $>$, $<$, or $=$.

31. $4 \times 3 \bullet 60 \div 5$ **32.** $14 + 2 \bullet 3 \times 4$ **33.** $15 + 6 \bullet 104 \div 4$

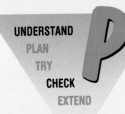

PROBLEM SOLVING

UNDERSTAND
PLAN
TRY
CHECK
EXTEND

Working Backward

Heidi has 3 times as many baseball cards as Michael. Jimmy has 5 more cards than Michael. Josh has 3 fewer cards than Jimmy. Josh has 9 cards. How many cards does Heidi have?

You can work backward to solve the problem. Start with what you know at the end and work back to the beginning.

You know that Josh has 9 baseball cards and that he has 3 fewer cards than Jimmy.

1. What can you do to find how many cards Jimmy has?

2. How many cards does Jimmy have? $9 + 3 = 12$

You also know that Jimmy has 5 more cards than Michael.

3. What can you do to find how many cards Michael has?

4. How many cards does Michael have? $12 - 5 = 7$

Finally you know that Heidi has 3 times as many cards as Michael.

5. What can you do to find how many cards Heidi has?

6. How many cards does Heidi have? $7 \times 3 = \blacksquare$

7. How can you check your answer?

PRACTICE

Work backward to solve the problem. Use mental math, a calculator, or paper and pencil.

8. Three friends are rehearsing a play. Josephine has 5 times as many lines as Loren. Claire has 7 more lines than Josephine. Claire has 47 lines. How many lines does Loren have?

9. Sue, Jean, and Cindy exercise after school. Sue exercises 10 minutes less than Jean, and Cindy exercises half as long as Sue. Cindy exercises for 15 minutes. How long does Jean exercise?

10. Mark, Jim, and Chang are at soccer practice. Jim kicks the ball twice as far as Mark. Chang kicks the ball 8 meters farther than Jim. Chang kicks the ball 32 meters. How far does Mark kick the ball?

11. Peter, Mary, Quentin, and Tom sell tickets to the game. Peter sells $\frac{1}{2}$ as many tickets as Mary, and Quentin sells 5 fewer tickets than Peter. Tom sells 3 times as many as Quentin. If Tom sells 21 tickets, how many does Mary sell?

Strategies and Skills Review

12. Marcus practices piano 2 hours every day of the year except on 6 days when he is sick. How many hours did he practice in the year?

13. Pam swims the breaststroke for 8 laps and the backstroke for 6 laps. The length of a lap is 25 yards. How many yards does Pam swim?

14. There are 6 more people in the first row of the stands than in the second row, and 5 fewer in the third row than in the second. There are twice as many people in the first row as in the fourth. There are 17 people in the fourth row. How many people are in each of the first three rows?

15. **Write a problem** that can be solved by working backward. Solve your problem. Ask others to solve it.

DEVELOPING A CONCEPT

Finding Equivalent Fractions

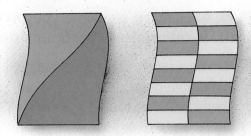

One-half of Jakki's beach towel is orange. Eight-sixteenths of Carla's towel is orange. Is the same part of each towel orange?

WORKING TOGETHER

Step 1 Fold a piece of paper in half and color one part.

1. What fraction shows how much you colored?

Step 2 Fold the paper in half again.

2. How many equal parts did you make by folding? How many are colored? What fraction is colored?

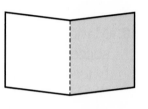

Step 3 Fold the paper one more time.

3. How many equal parts did you make now? How many are colored? What fraction is this?

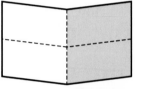

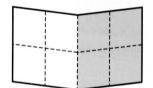

Step 4 Fold the paper again.

4. What fraction is colored?

$\frac{1}{2}$, $\frac{2}{4}$, $\frac{4}{8}$, and $\frac{8}{16}$ are **equivalent fractions.** Equivalent fractions name the same part of a whole or a set.

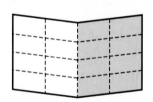

The same part of each towel is orange.

5. Fold another piece of paper to find fractions equivalent to $\frac{3}{4}$.

SHARING IDEAS

6. Look at the numerators and denominators in each set of equivalent fractions you found. What pattern do you see?

7. How can you use multiplication to find equivalent fractions? Why can you do this?

PRACTICE

Complete to name the equivalent fraction.

8.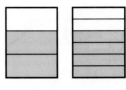

$$\frac{1}{3} = \frac{\blacksquare}{6}$$

9.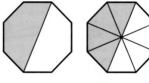

$$\frac{1}{2} = \frac{\blacksquare}{8}$$

10.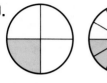

$$\frac{1}{4} = \frac{\blacksquare}{12}$$

11.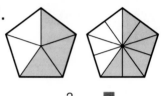

$$\frac{3}{5} = \frac{\blacksquare}{10}$$

12.

$$\frac{3}{4} = \frac{\blacksquare}{12}$$

13.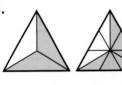

$$\frac{2}{3} = \frac{\blacksquare}{9}$$

14. $\frac{1}{2} = \frac{\blacksquare}{4}$ **15.** $\frac{1}{3} = \frac{\blacksquare}{9}$ **16.** $\frac{3}{8} = \frac{\blacksquare}{16}$ **17.** $\frac{5}{6} = \frac{\blacksquare}{12}$

Complete the pattern of equivalent fractions.

18. $\frac{1}{5} = \frac{\blacksquare}{10} = \frac{\blacksquare}{15} = \frac{4}{\blacksquare} = \frac{5}{\blacksquare}$ **19.** $\frac{2}{3} = \frac{\blacksquare}{6} = \frac{6}{\blacksquare} = \frac{\blacksquare}{12} = \frac{10}{\blacksquare}$

Critical Thinking

20. Brenda says $\frac{3}{6}$ of the beach balls are blue. Carlos says $\frac{2}{4}$ of them are blue. Who is correct? Why?

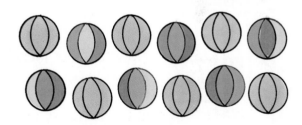

Mixed Applications

21. Bev and Sandra made 9 sand castles at the beach, but $\frac{2}{3}$ of them washed away. How many castles washed away?

22. Sol's beach towel is $\frac{8}{16}$ blue and $\frac{4}{8}$ green. Are the blue and green parts the same size? Why or why not?

Mixed Review

Find the answer. Which method did you use?

23. 24×7 **24.** $600 \div 6$ **25.** $\$36.59 + \$78.95$ **26.** $86 - 6$

| MENTAL MATH |
| CALCULATOR |
| PAPER/PENCIL |

MIND READING MADE EASY!

Investigating Patterns

A. Did you know that you can read minds? Here's a trick that will make your friends think you can. Tell a friend to follow these instructions carefully:

- Think of a number.
- Add 2 to your number.
- Double the amount you now have.
- Add 6.
- Divide by 2.
- Subtract your original number.

Now tell your friend to concentrate on the remaining number. Pretend you are reading your friend's mind. Then announce that the number is 5. You will be right!

Try the trick with a number of your own. Suppose you pick 10. Watch what happens.

10
$10 + 2 = 12$
$2 \times 12 = 24$
$24 + 6 = 30$
$30 \div 2 = 15$
$15 - 10 = 5$

Try it with another number. You will always get 5! Can you see how this trick works? Why don't you ever need to know your friend's starting number?

B. Let's draw pictures to see how this trick works. Let ??? stand for the mystery number. Use (○) to show how the number changes.

Think of a number.	???
Add 2 to your number.	??? ○ ○
Double the amount you now have.	??? ○ ○ ??? ○ ○
Add 6.	??? ○ ○ ○ ○ ○ ??? ○ ○ ○ ○ ○
Divide by 2.	??? ○ ○ ○ ○ ○
Subtract your original number.	○ ○ ○ ○ ○

No matter what ??? equals, the result is always 5.

1. Here is a chart for another trick. Copy the chart and fill in the second column.

Think of a number.	
Add 6 to your number.	
Triple the amount you now have.	
Add 9.	
Divide by 3.	
Subtract 2.	
Subtract your original number.	

2. What number is always the result?

3. Now invent your own trick. Draw pictures to check that your trick works.

Simplifying Fractions

Somati knows that 6 members of the band need new uniforms. There are 12 band members. What is the simplest fraction she can write to name the part of the band that needs new uniforms?

You can **simplify** a fraction by dividing the numerator and denominator by the same number. It is in **simplest form** when the numerator and denominator have no common factor greater than 1.

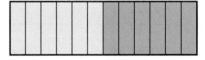

1. Complete: $\frac{6}{12} = \frac{6 \div 2}{12 \div 2} = \frac{\blacksquare}{\blacksquare}$

2. Is the fraction in simplest form? Can you divide again? Why or why not?

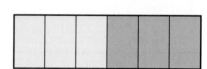

3. What is the simplest form of $\frac{6}{12}$?

Simplify the fraction if it is not in simplest form.

4. $\frac{4}{12}$ **5.** $\frac{5}{12}$ **6.** $\frac{2}{10}$ **7.** $\frac{2}{4}$

SHARING IDEAS

8. Why can you divide the numerator and denominator by the same number without changing the value of the fraction?

9. Are $\frac{6}{12}$, $\frac{3}{6}$, and $\frac{1}{2}$ equivalent fractions? How can you tell?

PRACTICE

Complete.

10.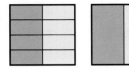

$$\frac{4}{8} = \frac{\blacksquare}{\blacksquare}$$

11.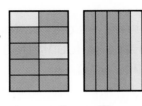

$$\frac{8}{10} = \frac{\blacksquare}{\blacksquare}$$

12.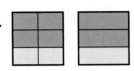

$$\frac{4}{6} = \frac{\blacksquare}{\blacksquare}$$

13. $\frac{6}{9} = \frac{6 \div 3}{9 \div 3} = \frac{\blacksquare}{\blacksquare}$

14. $\frac{8}{12} = \frac{8 \div 4}{12 \div 4} = \frac{\blacksquare}{\blacksquare}$

15. $\frac{12}{16} = \frac{12 \div 4}{16 \div 4} = \frac{\blacksquare}{\blacksquare}$

16. $\frac{6}{10} = \frac{\blacksquare}{5}$

17. $\frac{9}{12} = \frac{\blacksquare}{4}$

18. $\frac{6}{8} = \frac{\blacksquare}{4}$

19. $\frac{14}{16} = \frac{\blacksquare}{8}$

Is the fraction in simplest form? Write *yes* or *no*.

20. $\frac{1}{5}$

21. $\frac{5}{15}$

22. $\frac{5}{6}$

23. $\frac{4}{10}$

24. $\frac{8}{9}$

Write the fraction in simplest form.

25. $\frac{2}{8}$

26. $\frac{2}{6}$

27. $\frac{5}{10}$

28. $\frac{3}{9}$

29. $\frac{3}{12}$

30. $\frac{8}{16}$

31. $\frac{15}{18}$

32. $\frac{9}{15}$

33. $\frac{4}{16}$

34. $\frac{6}{15}$

Mixed Applications

35. The band spent 8 hours practicing in one week. Two of the hours were spent marching. Write this as a fraction in simplest form.

36. Kris has 18 tickets for the band concert. He wants to give them to 6 friends. How many tickets will each friend get?

37. Only 2 of the 12 band members carry a flag during a marching song. Write this as a fraction in simplest form.

38. Harry practiced 3 hours a day for 7 days. How many hours did he practice?

Mixed Review

Find the answer. Which method did you use?

39. $500 \div 5$
40. $153 - 92$
41. 4×34
42. $44 + 44$

43. $7{,}369 + 9{,}341 + 8{,}949$
44. 3×63
45. $651 \div 7$

MENTAL MATH
CALCULATOR
PAPER/PENCIL

UNDERSTANDING A CONCEPT

Whole Numbers and Mixed Numbers

Jana is on a hockey team. She uses circles to keep track of her practice time. She fills in one-fourth of each circle for every 15 minutes of practice time.

This is her time sheet for one week:

1. How many fourths did she fill in?

2. How much time is 4 fourths?

3. Use a whole number and a fraction to tell how many hours she practiced that week.

4. Why is $\frac{10}{4}$ hours the same as $2\frac{1}{2}$ hours?

You can rename a fraction greater than 1 as a **mixed number.**

Step 1	Step 2	Step 3
Divide the numerator by the denominator.	Write the quotient as the whole number. Write the remainder as a fraction.	Write the fraction in simplest form.
$$\frac{10}{4} \rightarrow 4\overline{)10}^{\,2\ R2}$$	quotient → $2\ \frac{2}{4}$ ← remainder ← divisor	$2\frac{2}{4} = 2\frac{1}{2}$ **Read:** two and one-half

5. Use this method to rename $\frac{5}{3}$ as a mixed number.

6. **What if** Jana filled in only 8 of the fourths? How much time would that be?

TRY OUT Write the letter of the correct answer. What is the whole number or the mixed number?

7. $\frac{8}{8}$ **a.** 8   **b.** 4   **c.** 2   **d.** 1

8. $\frac{8}{6}$ **a.** $\frac{3}{4}$ **b.** $1\frac{1}{3}$ **c.** $1\frac{2}{3}$ **d.** $1\frac{3}{4}$

9. $\frac{4}{2}$ **a.** $\frac{1}{2}$ **b.** 1   **c.** 2   **d.** 24

PRACTICE

Write the mixed number or the whole number.

10. 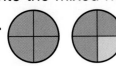 **11.** **12.** **13.**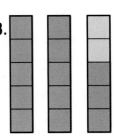

Write the fraction as a whole number or a mixed
number in simplest form.

14. $\frac{4}{3}$ **15.** $\frac{8}{2}$ **16.** $\frac{6}{5}$ **17.** $\frac{12}{6}$ **18.** $\frac{9}{4}$

19. $\frac{7}{3}$ **20.** $\frac{15}{3}$ **21.** $\frac{18}{5}$ **22.** $\frac{24}{4}$ **23.** $\frac{35}{8}$

24. $\frac{16}{10}$ **25.** $\frac{10}{4}$ **26.** $\frac{9}{6}$ **27.** $\frac{2}{2}$ **28.** $\frac{15}{6}$

Complete.

29. $5 = \frac{15}{\blacksquare}$ **30.** $\frac{12}{3} = \blacksquare$ **31.** $\frac{\blacksquare}{8} = 2$ **32.** $\frac{\blacksquare}{7} = 3$

33. $\frac{30}{\blacksquare} = 5$ **34.** $\frac{30}{\blacksquare} = 6$ **35.** $\frac{30}{\blacksquare} = 3$ **36.** $9 = \frac{\blacksquare}{3}$

Write the mixed number.

37. eight and three-fourths

38. four and one-half

39. two and three-tenths

40. five and two-fifths

Mixed Applications

41. Paul did 53 push-ups in 1 min. How many push-ups will he do in 6 min. at this rate?

42. One week Jana filled in 31 fourths on her practice sheet. How many hours is this?

43. The members of the skating team need $832 to go to the state contest. They can wash cars for $8 per car. How many cars will they need to wash to earn enough money?

44. Pineapples for an after-class party were each cut into 8 equal pieces. There were 92 pieces in all. How many pineapples was this?

UNDERSTANDING A CONCEPT

Comparing Fractions and Mixed Numbers

A. The school choir gives a concert on Saturday that lasts $1\frac{3}{4}$ hours. It gives a concert that lasts $1\frac{1}{2}$ hours on Sunday. Which concert lasts longer?

You can compare fractions or mixed numbers on a number line.

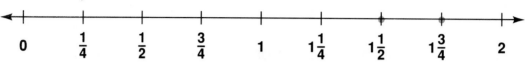

0	$\frac{1}{4}$	$\frac{1}{2}$	$\frac{3}{4}$	1	$1\frac{1}{4}$	$1\frac{1}{2}$	$1\frac{3}{4}$	2

Think: $1\frac{1}{2}$ comes **before** $1\frac{3}{4}$.

Read: $1\frac{1}{2}$ is **less than** $1\frac{3}{4}$.

Write: $1\frac{1}{2} < 1\frac{3}{4}$

Think: $1\frac{3}{4}$ comes **after** $1\frac{1}{2}$.

Read: $1\frac{3}{4}$ is **greater than** $1\frac{1}{2}$.

Write: $1\frac{3}{4} > 1\frac{1}{2}$

The Saturday concert lasts longer.

B. You can compare without a number line.

Compare: $1\frac{1}{2}$ and $1\frac{3}{4}$

Step 1	Step 2	Step 3
Compare the whole numbers.	Compare the fractions. Write equivalent fractions with like denominators.	Compare the numerators.
$1\frac{1}{2}$ **Think:** $1 = 1$ $1\frac{3}{4}$	$\frac{1}{2} = \frac{1 \times 2}{2 \times 2} = \frac{2}{4}$ $\frac{3}{4}$ $\qquad$ $\frac{3}{4}$	$2 < 3$ $\frac{2}{4} < \frac{3}{4}$ $\quad$ $\frac{1}{2} < \frac{3}{4}$

So $1\frac{1}{2} < 1\frac{3}{4}$.

C. You can order $\frac{1}{3}$, $\frac{5}{6}$, and $\frac{7}{12}$ by comparing.

Step 1	Step 2	Step 3
Write equivalent fractions with like denominators.	Compare the fractions.	Order the fractions.
$\frac{1}{3} = \frac{1 \times 4}{3 \times 4} = \frac{4}{12}$ $\frac{5}{6} = \frac{5 \times 2}{6 \times 2} = \frac{10}{12}$ $\frac{7}{12} \qquad \frac{7}{12}$	$\frac{4}{12} < \frac{7}{12}$ $\frac{4}{12} < \frac{10}{12}$ $\frac{7}{12} < \frac{10}{12}$	From least to greatest: $\frac{1}{3}, \frac{7}{12}, \frac{5}{6}$ From greatest to least: $\frac{5}{6}, \frac{7}{12}, \frac{1}{3}$

TRY OUT

Compare. Write $>$, $<$, or $=$.

1. $\frac{1}{4}$ ● $\frac{2}{4}$ **2.** $\frac{2}{3}$ ● $\frac{4}{6}$ **3.** $1\frac{3}{5}$ ● $1\frac{1}{10}$ **4.** $2\frac{1}{8}$ ● $1\frac{1}{3}$

Write in order from least to greatest.

5. $\frac{3}{4}, \frac{3}{8}, \frac{5}{16}$ **6.** $1\frac{2}{3}, 2\frac{5}{9}, 1\frac{5}{18}$ **7.** $1\frac{3}{8}, 1\frac{2}{3}, 1\frac{7}{12}$

PRACTICE

Use the number line to compare. Write $>$, $<$, or $=$.

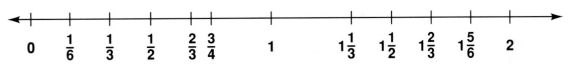

8. $\frac{1}{2}$ ● $\frac{1}{6}$ **9.** $\frac{1}{3}$ ● $\frac{1}{2}$ **10.** $\frac{3}{4}$ ● $\frac{2}{3}$ **11.** $1\frac{1}{2}$ ● $1\frac{2}{3}$

Compare. Write $>$, $<$, or $=$.

12. $\frac{3}{4}$ ● $\frac{1}{4}$ **13.** $\frac{2}{5}$ ● $\frac{4}{5}$ **14.** $\frac{5}{8}$ ● $\frac{3}{8}$ **15.** $\frac{4}{7}$ ● $\frac{3}{7}$

16. $1\frac{5}{6}$ ● $2\frac{1}{6}$ **17.** $1\frac{4}{9}$ ● $1\frac{7}{9}$ **18.** $\frac{33}{10}$ ● $\frac{37}{10}$ **19.** $6\frac{1}{3}$ ● $6\frac{2}{3}$

20. $7\frac{5}{8}$ ● $6\frac{3}{4}$ **21.** $4\frac{7}{9}$ ● $4\frac{2}{3}$ **22.** $\frac{27}{12}$ ● $2\frac{5}{6}$ **23.** $1\frac{4}{16}$ ● $1\frac{1}{4}$

Write in order from least to greatest.

24. $\frac{1}{4}, \frac{1}{2}, \frac{1}{8}$ **25.** $\frac{11}{18}, \frac{5}{9}, \frac{2}{3}$ **26.** $4\frac{1}{2}, 4\frac{5}{12}, 4\frac{5}{6}$ **27.** $1\frac{3}{8}, 2, 1\frac{5}{6}$

Write in order from greatest to least.

28. $\frac{3}{4}, \frac{13}{16}, \frac{7}{8}$ **29.** $\frac{2}{3}, \frac{5}{6}, \frac{11}{12}$ **30.** $4\frac{7}{16}, 5\frac{1}{4}, 4\frac{5}{8}$ **31.** $1\frac{1}{2}, 1\frac{2}{3}, 1\frac{4}{9}$

Mixed Applications

32. The Poetry Club meets for $1\frac{5}{8}$ hours. The French Club meets for $1\frac{3}{4}$ hours. Which club had the longer meeting?

33. *Write a problem* about comparing fractions or mixed numbers. Solve your problem. Ask others to solve it.

Measuring Length: Customary Units

An **inch (in.)** is a customary unit of length used to measure short lengths.

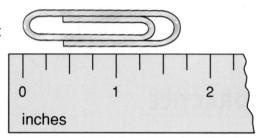

To measure something to the nearest inch:

Step 1 Line up the edge of the object with the left end of the scale.

Step 2 Look at the right end of the object. Find the closest inch mark.

1. Find the length of the paper clip to the nearest inch.

You can also measure length to the nearest **half inch ($\frac{1}{2}$ in.)** or **quarter inch ($\frac{1}{4}$ in.).**

2. Find the length of the paper clip to the nearest $\frac{1}{2}$ inch, and $\frac{1}{4}$ inch.

3. Work with a partner. Estimate, then measure, the length of each other's foot to the nearest inch, the nearest $\frac{1}{2}$ in., and the nearest $\frac{1}{4}$ in. Record your measurements.

SHARING IDEAS

4. Compare your three groups of measurements. When are they the same?

5. When do you think you may need to measure something using a smaller unit?

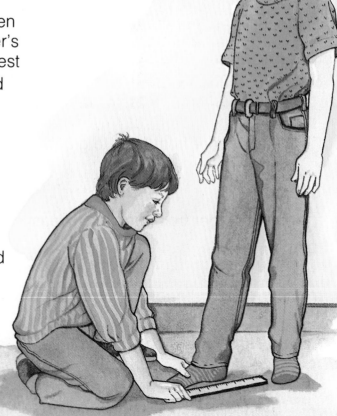

PRACTICE

Measure to the nearest inch, $\frac{1}{2}$ in., and $\frac{1}{4}$ in.

6.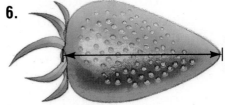

7.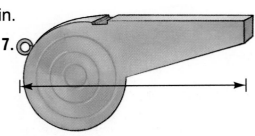

Complete the chart. Estimate, then use your inch ruler to measure.

	Object	Estimate	Measurement
8.	pencil	■	■
9.	width of a calculator	■	■
10.	length of your math book	■	■
11.	length of your arm	■	■

12. Choose other objects to measure. Record your work on your chart.

Use your inch ruler. Draw a line segment of the given length.

13. $3\frac{1}{4}$ in.   **14.** $2\frac{1}{2}$ in.   **15.** 4 in.

Mixed Applications

16. Ruth wants to buy a hat. Size $6\frac{5}{8}$ is too small. Which size should she try on next, $6\frac{3}{4}$ or $6\frac{3}{8}$?

17. Roger measured the length of his book to the nearest quarter inch as $13\frac{3}{4}$ in. What would it measure to the nearest inch?

18. Claire's clarinet is 25 in. long. She puts in a mouthpiece that is 3 in. long. How long is the clarinet with the mouthpiece?

19. Hank's electric guitar is 14 in. wide. This is 2 in. wider than June's acoustic guitar. How wide is June's guitar?

UNDERSTANDING A CONCEPT

Estimating Length: Customary Units

The inch (in.), the **foot (ft),** and the **yard (yd)** are customary units used to measure length. The **mile (mi)** is the measure for a long distance.

The length of a football is about 1 foot.

The length of a guitar is about 1 yard.

A mile is about how far you can walk in 20 minutes.

This chart shows you how the units are related.

12 inches (in.)	= 1 foot (ft)
3 feet (ft)	= 1 yard (yd)
1,760 yards (yd)	= 1 mile (mi)

Tell whether you would use *inches, feet, yards,* **or** *miles* **to measure.**

1. the length of the playground

2. the width of a desk

3. the height of a cup

4. the height of a woman

TRY OUT Write the letter of the correct answer. Would you use inch, foot, yard, or mile to measure?

5. the width of a door   **a.** in.   **b.** ft   **c.** yd   **d.** mi

6. the length of a pencil   **a.** in.   **b.** ft   **c.** yd   **d.** mi

7. the distance across a state   **a.** in.   **b.** ft   **c.** yd   **d.** mi

8. the height of a telephone pole   **a.** in.   **b.** ft   **c.** yd   **d.** mi

PRACTICE

Tell whether you would use *inches, feet, yards,* or *miles* to measure.

9. the length of a baseball bat

10. the distance around a lake

11. the width of a piece of paper

12. the length of a crayon

Write the letter of the best estimate.

13. length of a car   **a.** 15 in.   **b.** 15 ft   **c.** 15 yd   **d.** 15 mi

14. distance a plane travels   **a.** 9 in.   **b.** 90 ft   **c.** 90 yd   **d.** 900 mi

15. height of a flagpole   **a.** 80 in.   **b.** 80 ft   **c.** 80 yd   **d.** 80 mi

16. width of a teacup   **a.** 3 in.   **b.** 3 ft   **c.** 3 yd   **d.** 3 mi

Complete the chart. Use the units given.

	Object	Estimate	Actual Measure
17.	height of your desk (ft)	■	■
18.	width of the door (in.)	■	■
19.	length of your classroom (yd)	■	■

Mixed Applications

Solve. You may need to use the Databank on page 521.

20. Peter throws the football 30 yd. John catches the ball and runs 35 yd before he is tackled. How many yards were gained on the play?

21. What is the tallest species of tree found in the United States? What is its height in yards?

22. How much taller is the Coast Douglas Fir than the Western Hemlock?

23. **Write a problem** using the information in the Databank. Solve it. Then trade your problem with others.

PROBLEM SOLVING

Strategy: Using Number Sense

Tom's goal is to run more than 9 miles on the weekend. He runs $4\frac{3}{4}$ miles on Saturday. If he runs $4\frac{3}{4}$ miles on Sunday, will he have run more than 9 miles on the two days?

Tom uses number sense to find out whether he will meet his goal.

1. He thinks: "I know that I will have run more than 8 miles on the weekend." Do you agree? Why or why not?

2. Tom thinks: "Since $\frac{3}{4}$ is greater then $\frac{1}{2}$, I will run at least another mile." Do you agree? Why or why not?

3. Will Tom run more than 9 miles? How can you tell?

4. ***What if*** Tom runs $4\frac{1}{8}$ miles on Saturday and $4\frac{3}{8}$ miles on Sunday. Will he run more than 9 miles on the weekend?

5. Talk about how Tom used number sense to solve the problem.

PRACTICE

Use number sense to solve the problem.

6. Grace wants to exercise at least 4 hours this weekend. She exercises $1\frac{1}{4}$ hours on Saturday and $2\frac{1}{4}$ hours on Sunday. Has she met her goal?

7. Karen and Dave want to paint 3 walls of the kitchen by lunchtime. They each paint $1\frac{2}{3}$ walls. Have they met their goal?

8. Drew needs 6 pounds of fruit to make a fruit salad. He picks $2\frac{5}{16}$ pounds of apples and $3\frac{3}{16}$ pounds of pears. Does he have as much fruit as he needs?

9. Ruben wants to lose 4 pounds in 2 weeks. He loses $2\frac{3}{4}$ pounds the first week and $1\frac{3}{4}$ pounds the next week. Has he met his goal?

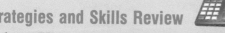

Strategies and Skills Review

Solve. Use mental math, a calculator, or paper and pencil.

10. Dave is packing 137 pounds of cans into boxes for recycling. Each box holds 4 pounds of cans. How many full boxes will he pack?

11. The Tigers won the football game by a score of 35 to 17. How many more points did the Tigers score than their opponent?

12. Sandy wants to buy a sketch pad for $3.73 and pastels for $6.57. She has $10.00. Does she have enough money?

13. A football field is 100 yards long. There is a line across the field every 10 yards. Including the end lines, how many lines are there?

14. In the skate-boarding competition, Ed receives scores of 46 points, 53 points and 61 points. The scorer records Ed's total as 150 points. Did the scorer record his total correctly?

15. *Write a problem* that you can solve using number sense. Ask others to solve your problem. Talk about how you used number sense.

DECISION MAKING

Problem Solving: Choosing a Musical Instrument to Play

SITUATION

For an after-school activity, Luis Allende wants to learn to play a musical instrument and join the school orchestra.

PROBLEM

Which instrument should Luis choose?

DATA

Whitman Elementary School Orchestra

Information Sheet for New Members

Learn to play one of these musical instruments:

Instrument	PIANO	TRUMPET	VIOLIN	DRUMS	FLUTE
Lessons (starting at 3:00 P.M.)	Monday	Wednesday & Friday	Monday & Tuesday	Tuesday & Wednesday	Wednesday
	1 hour	1 hour	1 hour	$\frac{1}{2}$ hour	1 hour
Minimum Daily Practice	$\frac{1}{2}$ hour	$\frac{1}{2}$ hour	$\frac{1}{2}$ hour	$\frac{1}{2}$ hour	$\frac{1}{2}$ hour
	may use school piano	Rental available	Rental available	may use school drums	Rental available

Orchestra rehearsal: Thursday, 3:00 P.M.-4:30 P.M.

School piano and drums available: Monday — Friday 7:30 A.M.-8:30 A.M.
3:00 P.M.-4:30 P.M.
Saturday & Sunday 12:00 P.M.-3:00 P.M.

USING THE DATA

1. How many hours of practice do the musical instruments require each week?

How many hours would Luis spend on each instrument in one week, including practice, lessons, and orchestra rehearsal?

2. piano   3. trumpet   4. violin

5. drums   6. flute

MAKING DECISIONS

7. What are the things Luis should think about when deciding which musical instrument to play?

8. *What if* Luis wants to play in the high school marching band? Which musical instrument should he choose?

9. *What if* Luis has a Cub Scout troop meeting after school on Monday? Are there instruments that he should not choose? Why?

10. How would choosing the piano or the drums affect Luis' after-school activities?

11. *What if* Luis' mom offers to buy the instrument he chooses to play? Luis lives in a small apartment. How will that affect his choice?

12. Which musical instrument would you choose? Why?

Math and Music

In music, musical notes are grouped in **measures.** Rhythm in each measure is counted in units called **beats.** Here are three measures of music. They each have the same number of beats.

Quarter notes   Half notes   Whole note

MEASURE   MEASURE   MEASURE

The numbers at the left are known as the **meter signature.** The top number of the meter signature tells how many beats are in each measure. In $\frac{4}{4}$ meter, there are four beats in a measure. The bottom number tells what kind of note sounds for one beat. The 4 means a quarter note ($\quarternote$) sounds for one beat in $\frac{4}{4}$ meter. So each quarter note is $\frac{1}{4}$ of a measure. A half note ($\halfnote$) sounds for 2 beats. A whole note ($\circ$) sounds for 4 beats.

In $\frac{4}{4}$ meter, what fraction of a measure is a half note?

Think: There are 4 beats in a measure. A half note gets 2 beats. 2 is $\frac{1}{2}$ of 4.

So a half note is $\frac{1}{2}$ of a measure in $\frac{4}{4}$ meter.

ACTIVITIES

1. Write as many different combinations of notes for one measure in $\frac{4}{4}$ meter as you can. Tap out the rhythm for each measure you write. Remember to tap only once for each note.

2. Suppose the meter signature is $\frac{3}{4}$. How many beats are in a measure? What kind of note sounds for 1 beat?

Computer Applications: Bar Graphs

For the Spring Fair your class is going to sell fruit punch. Your teacher wants you to draw a bar graph to show how much of each ingredient will be needed for the punch. You can input the data into a computer program and the computer will round the numbers and draw the bar graph for you.

FRUIT PUNCH

Ingredient	Amount (in cups)	Rounded to Nearest Whole Number
Apple juice	$5\frac{1}{4}$	
Grapefruit juice	$7\frac{1}{3}$	
Lemonade	$9\frac{1}{2}$	
Orange juice	$6\frac{2}{3}$	
Pink lemonade	$8\frac{3}{4}$	
Club soda	$10\frac{1}{4}$	

DATA

Copy the table above. Round the mixed numbers to the nearest whole number and complete the table. Draw a bar graph showing the amount of each ingredient in the punch.

THINKING ABOUT COMPUTERS

1. Why is the last column in the table helpful in creating a bar graph? Would this column be necessary if you were using a computer?

2. **What if** you were going to make only half the amount of punch? How would this change the bar graph? Why would a computer be useful in making this change?

3. The sale of fruit punch was so successful that your teacher suggests the class sell some at the County Fair. Why would a computer be helpful in drawing a bar graph for this new situation?

EXTRA PRACTICE

The Meaning of Fractions, page 345 ....

Write a fraction for the shaded part.

1.

2.

3.

4.

5.

6.

Parts of a Whole, page 347 ....

Write a fraction for the part of each figure that is shaded.

1.

2.

3.

Write the fraction.

4. two-thirds

5. two-sixths

6. five-eighths

7. nine-tenths

8. four-fifths

9. two-fifths

Parts of a Set, page 349 ....

Write a fraction that shows the part of each set
that is shaded.

1.

2.

3.

Write the fractions that are blue.

4. Seven of ten are blue.

5. Three of the five are blue.

6. Four of the nine are blue.

7. None of the six are blue.

8. All five are blue.

9. Six of the seven are blue.

EXTRA PRACTICE

Finding Parts of a Set, page 351 ....

Complete.

1. $\frac{1}{6}$ of 24 = ■

2. $\frac{1}{4}$ of 12 = ■

3. $\frac{1}{9}$ of 18 = ■

4. $\frac{1}{4}$ of 8 = ■

5. $\frac{1}{5}$ of 25 = ■

6. $\frac{1}{2}$ of 8 = ■

Use division and multiplication to find the part of the set.

7. $\frac{3}{4}$ of 24

8. $\frac{3}{4}$ of 12

9. $\frac{2}{6}$ of 12

10. $\frac{2}{5}$ of 10

11. $\frac{2}{4}$ of 8

12. $\frac{1}{4}$ of 24

Problem Solving Strategy: Working Backward, page 353 ....

Work backward to solve the problem.

1. Jay is 3 years older than Jeff. Jeff is 4 years younger than Pat. Pat is 12 years old. How old is Jay?

2. Meg is five times older than Evan. Evan is a year older than Emily. Emily is 3 years old. How old is Meg?

3. Sam is twice as old as Jeff. Jeff is 4 years older than Sadie. Rifka is 2 years younger than Sadie. Rifka is 8 years old. How old is Sam?

Finding Equivalent Fractions, page 355 ....

Complete to show the equivalent fraction.

1.

$$\frac{4}{8} = \frac{1}{■}$$

2.

$$\frac{2}{8} = \frac{1}{■}$$

3.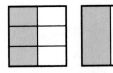

$$\frac{3}{6} = \frac{1}{■}$$

4. $\frac{3}{4} = \frac{■}{8}$

5. $\frac{2}{4} = \frac{■}{8}$

6. $\frac{1}{6} = \frac{■}{12}$

7. $\frac{1}{2} = \frac{■}{10}$

8. $\frac{3}{4} = \frac{9}{■}$

9. $\frac{2}{5} = \frac{10}{■}$

10. $\frac{2}{3} = \frac{12}{■}$

11. $\frac{2}{6} = \frac{6}{■}$

EXTRA PRACTICE

Simplifying Fractions, page 359 ....

Simplify the fraction.

1. $\frac{6}{8} = \frac{\blacksquare}{4}$
2. $\frac{9}{12} = \frac{\blacksquare}{4}$
3. $\frac{10}{18} = \frac{\blacksquare}{9}$

4. $\frac{6}{10} = \frac{\blacksquare}{5}$
5. $\frac{10}{15} = \frac{\blacksquare}{3}$
6. $\frac{12}{14} = \frac{\blacksquare}{7}$

Write the fraction in simplest form.

7. $\frac{8}{12}$
8. $\frac{4}{10}$
9. $\frac{6}{9}$
10. $\frac{4}{8}$

11. $\frac{4}{6}$
12. $\frac{9}{12}$
13. $\frac{10}{18}$
14. $\frac{12}{15}$

Whole Numbers and Mixed Numbers, page 361 ....

Write the mixed number or whole number.

1.
2.
3.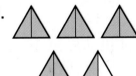

Find the mixed number or whole number. Write fractions in simplest form.

4. $\frac{24}{4}$
5. $\frac{15}{3}$
6. $\frac{7}{3}$
7. $\frac{5}{2}$

8. $\frac{9}{4}$
9. $\frac{3}{2}$
10. $\frac{4}{4}$
11. $\frac{11}{10}$

Comparing Fractions and Mixed Numbers, page 363 ....

Compare. Write > or <.

1. $\frac{2}{3} \bullet \frac{3}{4}$
2. $4\frac{1}{2} \bullet \frac{4}{8}$
3. $1\frac{5}{6} \bullet 6\frac{1}{6}$

Write in order from least to greatest.

4. $\frac{1}{3}, \frac{5}{8}, \frac{1}{2}$
5. $\frac{9}{10}, \frac{1}{5}, \frac{3}{4}$
6. $\frac{3}{4}, \frac{5}{6}, \frac{1}{8}$

Measuring Length: Customary Units, page 365 . . . .

Write each mixed number measurement to the nearest whole number.

1. $4\frac{5}{8}$ in.

2. $1\frac{1}{6}$ in.

3. $6\frac{7}{8}$ yd

4. $5\frac{1}{8}$ ft

5. $9\frac{1}{10}$ mi

6. $2\frac{3}{4}$ ft

Estimating Length: Customary Units, page 367 . . . .

Tell whether you would use in., ft, yd, or mi to measure.

1. the height of a house

2. the length of a pencil

3. the length of a field

4. the length of a shoe

5. the distance from New York to Chicago.

Write the letter of the best estimate.

6.	the height of a refrigerator	**a.** 4 ft	**b.** 4 yd	**c.** 2 ft
7.	the length of a river	**a.** 10 ft	**b.** 10 yd	**c.** 10 mi
8.	the width of the classroom	**a.** 35 in.	**b.** 35 ft	**c.** 35 mi

Problem Solving Strategy: Using Number Sense, page 369 . . . .

Use number sense to solve the problem.

1. Gary lives 6 miles from school. His friend Jay lives 2 miles from school. Phil lives half way between them. About how far does Phil live from school?

2. Lou travels 12 hours a week. Diane travels one-half that number. Scott travels one hour more than Diane. How many hours does Scott travel?

3. Dotty got on the elevator on her own floor. She went up 11 floors. Then she went down 15 floors to the lobby. What floor does she live on?

4. Sandy started on the 5th floor. She went down 4 floors. Then she went up 9 floors. What floor is she on now?

Practice PLUS

KEY SKILL: Simplifying Fractions (Use after page 359.)

Level A

Is each fraction in simplest form? Write *yes* or *no*.

1. $\frac{4}{8}$ **2.** $\frac{1}{2}$ **3.** $\frac{2}{3}$ **4.** $\frac{6}{12}$ **5.** $\frac{9}{18}$ **6.** $\frac{9}{10}$

Write each fraction in simplest form.

7. $\frac{5}{10}$ **8.** $\frac{9}{12}$ **9.** $\frac{15}{18}$ **10.** $\frac{3}{9}$ **11.** $\frac{8}{16}$ **12.** $\frac{4}{12}$

13. Stephen had 12 concert tapes. He gave 6 away to his friends. Write this fraction in simplest form.

Level B

Is the fraction in simplest form? Write *yes* or *no*.

14. $\frac{11}{13}$ **15.** $\frac{9}{15}$ **16.** $\frac{1}{8}$ **17.** $\frac{8}{12}$ **18.** $\frac{4}{9}$ **19.** $\frac{3}{6}$

Write the fraction in simplest form.

20. $\frac{12}{15}$ **21.** $\frac{5}{10}$ **22.** $\frac{10}{25}$ **23.** $\frac{5}{15}$ **24.** $\frac{3}{9}$ **25.** $\frac{9}{18}$

26. $\frac{6}{12}$ **27.** $\frac{4}{8}$ **28.** $\frac{8}{12}$ **29.** $\frac{27}{3}$ **30.** $\frac{18}{21}$ **31.** $\frac{11}{22}$

32. Tom has 21 copies of the school song. He plans to give 7 of them away. Write this fraction in simplest form.

Level C

Write the fraction in simplest form.

33. $\frac{4}{18}$ **34.** $\frac{9}{12}$ **35.** $\frac{12}{24}$ **36.** $\frac{6}{18}$ **37.** $\frac{9}{12}$ **38.** $\frac{4}{16}$

39. $\frac{8}{12}$ **40.** $\frac{6}{18}$ **41.** $\frac{24}{48}$ **42.** $\frac{5}{25}$ **43.** $\frac{13}{26}$ **44.** $\frac{7}{35}$

45. There are 30 members of the school chorus. Ten of them are in the fourth grade. Write this fraction in simplest form.

PRACTICE PLUS

KEY SKILL: Whole Numbers and Mixed Numbers (Use after page 361.)

Level A

Write the whole number or mixed number in simplest form.

1. $\frac{5}{3}$ **2.** $\frac{7}{4}$ **3.** $\frac{10}{4}$ **4.** $\frac{11}{10}$ **5.** $\frac{8}{4}$

Complete.

6. $\frac{\blacksquare}{2} = 2$ **7.** $\frac{\blacksquare}{7} = 2$ **8.** $\frac{\blacksquare}{2} = 1$ **9.** $\frac{\blacksquare}{3} = 3$

10. Taryn had 7 halves of apple. How many apples did she have?

Level B

Write the whole number or mixed number in simplest form.

11. $\frac{14}{10}$ **12.** $\frac{10}{3}$ **13.** $\frac{6}{2}$ **14.** $\frac{13}{3}$ **15.** $\frac{19}{6}$

Complete.

16. $\frac{\blacksquare}{3} = 3$ **17.** $\frac{\blacksquare}{6} = 3$ **18.** $\frac{\blacksquare}{5} = 5$ **19.** $\frac{\blacksquare}{4} = 1$

20. Robin had 9 orange quarters to share. How many oranges did she have?

Level C

Write the whole number and mixed number in simplest form.

21. $\frac{27}{3}$ **22.** $\frac{14}{5}$ **23.** $\frac{5}{2}$ **24.** $\frac{25}{4}$ **25.** $\frac{14}{3}$

Complete.

26. $\frac{\blacksquare}{4} = 1\frac{1}{4}$ **27.** $\frac{\blacksquare}{3} = 3$ **28.** $\frac{\blacksquare}{10} = 1\frac{1}{10}$ **29.** $\frac{7}{\blacksquare} = 2\frac{1}{3}$

30. It is equivalent to $\frac{6}{10}$. The denominator is 5.

31. Cathy practiced the piano for 17 fourths of an hour. How many hours did she practice?

Understanding Fractions and Mixed Numbers **379**

CHAPTER REVIEW

LANGUAGE AND MATHEMATICS

Complete the sentences. Use the words in the chart.

1. A ■ names each of the equal parts of an object. *(page 344)*

2. The top part of a fraction is called the ■. *(page 346)*

3. A ■ is a measure for long distances. *(page 366)*

4. ■ fractions name the same part of a whole. *(pages 354–355)*

5. ***Write a definition*** or give an example of the words you did not use from the chart.

CONCEPTS AND SKILLS

There are 3 red blocks and 1 blue block. *(page 344)*

6. How many are in the set?

7. What fraction shows how many are red? blue?

Write the fraction. *(page 346)*

8. one-tenth   9. three-fifths   10. seven-twelfths   11. two-ninths

Find the part. *(page 350)*

12. $\frac{1}{5}$ of 15   13. $\frac{2}{3}$ of 12   14. $\frac{1}{6}$ of 24

Write a fraction that names the shaded part. *(page 348)*

15.  16.   17.   18.

19. Tell which of these figures shows fourths. How do you know? *(page 346)*

a.   b.

Complete to show the equivalent fraction. *(page 354)*

20. $\frac{1}{4} = \frac{2}{\blacksquare}$
21. $\frac{3}{5} = \frac{9}{\blacksquare}$
22. $\frac{1}{6} = \frac{\blacksquare}{24}$
23. $\frac{2}{3} = \frac{\blacksquare}{9}$

Write each fraction in simplest form. *(pages 358, 360)*

24. $\frac{8}{20}$
25. $\frac{3}{9}$
26. $\frac{12}{4}$
27. $\frac{8}{3}$
28. $\frac{25}{40}$
29. $\frac{16}{5}$

Order from greatest to least. *(page 362)*

30. $\frac{2}{16}, \frac{3}{8}, \frac{5}{8}$
31. $\frac{13}{18}, \frac{4}{9}, \frac{2}{3}$
32. $3\frac{1}{2}, 3\frac{7}{12}, 3\frac{5}{6}$

Use your inch ruler. Draw a line to the nearest $\frac{1}{4}$ in. *(page 364)*

33. $3\frac{3}{4}$ in.
34. $3\frac{1}{2}$ in.
35. $3\frac{15}{16}$ in.

Write the letter of the best estimate. *(page 366)*

36. length of a notebook   **a.** 12 ft   **b.** 12 yd   **c.** 12 mi   **d.** 12 in.

37. width of a movie screen   **a.** 21 ft   **b.** 21 yd   **c.** 21 mi   **d.** 21 in.

CRITICAL THINKING

38. Zeke cuts a piece of paper into 8 equal parts. He colors half of them. He gives Clem half of the colored parts. How many parts does he give Clem? *(page 350)*

39. Zeke says each part is 2 in. long. Clem says that each is $2\frac{1}{4}$ in. long. Can they both be right? Why or why not? *(page 364)*

MIXED APPLICATION

40. There are 36 pieces in Pam's dinnerware set. She places $\frac{1}{3}$ of the pieces on the table. How many pieces are on the table? *(page 346)*

41. Jack wants to hike 8 miles today. He hikes $4\frac{7}{8}$ miles before lunch and $3\frac{7}{8}$ miles after lunch. Has he met his goal? *(page 368)*

42. Kim, Beth, and Eva are making ornaments. Kim has made 8 less than Eva. Eva has made twice as many as Beth. Beth has made 12. How many has Kim made? *(page 352)*

CHAPTER TEST

What fraction is shaded?

1. **2.** ⬤⬤◯◯◯ **3.** **4.** [figure]

Complete to show the equivalent fraction.

4. $\frac{4}{5} = \frac{\blacksquare}{15}$

6. $\frac{1}{8} = \frac{4}{\blacksquare}$

Write the fraction in simplest form.

7. $\frac{8}{16}$

8. $\frac{9}{12}$

Write the whole number or mixed number in simplest form.

9. $\frac{5}{4}$

10. $\frac{9}{9}$

11. $\frac{25}{5}$

12. $\frac{14}{6}$

Compare. Write $>$, $<$, or $=$.

13. $\frac{2}{5}$ ● $\frac{3}{4}$

14. $3\frac{1}{2}$ ● $3\frac{1}{3}$

Order from least to greatest.

15. $\frac{3}{4}, \frac{1}{3}, \frac{5}{6}$

Write the letter of the best estimate.

16. width of a driveway   **a.** 15 inches   **b.** 15 feet   **c.** 15 yards

17. length of a caterpillar   **a.** 2 inches   **b.** 2 feet   **c.** 2 yards

Solve.

18. Robin, Jay, and Tara volunteer at the library. Robin works twice as long as Jay. Jay works 4 hours more than Tara. Tara works for 7 hours. How long does Robin work?

19. Erin reads twice as many books as Fran and Al together. Al reads 7 books. Fran reads twice as many as Al. How many books does Erin read?

20. Fay wants to bolt something to the wall. A $1\frac{1}{2}$-in. bolt is too short. Should she try a $1\frac{1}{4}$-in. bolt or a $1\frac{7}{8}$-in. bolt?

382 Chapter 9

ENRICHMENT FOR ALL

RATIOS

Charlie and Jeannie helped make chili for a class supper. This is the basic recipe they used.

CHILI
3 cups ground meat
1 cup tomato sauce
4 cups pinto beans
1 cup chopped onions
Mix together. Cook for 30 minutes.

1. How many cups of meat did they use for the recipe?

2. How many cups of beans did they use for the recipe?

You can describe how these parts of the recipe compare to each other by writing a **ratio.** A ratio is a way to compare two numbers.

The ratio of meat to beans is 3 to 4. You can also write this ratio as 3:4 or 3/4.

3. What is the ratio of onions to meat in the recipe?

4. What is the ratio of sauce to beans in the recipe?

5. **What if** Charlie and Jeannie decide to double the recipe? How much of each ingredient would they use now?

6. What would be the ratio of meat to beans if the recipe were doubled? What would be the ratio of onions to meat? of sauce to beans?

7. You can simplify ratios the same way you simplify fractions. Give the ratios for Problem 6 in simplest form. What do you notice?

8. **What if** Charlie and Jeannie decide to make more of the chili? They use 9 cups of ground meat. How much of the other ingredients should they use? How do you know?

CUMULATIVE REVIEW

Choose the letter of the correct answer.

1. Write $\frac{8}{32}$ in simplest form.

   a. $\frac{1}{3}$ c. $\frac{1}{4}$

   b. $\frac{2}{8}$ d. not given

2. Find $\frac{3}{4}$ of 12.

   a. 6   c. 12

   b. 9   d. not given

3. Rename $\frac{21}{18}$.

   a. $1\frac{1}{6}$ c. $1\frac{3}{8}$

   b. $1\frac{1}{5}$ d. not given

4. Name the shaded part.

   a. $\frac{3}{8}$ c. $\frac{5}{8}$

   b. $\frac{3}{5}$ d. $\frac{5}{3}$

5. $\frac{4}{9} = \frac{12}{\blacksquare}$

   a. 3   c. 17

   b. 7   d. not given

6. Which is a space figure?

   a. triangle   c. square

   b. cylinder   d. hexagon

7. Compare: $\frac{36}{12}$ ● 3

   a. <   c. =

   b. >   d. not given

8. What is 4 × 968?

   a. 3,622   c. 3,822

   b. 3,672   d. not given

9. Estimate the width of a mug.

   a. 3 in.   c. 3 yd

   b. 3 ft   d. 3 mi

10.

   The figures are:

   a. parallel   c. similar

   b. congruent   d. not given

11. Divide 189 by 5.

   a. 36 R1   c. 37 R4

   b. 36 R4   d. not given

12. Find the product of 8 × $.62.

   a. $.49   c. $4.96

   b. $.50   d. not given

13. Estimate: 5 × 887

   a. 400   c. 3,500

   b. 450   d. 4,500

14. 3,000 ÷ 6

   a. 50   c. 600

   b. 500   d. not given

Using Fractions

CHAPTER

10

MATH CONNECTIONS: PERIMETER • AREA • CAPACITY, WEIGHT, AND TEMPERATURE • PROBLEM SOLVING

Backpack
$4\frac{1}{2}$ pounds

2-Person Tent
$7\frac{3}{4}$ pounds

Sleeping Bag
$3\frac{3}{4}$ pounds

1. What information do you see in this photo?

2. How would you use this information if you were planning a camping trip?

3. Write a problem about a camping trip.

Cara walks $\frac{1}{8}$ mi from the park entrance to the information center. She then walks $\frac{5}{8}$ mi to the scenic view. How far does Cara walk?

WORKING TOGETHER

You can make a model with fraction strips to solve the problem.

Show $\frac{1}{8}$ with fraction strips. Then add $\frac{5}{8}$ next to it to make one long strip.

1. How many eighths are there in all? How far does Cara walk?

2. Write an addition sentence to show how you added the two fractions.

3. How does the sum compare to $\frac{1}{2}$? to 1? How can you tell?

4. What if Cara walks another $\frac{5}{8}$ mi? How far does she walk?

5. How can you write this fraction as a mixed number?

6. How does the sum compare to 1? How can you tell?

Use fraction strips to show each sum.
Tell how the sum compares to $\frac{1}{2}$ and to 1.

7. $\frac{1}{3} + \frac{1}{3}$

8. $\frac{1}{12} + \frac{4}{12}$

9. $\frac{1}{8} + \frac{7}{8}$

10. $\frac{5}{6} + \frac{5}{6}$

11. $\frac{3}{4} + \frac{2}{4}$

12. $\frac{7}{8} + \frac{3}{8}$

13. $\frac{1}{6} + \frac{2}{6}$

14. $\frac{2}{8} + \frac{1}{8}$

15. $\frac{2}{12} + \frac{6}{12}$

SHARING IDEAS

16. Look at the numerators of the addends and the sums. What do you notice?

17. Look at the denominators of the addends and the sums. What do you notice?

18. Tell how you can add two fractions with the same denominator.

ON YOUR OWN

Complete the table. Use fraction strips if needed.

19. Rule: Add $\frac{1}{8}$.

$\frac{0}{8}$	$\frac{1}{8}$	$\frac{2}{8}$	$\frac{3}{8}$	$\frac{4}{8}$	$\frac{5}{8}$	$\frac{6}{8}$	$\frac{7}{8}$
■	■	■	■	■	■	■	■

20. Rule: Add $\frac{1}{6}$.

$\frac{0}{6}$	$\frac{1}{6}$	$\frac{2}{6}$	$\frac{3}{6}$	$\frac{4}{6}$	$\frac{5}{6}$
■	■	■	■	■	■

Solve. Use fraction strips if needed.

21. Cara hikes $\frac{3}{8}$ mi from Rocky Ridge to the waterfall. Then she hikes $\frac{4}{8}$ mi back to her campsite. How far does Cara hike?

22. It takes Ben $\frac{3}{4}$ hour to walk to the park. It takes another $\frac{1}{4}$ hour to walk to the pond. How long does Ben walk?

23. *Write a problem* involving adding fractions with like denominators. Ask others to solve it.

Finding Sums

A. Joanna is testing how clean the lake water is. She fills $\frac{1}{8}$ of a jar with lake water. She fills another $\frac{1}{8}$ of the jar with test chemicals. What part of the jar has she filled? Has she filled over $\frac{1}{2}$ of the jar?

You can use fraction strips to answer these questions.

 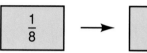 

1. How many eighths are there in all?

2. How many fourths is this equal to?

3. What part of the jar has she filled?

4. How does this compare to $\frac{1}{2}$? to 1?

B. You can also use this method to add fractions with like denominators.

Step 1	Step 2	Step 3
Add the numerators.	**Use the common denominator.**	**Write the sum in simplest form.**
$\frac{1}{8} + \frac{1}{8} = 2$	$\frac{1}{8} + \frac{1}{8} = \frac{2}{8}$	$\frac{2}{8} = \frac{1}{4}$ *Think:* $\frac{2 \div 2}{8 \div 2} = \frac{1}{4}$

5. **What if** she had filled $\frac{4}{8}$ of the jar with water and $\frac{1}{8}$ of the jar with chemicals? What part of the jar would be filled?

SHARING IDEAS

6. Why does the denominator stay the same when you add fractions?

7. Do you always need to simplify the sum of two fractions? Why or why not?

PRACTICE

Add. Write the sum in simplest form.

8. $\frac{1}{5} + \frac{1}{5}$ **9.** $\frac{1}{3} + \frac{1}{3}$ **10.** $\frac{1}{6} + \frac{4}{6}$ **11.** $\frac{4}{9} + \frac{1}{9}$ **12.** $\frac{2}{8} + \frac{3}{8}$

13. $\frac{2}{4} + \frac{1}{4}$ **14.** $\frac{3}{7} + \frac{2}{7}$ **15.** $\frac{5}{12} + \frac{2}{12}$ **16.** $\frac{1}{6} + \frac{1}{6}$ **17.** $\frac{1}{4} + \frac{1}{4}$

18. $\frac{1}{10} + \frac{3}{10}$ **19.** $\frac{5}{8} + \frac{1}{8}$ **20.** $\frac{2}{9} + \frac{4}{9}$ **21.** $\frac{3}{16} + \frac{5}{16}$ **22.** $\frac{7}{12} + \frac{2}{12}$

23. $\frac{1}{7}$ **24.** $\frac{1}{8}$ **25.** $\frac{3}{9}$ **26.** $\frac{5}{16}$ **27.** $\frac{3}{10}$ **28.** $\frac{0}{4}$
    $+\frac{1}{7}$ $+\frac{2}{8}$ $+\frac{2}{9}$ $+\frac{6}{16}$ $+\frac{4}{10}$ $+\frac{3}{4}$

29. $\frac{1}{6}$ **30.** $\frac{4}{10}$ **31.** $\frac{1}{9}$ **32.** $\frac{4}{8}$ **33.** $\frac{11}{16}$ **34.** $\frac{5}{12}$
    $+\frac{2}{6}$ $+\frac{2}{10}$ $+\frac{5}{9}$ $+\frac{0}{8}$ $+\frac{3}{16}$ $+\frac{3}{12}$

Find the missing numerator.

35. $\frac{\blacksquare}{8} + \frac{3}{8} = \frac{3}{8}$ **36.** $\frac{5}{7} + \frac{1}{7} = \frac{\blacksquare}{7} + \frac{5}{7}$ **37.** $\frac{4}{12} + \frac{3}{12} + \frac{1}{12} = \frac{1}{12} + \frac{3}{12} + \frac{\blacksquare}{12}$

Mixed Applications

Solve. You may need to use the Databank on page 521.

38. Joanna walked $\frac{4}{8}$ mi on the park trails to the waterfall. Then she walked to the lake. How far did she walk?

39. There are 644 markers on 7 trails. Each trail has the same number of markers. How many markers are on each trail?

40. The trail markers come 25 to a box. If there are 4 boxes, how many trail markers are there?

41. How far is it from the picnic grounds to the lake and then to the waterfall?

Mixed Review

Find the answer. Which method did you use?

42. $7,007 - 889$ **43.** 8×40 **44.** $373 \div 4$

45. $\frac{3}{16} + \frac{5}{16}$ **46.** $400 \div 5$ **47.** 6×37

MENTAL/MATH
CALCULATOR
PAPER/PENCIL

UNDERSTANDING A CONCEPT

Finding Greater Sums

Lars swam in the lake for $\frac{3}{4}$ hour.
He rowed a boat for another $\frac{3}{4}$ hour.
How much time did this take
in all?

You can add to find the answer.

Add: $\frac{3}{4} + \frac{3}{4}$

Step 1	Step 2	Step 3
Add the numerators.	**Use the common denominator.**	**Write the sum in simplest form.**
$\frac{3}{4} + \frac{3}{4} = 6$	$\frac{3}{4} + \frac{3}{4} = \frac{6}{4}$	$\frac{6}{4} = 1\frac{2}{4} = 1\frac{1}{2}$
		Think: $4\overline{)6}$ $\quad$ 1 R2

It took $1\frac{1}{2}$ hours in all.

1. **What if** Lars swam for $\frac{3}{4}$ hour, but rowed the boat for
   only $\frac{1}{4}$ hour? How much time would this have taken?

TRY OUT
Write the letter of the correct answer.
Find the sum in simplest form.

2. $\frac{1}{2} + \frac{1}{2}$ **a.** $\frac{2}{4}$ **b.** $\frac{1}{2}$ **c.** $\frac{2}{2}$ **d.** 1

3. $\frac{3}{5} + \frac{4}{5}$ **a.** $\frac{7}{10}$ **b.** $\frac{1}{2}$ **c.** $1\frac{2}{5}$ **d.** $\frac{7}{5}$

4. $\frac{5}{8} + \frac{7}{8}$ **a.** $1\frac{2}{8}$ **b.** $1\frac{1}{2}$ **c.** $\frac{12}{16}$ **d.** $\frac{3}{4}$

PRACTICE

Add.

5. $\frac{1}{8} + \frac{7}{8}$　　**6.** $\frac{4}{5} + \frac{1}{5}$　　**7.** $\frac{2}{6} + \frac{5}{6}$　　**8.** $\frac{3}{4} + \frac{2}{4}$　　**9.** $\frac{4}{7} + \frac{4}{7}$

10. $\frac{3}{5} + \frac{4}{5}$　　**11.** $\frac{6}{9} + \frac{7}{9}$　　**12.** $\frac{5}{16} + \frac{13}{16}$　　**13.** $\frac{7}{10} + \frac{7}{10}$　　**14.** $\frac{6}{8} + \frac{5}{8}$

15. $\begin{array}{r} \frac{9}{10} \\ + \frac{1}{10} \\ \hline \end{array}$　　**16.** $\begin{array}{r} \frac{2}{3} \\ + \frac{2}{3} \\ \hline \end{array}$　　**17.** $\begin{array}{r} \frac{7}{12} \\ + \frac{5}{12} \\ \hline \end{array}$　　**18.** $\begin{array}{r} \frac{9}{16} \\ + \frac{8}{16} \\ \hline \end{array}$　　**19.** $\begin{array}{r} \frac{4}{10} \\ + \frac{9}{10} \\ \hline \end{array}$

20. $\begin{array}{r} \frac{5}{6} \\ + \frac{5}{6} \\ \hline \end{array}$　　**21.** $\begin{array}{r} \frac{9}{10} \\ + \frac{7}{10} \\ \hline \end{array}$　　**22.** $\begin{array}{r} \frac{8}{9} \\ + \frac{7}{9} \\ \hline \end{array}$　　**23.** $\begin{array}{r} \frac{7}{8} \\ + \frac{7}{8} \\ \hline \end{array}$　　**24.** $\begin{array}{r} \frac{14}{16} \\ + \frac{6}{16} \\ \hline \end{array}$　　**25.** $\begin{array}{r} \frac{5}{12} \\ + \frac{11}{12} \\ \hline \end{array}$

Critical Thinking　Give examples. Write *not possible* if you cannot give an example.

26. Give two fractions that are both less than $\frac{1}{2}$ whose sum is less than 1, is equal to 1, is greater than 1.

27 Give two fractions that are both greater than $\frac{1}{2}$ whose sum is less than 1, is equal to 1, is greater than 1.

Mixed Applications　Solve. Which method did you use?

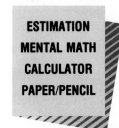

ESTIMATION
MENTAL MATH
CALCULATOR
PAPER/PENCIL

28. Last year 2,357 people rented boats on the lake. This year 3,004 people rented boats. How many more people rented boats this year?

29. Tamla sailed her boat $\frac{7}{8}$ mi to the buoy. Then she sailed $\frac{3}{8}$ mi to the dock. How far did she sail?

30. A swimming pass costs $2.10. Jo and Bo need 2 passes. They have $5.00. Do they have enough to buy the passes?

LOGICAL REASONING

Complete the statements. The first one is done for you.

1. If ## is 1, then ###### is 3.　　**2.** If @@ is 4, then @ is ■.

3. If ∗ is 2, then ∗∗∗∗∗ is ■.　　**4.** If §§§ is 9, then § is ■.

PROBLEM SOLVING

Strategy: Using Different Strategies

John has a collection of wildlife cards. He has 10 fewer reptile cards than insect cards. He has twice as many bird cards as reptile cards. He has 50 bird cards. How many insect cards does he have?

You can often use more than one strategy to solve a problem.

Maria is solving the problem by working backward.

She knows John has 50 bird cards and twice as many bird cards as reptile cards. So she divides to find how many reptile cards John has.

50 ÷ 2 = 25

She also knows that John has 10 fewer reptile cards than insect cards. She adds to find how many insect cards John has.

25 + 10 = ■

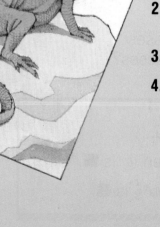

1. How many insect cards does John have?

Vincent is using the Guess, Test, and Revise strategy to check Maria's answer.

His first guess was too high. So he revised it lower.

First Guess
40 insect cards

Test
40 − 10 = 30
30 × 2 = 60

2. Is his revised guess correct? Why or why not?

Revised Guess
35 insect cards

3. Does his answer match Maria's?

Test
35 − 10 = 25
25 × 2 = ■

4. Which strategy do you prefer? Why?

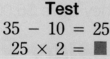

PRACTICE

Solve the problem. Then use a different strategy to check your answer. Tell which strategies you used. Use mental math, a calculator, or paper and pencil.

5. Animals drink at a water hole in the morning, at noon, in the afternoon, or in the evening. Lions drink in the afternoon. Giraffes do not drink at noon. Warthogs do not drink in the evening. Hyenas drink before warthogs. In what order do the animals drink?

6. The exhibits at the Nature Museum are in a row on one side of the building. The insects are between the mammals and the reptiles. The birds are to the right of the reptiles, and the reptiles to the left of the mammals. In what order are the exhibits?

7. There are 4 times as many insects as mammals. There are 3 fewer birds than reptiles and half as many reptiles as insects. If there are 4 mammals, how many birds are there?

8. Mr. Lund is buying equipment for a hiking trip. He buys a pack for $69, a sweater for $32.75, boots for $97.50, and field glasses for $78. Will $300 be enough to pay for these items?

Strategies and Skills Review

9. Irene spent $18 for two ecology books. One book cost $4 more than the other. How much did the more expensive book cost?

10. If 7 girls and 6 boys each plant 8 trees, how many trees will they plant in all?

11. Oren wants to buy tomato seeds for $2.69 and lettuce seeds for $1.89. Will $5 be enough to pay for both items? How can you tell?

12. *Write a problem* that can be solved by more than one strategy. Solve the problem. Ask others to use a different strategy to check your answer.

POUR MORE

Measuring

A. Suppose you want to fill a pitcher with 4 ounces of water, but you have only these two glasses:

 3 ounces   5 ounces

How can you get exactly 4 ounces in the pitcher?

Think about the amounts you can measure with these glasses. You can measure 3 or 5 ounces easily.

1. How can you measure 8 ounces easily?

2. Here is a simple way to measure 2 ounces. Fill the 5-ounce glass. Then pour from that glass into the 3-ounce glass until it is full. How many ounces will be left in the 5-ounce glass? Will this help you measure 4 ounces? Tell how.

3. Some measuring problems will require many steps. You can use a table to keep track. This table shows how you measure 4 ounces. Tell what happens at each of the other steps.

3-Ounce Glass	5-Ounce Glass	Pitcher	
0	5	0	← Fill the 5-ounce glass.
3	2	0	← Fill the 3-ounce glass from the 5-ounce glass.
3	0	2	← Pour what remains in the 5-ounce glass into the pitcher.
0	0	2	← Empty the 3-ounce glass.
0	5	2	← ?
3	2	2	← ?
3	0	4	← ?

B. Can you tell how to fill the pitcher with 6, 7, 8, 9, or 10 ounces using a 3-ounce and a 5-ounce glass?

4. Use tables to show the steps for each amount.

5. Which amount takes the least number of steps?

6. Now imagine that you have three glasses:

 2 ounces   5 ounces   8 ounces

What is the easiest way you could fill a pitcher with each of these amounts of water?

a. 6 ounces   **b.** 9 ounces   **c.** 1 ounce

Remember, you can pour water back from the pitcher into any glass.

Subtracting Fractions

Jerry is trimming the bushes in his front yard. He cuts off $\frac{3}{8}$ yd from a bush that was $\frac{7}{8}$ yd tall. How tall is the bush after he trims it?

WORKING TOGETHER

You can make a model with fraction strips to solve the problem.

Show $\frac{7}{8}$ with fraction strips. Take away $\frac{3}{8}$ by taking away three of the strips.

1. How many eighths are left? How tall is the trimmed bush?

2. How does this compare to $\frac{1}{2}$? How can you tell?

3. Write a subtraction sentence to show how you subtracted the two fractions.

4. **What if** Jerry had cut $\frac{5}{8}$ yd from the bush? How tall would it be?

5. How does this difference compare to $\frac{1}{2}$? How can you tell?

Use fraction strips to find the difference. Tell how it compares to $\frac{1}{2}$.

6. $\frac{2}{3} - \frac{1}{3}$

7. $\frac{5}{6} - \frac{1}{6}$

8. $\frac{5}{8} - \frac{3}{8}$

9. $\frac{9}{12} - \frac{3}{12}$

10. $\frac{3}{4} - \frac{3}{4}$

11. $\frac{10}{12} - \frac{3}{12}$

12. $\frac{3}{4} - \frac{2}{4}$

13. $\frac{5}{6} - \frac{2}{6}$

14. $\frac{8}{12} - \frac{1}{12}$

15. Look at the numerators of the fractions you subtracted. Then look at the numerators of the differences. What do you notice?

16. Look at the denominators of the numbers you subtracted. Then look at the denominators of the differences. What do you notice?

17. Tell how you can subtract two fractions with the same denominator.

ON YOUR OWN

Complete the table. Use fraction strips if needed.

18. Rule: Subtract $\frac{1}{8}$.

$\frac{8}{8}$	$\frac{7}{8}$	$\frac{6}{8}$	$\frac{5}{8}$	$\frac{4}{8}$	$\frac{3}{8}$	$\frac{2}{8}$	$\frac{1}{8}$
■	■	■	■	■	■	■	■

19. Rule: Subtract $\frac{1}{6}$.

$\frac{6}{6}$	$\frac{5}{6}$	$\frac{4}{6}$	$\frac{3}{6}$	$\frac{2}{6}$	$\frac{1}{6}$
■	■	■	■	■	■

Solve. Use fraction strips if needed.

20. Carlos has a piece of string $\frac{10}{12}$ ft long. He cuts off a piece $\frac{5}{12}$ ft long. How long is the rope now?

21. Aisha has $\frac{3}{4}$ hour to wait before dinner. She reads a story for $\frac{1}{4}$ hour. How long is it until dinner?

22. *Write a problem* involving subtracting fractions with like denominators. Ask others to solve it.

DEVELOPING A CONCEPT

Finding Differences

A. Martha bakes a loaf of oatmeal bread $\frac{5}{6}$ ft long. She cuts a piece $\frac{1}{6}$ ft long to give to Sara. How long is the loaf of bread now?

You can use fraction strips to answer these questions.

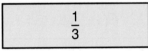

1. How many sixths are left?

2. How many thirds is this equal to?

3. How long is the loaf of bread now?

4. How does this difference compare to $\frac{1}{2}$?

B. You can also use this method to subtract fractions with like denominators.

Step 1	Step 2	Step 3
Subtract the numerators.	Use the common denominator.	Write the sum in simplest form.
$\frac{5}{6} - \frac{1}{6} = {}^4$	$\frac{5}{6} - \frac{1}{6} = \frac{4}{6}$	$\frac{4}{6} = \frac{2}{3}$ *Think:* $\frac{4 \div 2}{6 \div 2} = \frac{2}{3}$

5. **What if** Martha bakes another loaf $\frac{5}{8}$ ft long and cuts a piece $\frac{4}{8}$ ft long? How long would the loaf be then?

SHARING IDEAS

6. Why does the denominator stay the same when you subtract fractions?

7. Do you always need to simplify the difference of two fractions? Why or why not?

8. How can you check your answers?

PRACTICE

Subtract. Write the difference in simplest form.

9. $\frac{4}{5} - \frac{1}{5}$ **10.** $\frac{2}{3} - \frac{1}{3}$ **11.** $\frac{5}{6} - \frac{4}{6}$ **12.** $\frac{8}{9} - \frac{7}{9}$ **13.** $\frac{7}{8} - \frac{4}{8}$

14. $\frac{3}{4} - \frac{1}{4}$ **15.** $\frac{3}{7} - \frac{0}{7}$ **16.** $\frac{8}{12} - \frac{5}{12}$ **17.** $\frac{5}{6} - \frac{1}{6}$ **18.** $\frac{8}{16} - \frac{4}{16}$

19. $\frac{9}{10} - \frac{3}{10}$ **20.** $\frac{5}{8} - \frac{3}{8}$ **21.** $\frac{7}{9} - \frac{4}{9}$ **22.** $\frac{14}{16} - \frac{4}{16}$ **23.** $\frac{10}{12} - \frac{2}{12}$

24. $\begin{array}{r}\frac{5}{7}\\-\frac{2}{7}\\\hline\end{array}$ **25.** $\begin{array}{r}\frac{6}{8}\\-\frac{3}{8}\\\hline\end{array}$ **26.** $\begin{array}{r}\frac{8}{9}\\-\frac{3}{9}\\\hline\end{array}$ **27.** $\begin{array}{r}\frac{9}{16}\\-\frac{6}{16}\\\hline\end{array}$ **28.** $\begin{array}{r}\frac{8}{10}\\-\frac{5}{10}\\\hline\end{array}$ **29.** $\begin{array}{r}\frac{3}{4}\\-\frac{0}{4}\\\hline\end{array}$

30. $\begin{array}{r}\frac{5}{6}\\-\frac{3}{6}\\\hline\end{array}$ **31.** $\begin{array}{r}\frac{9}{10}\\-\frac{4}{10}\\\hline\end{array}$ **32.** $\begin{array}{r}\frac{5}{8}\\-\frac{3}{8}\\\hline\end{array}$ **33.** $\begin{array}{r}\frac{7}{9}\\-\frac{1}{9}\\\hline\end{array}$ **34.** $\begin{array}{r}\frac{15}{16}\\-\frac{15}{16}\\\hline\end{array}$ **35.** $\begin{array}{r}\frac{11}{12}\\-\frac{7}{12}\\\hline\end{array}$

Find the missing numerator.

36. $\frac{4}{12} - \frac{\blacksquare}{12} = \frac{4}{12}$ **37.** $\frac{8}{16} - \frac{\blacksquare}{16} = \frac{0}{16}$ **38.** $\frac{4}{7} + \frac{\blacksquare}{7} = 1$

Mixed Applications

Solve. You may need to use the Databank on page 521.

39. Which is farther from the parking lot, the picnic grounds or the waterfall? How much farther?

40. Geoff brings 8 muffins to the picnic. He gives $\frac{1}{4}$ of the muffins to Sam. How many muffins does he give Sam?

41. Khoury brings 4 bags of rolls to the picnic. Each bag holds 12 rolls. How many rolls did Khoury bring?

42. Sara mixes batter for $\frac{1}{4}$ hour. The cake bakes for $\frac{3}{4}$ hour. How long does it take Sara to make a cake?

Mixed Review

Find the answer. Which method did you use?

43. $4,379 + 7,637$ **44.** $50 \div 5$ **45.** 48×6

46. $\frac{7}{12} - \frac{3}{12}$ **47.** $737 \div 4$ **48.** $5,999 - 4,999$

| MENTAL MATH |
| CALCULATOR |
| PAPER/PENCIL |

EXTRA Practice, page 421; Practice **PLUS**, page 425

Add and Subtract Mixed Numbers

The ecology club is helping to plant trees on a bare hillside not far from their school. They plant $1\frac{7}{12}$ boxes of seedlings the first day and $2\frac{7}{12}$ boxes the next day. How many boxes did they plant?

WORKING TOGETHER

You can use fraction strips to make a model to solve the problem.

Step 1 Show $1\frac{7}{12}$ with fraction strips. Then show $2\frac{7}{12}$.

Step 2 Combine the twelfths.

1. How many twelfths are there in all? Can you regroup to form a whole?

Step 3 Combine the whole strips.

2. How many twelfths are there? How many whole strips? How many boxes did they plant?

3. How can you write the sum in simplest form?

What if $2\frac{10}{12}$ boxes of the seedlings are pine trees? How many boxes are not pine tree seedlings?

Step 4 Take away $2\frac{10}{12}$ from your model of the total. Start with the twelfths.

4. Do you need to regroup a whole strip as twelfths before you can take away? Why?

5. How many twelfths are left? How many whole strips? How many boxes are not pine tree seedlings?

6. How can you write the difference in simplest form?

Use fraction strips to add or subtract. Write the answer in simplest form.

7. $3\frac{1}{8}$
   $+2\frac{4}{8}$

8. $5\frac{11}{12}$
   $-3\frac{4}{12}$

9. $4\frac{2}{3}$
   $+1\frac{2}{3}$

10. $6\frac{1}{4}$
    $-2\frac{2}{4}$

11. $2\frac{3}{6}$
    $+1\frac{5}{6}$

12. $4\frac{1}{4}$
    $-\frac{3}{4}$

13. $4\frac{1}{2} + 2\frac{1}{2}$

14. $2\frac{5}{12} - 1\frac{7}{12}$

15. $3\frac{5}{8} - 2\frac{5}{8}$

SHARING IDEAS

Compare your models with those of others.

16. When do you need to regroup the fraction part when you add mixed numbers?

17. When do you need to regroup the whole number part when you subtract mixed numbers?

18. Write a method to tell how to add mixed numbers.

19. Write a method to tell how to subtract mixed numbers.

ON YOUR OWN

20. In the morning Lyn uses $2\frac{3}{8}$ bags of fertilizer. She uses $1\frac{5}{8}$ bags that afternoon. How much fertilizer does Lyn use?

21. How much more fertilizer did Lyn use in the morning than in the afternoon?

22. **Write a problem** involving adding or subtracting mixed numbers. Ask others to solve it.

UNDERSTAND
PLAN
TRY
CHECK
EXTEND

PROBLEM SOLVING

Strategy: Solving a Simpler Problem

Dick went on a hike in the wildlife preserve. It took him $\frac{7}{8}$ hour to get to the wildflower meadow and $\frac{5}{8}$ hour to get back to his campsite. How long was he on his hike?

Dick cannot decide what he can do to solve the problem. He plans to solve a simpler problem by using whole numbers.

Dick thinks: If I took 3 hours to get to the meadow and 2 hours to get back to camp, I could add to find the amount of time I hiked.

3 + 2 = 5

Dick can solve the simpler problem by adding the hours.

1. What can he do to solve the original problem?

   $\frac{7}{8} + \frac{5}{8} = $ ■

2. How long was he on his hike?

3. **What if** Dick wants to know how much longer it took him to get to the meadow than to get back to his campsite? Write a simpler problem by using whole numbers. Then solve the original problem.

4. How can solving a simpler problem help you solve more difficult problems?

PRACTICE

Solve the problem. Solve a simpler problem first if you need to. Use mental math, a calculator, or paper and pencil.

5. A tour train has 6 cars. It can seat 168 people. How many people will 1 car seat?

6. The park has 8 campgrounds. Each campground has 75 campsites. How many campsites does the park have?

7. A mountain stream is $\frac{1}{4}$ ft deep. It empties into a pool of water that is $\frac{3}{4}$ ft deep. How much deeper is the pool of water than the stream?

8. On Monday the park ranger patrolled 115 miles of park land. She patrolled 102 miles on Tuesday and 111 miles on Wednesday. How many miles did she patrol in all?

Strategies and Skills Review

9. The Kanes are conserving water. They used 380 gallons of water in June, 366 gallons in July, and 352 gallons in August. If they continue this pattern, how many gallons will they use in September?

10. The wildlife preserve has biking and hiking trails. There are 16 hiking trails. What else do you need to know to find how many trails there are in all?

11. Alexis bought 2 animal posters for $3.00 each and 1 postcard for $.50. How much change will she get from $10.00?

12. The wildlife preserve has 125 more acres of forest than of prairie. It has 385 acres of forest. How many acres of prairie does it have?

13. A baby chimp drinks $\frac{3}{4}$ bottle of milk and $\frac{3}{4}$ bottle of sugared water at a feeding. How much liquid does it drink at a feeding?

14. **Write a problem.** Solve it. Ask others to solve your problem by first writing a simpler problem.

DEVELOPING A CONCEPT

Perimeter and Area

Felipe is helping to build a new corral for the burros used on the mountain trails. The corral is 20 feet long on 2 sides and 9 feet long on the other 2 sides. What is the perimeter of the corral? What is its area?

Think: *Perimeter* is the distance around an object.
Area is the number of square units it takes to cover the surface of the object.

WORKING TOGETHER

1. What is the perimeter of the corral? How did you find it?

2. What is the area of the corral? How did you find it?

3. Estimate, then measure the perimeter and area of objects around you. Make a table.

4. Use graph paper to draw different figures with an area of 12 square inches. Find the perimeter of each shape. What do you notice?

SHARING IDEAS

5. Why can you add the length and the width and multiply by 2 to find the perimeter of a rectangle?

6. Can you always multiply to find the area of a rectangle? Why or why not?

7. Do figures with the same area always have the same perimeter? Give examples to support your answer.

PRACTICE

Measure. Then find the perimeter and area.

8.

9.

Find the perimeter and area.

10.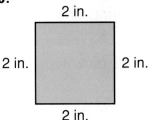
2 in. (top) 2 in. (left) 2 in. (right) 2 in. (bottom)

11.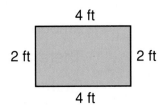
4 ft (top) 2 ft (left) 2 ft (right) 4 ft (bottom)

12.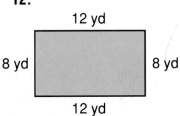
12 yd (top) 8 yd (left) 8 yd (right) 12 yd (bottom)

13. length = 10 ft
width = 7 ft

14. length = 42 in.
width = 8 in.

15. length = 15 yd
width = 5 yd

Find the missing length.

16.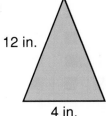
12 in.
4 in.
Perimeter: 24 in.

17.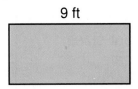
9 ft
Perimeter: 32 ft

18.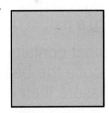
Perimeter: 56 yd

Mixed Applications

19. The bottom of Hannah's fish tank is a square $\frac{1}{4}$ ft on each side. What is the perimeter of the fish tank?

20. The floor of Archie's storage shed is a rectangle 3 yd wide by 4 yd long. What is the area of the floor?

21. The area of a hog pen is 178 sq ft. The area of a sheep pen is 532 sq ft. How much larger is the sheep pen?

22. *Draw a figure.* Find its perimeter and area. Ask others to find them also.

Capacity

1 cup

1 pint

1 quart

1 gallon

Julie brings along a canteen of water on a hike. How much water does it hold?

The amount a container can hold is its **capacity.** The customary units of capacity are **cup (c), pint (pt), quart (qt),** and **gallon (gal).**

Cups and pints are customary units used for measuring small amounts of liquids.

There are 2 cups in 1 pint.

Quarts and gallons are customary units used to measure larger amounts.

There are 2 pints in a quart.
There are 4 quarts in a gallon.

1. Do you think Julie's canteen holds a cup, a pint, a quart, or a gallon? Why?

2. Collect containers of different sizes. Choose the best unit. Estimate, then measure their capacities. Record your measures in a table. Compare them with those of others.

CAPACITY

Object	c	pt	qt	gal
Cereal bowl	■	■	■	■
Small glass	■	■	■	■
Large glass	■	■	■	■
Bucket	■	■	■	■

SHARING IDEAS

3. Why did you pick the units you chose to measure the different objects?

4. Why would you not want to tell someone the capacity of a container by the number of glasses it holds?

5. **Write a list** of the types of containers you use every day and how their capacities are labeled. Compare your list to those of others.

PRACTICE

Choose the best estimate of capacity.

6. glass of juice   **a.** 1 c   **b.** 1 pt   **c.** 1 qt   **d.** 1 gal

7. large jug of water   **a.** 1 c   **b.** 1 pt   **c.** 1 qt   **d.** 1 gal

8. container of cream   **a.** 1 c   **b.** 1 pt   **c.** 1 qt   **d.** 1 gal

9. large container of juice   **a.** 2 c   **b.** 2 pt   **c.** 2 qt   **d.** 2 gal

10. large bucket   **a.** 5 c   **b.** 5 pt   **c.** 5 qt   **d.** 5 gal

11. cocoa mug   **a.** 2 c   **b.** 2 pt   **c.** 2 qt   **d.** 2 gal

Choose the best unit to use to measure the capacity. Write
cup, pint, quart, or *gallon.*

12. kitchen sink   **13.** bud vase   **14.** juice pitcher

15. large glass   **16.** bath tub   **17.** large can of paint

Critical Thinking

Do these sentences make sense? Why or why not?

18. Barry drank a gallon of milk with dinner.

19. Kevin used 3 cups of water in his bath.

Mixed Applications

20. Dale brings enough drinking water to last for 2 days. Does she bring 5 cups or 5 quarts?

21. Hank pours $\frac{1}{2}$ pint of apple juice and $\frac{1}{2}$ pint of grape juice into the same bottle. How much juice is in the bottle?

22. Yoko finds that each person drinks about 4 cups of juice each day. If there are 24 people on the camping trip, how much juice do they drink daily?

23. The pool at the camp usually holds 5,000 gallons of water. After a rain there are 6,243 gallons in it. How much rainwater went in the pool?

DEVELOPING A CONCEPT
Weight

Xavier's class is collecting aluminum cans as part of their recycling project. They can earn money based on the weight of the cans they collect.

The customary units of weight are **ounces (oz)** and **pounds (lb).**

Ounces are customary units used to measure light objects.

A slice of bread weighs about 1 oz. A small bowl of cereal weighs about 4 oz.

1. Name other things that could be measured using ounces.

Pounds are customary units used to measure heavier objects.

There are 16 ounces in 1 pound.

A loaf of bread weighs about 1 pound. A large box of cereal weighs about 1 pound.

2. Name other things that could be measured using pounds.

3. Collect objects in your classroom. Decide which units you would use to measure their weights. Record your decisions in a chart like this.

1 ounce

1 pound

WEIGHT

Object	oz	lb
■	■	■
■	■	■

SHARING IDEAS

4. How did you choose which units to use to measure the different objects?

5. Is the larger object always heavier than the smaller object? Why or why not?

PRACTICE

Choose the best estimate of weight.

6. telephone   **a.** 1 oz   **b.** 10 oz   **c.** 1 lb   **d.** 10 lb

7. letter   **a.** 1 oz   **b.** 10 oz   **c.** 1 lb   **d.** 10 lb

8. textbook   **a.** 2 oz   **b.** 12 oz   **c.** 2 lb   **d.** 20 lb

9. sneaker   **a.** 3 oz   **b.** 13 oz   **c.** 3 lb   **d.** 13 lb

10. large cat   **a.** 4 oz   **b.** 14 oz   **c.** 4 lb   **d.** 14 lb

11. scissors   **a.** 5 oz   **b.** 15 oz   **c.** 5 lb   **d.** 15 lb

Choose the object closest to the given weight.

12. 2 oz   **a.** paper clip   **b.** eraser   **c.** desk

13. 8 lb   **a.** compass   **b.** blanket   **c.** hiking boots

14. 20 lb   **a.** record   **b.** television   **c.** car

Mixed Applications

15. Kathy wrote that her large bag of cans weighs 25. Does she mean ounces or pounds?

16. If there are 16 cans in 1 lb, how many cans are in 8 lb?

17. So far Dan has collected $\frac{7}{8}$ lb of cans. Stan has collected $\frac{5}{8}$ lb. How much more has Dan collected?

18. *Write a problem* that involves weight. Solve your problem. Ask others to solve it.

CHALLENGE

Gravity on the moon is $\frac{1}{6}$ the gravity on earth. So objects on the moon weigh $\frac{1}{6}$ what they would weigh on earth.

Tell how much the object would weigh on the moon.

1. a 24-oz book   **2.** a 168-lb astronaut   **3.** a 36-lb telescope

Renaming Customary Units

Niko needs to buy special paper for a banner she is making. The banner is going to be 4 yards long. At the store Niko learns that the special paper is measured in feet. How many feet of paper will she need?

You can solve the problem by finding patterns and using them to rename the units.

WORKING TOGETHER

1. Look at a yardstick. How many feet are equal to 1 yard?

2. Use the yardstick to mark off 2 yards on the chalkboard. Then use the yardstick to find this length in feet.

3. What if you were to mark off 3 yd? What would be its length in feet?

4. Complete the table.

yd	1	2	3	4
ft	■	■	■	■

5. How many feet of special paper will Niko need?

6. You know that there are 12 inches in 1 foot. How many inches are in 2 feet?

7. Complete the table.

ft	1	2	3	4
in.	■	■	■	■

SHARING IDEAS

8. As the size of the unit decreases what happens to the number of units needed to measure the same length?

9. What patterns do you see in the two tables? How can they help you rename measures in yards as feet and measures in feet as inches?

ON YOUR OWN

10. You know that there are 4 quarts in a gallon. If you know the capacity of a container in gallons, how can you find the capacity in quarts?

11. You know there are 16 ounces in a pound. If you know the weight of an object in pounds, how can you find the weight in ounces?

Complete.

12. 5 yd = ■ ft

13. 7 ft = ■ in.

14. 3 gal = ■ qt

15. 8 qt = ■ gal

16. 10 lb = ■ oz

17. 64 oz = ■ lb

Complete the table.

18.

qt	1	2	3	4
pt	2	■	■	■

19.

pt	1	2	3	4
c	2	■	■	■

Solve.

20. How long is Niko's banner in inches?

21. Niko's printer weighs 6 lb. How many ounces does it weigh?

22. Carl's fish tank holds 10 gal of water. How many quarts does it hold?

23. *Write a problem* that involves renaming units of measure. Solve your problem. Then ask others to solve it.

UNDERSTANDING A CONCEPT

Temperature

To measure temperature in customary units, use a **Fahrenheit thermometer.**

To read the temperature on a thermometer, look at the mark or number next to the top of the red column. You read the temperature in **degrees Fahrenheit (°F).**

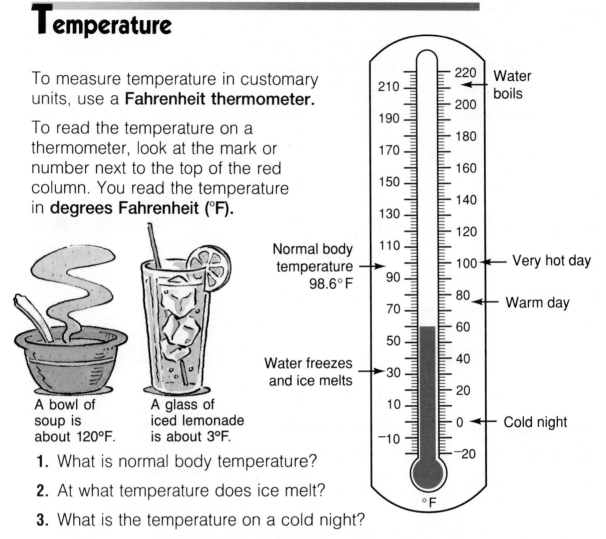

A bowl of soup is about 120°F.

A glass of iced lemonade is about 3°F.

Normal body temperature → 98.6°F

Water freezes and ice melts →

Water boils

Very hot day ←
Warm day ←
Cold night ←

°F

1. What is normal body temperature?

2. At what temperature does ice melt?

3. What is the temperature on a cold night?

The coldest day ever recorded in Florida so far was ⁻2°F. Read this as "minus 2 degrees Fahrenheit" or "2 degrees below 0, Fahrenheit."

4. Write the temperature for minus 12 degrees Fahrenheit.

TRY OUT
Write the letter of the correct answer. Choose the most reasonable temperature.

5. Arctic winter   **a.** ⁻23°F   **b.** 55°F   **c.** 96°F   **d.** 143°F

6. summer day   **a.** ⁻3°F   **b.** 45°F   **c.** 93°F   **d.** 178°F

7. bowl of oatmeal   **b.** 16°F   **b.** 32°F   **c.** 76°F   **d.** 98°F

8. icicle   **a.** ⁻15°F   **b.** 0°F   **c.** 42°F   **d.** 212°F

PRACTICE

Write the temperature in degrees Fahrenheit.

9.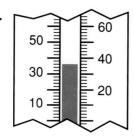

10.

11.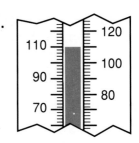

Choose the most reasonable temperature.

12. ice-skating   **a.** ⁻42°F   **b.** 28°F   **c.** 43°F   **d.** 76°F

13. having a beach party   **a.** 13°F   **b.** 45°F   **c.** 85°F   **d.** 201°F

14. sleigh-riding   **a.** 24°F   **b.** 52°F   **c.** 73°F   **d.** 88°F

Choose the object closest to the given temperature.

15. 123°F   **a.** ice water   **b.** stew   **c.** yogurt

16. 30°F   **a.** ice cube   **b.** taco   **c.** shower

17. 75°F   **a.** igloo   **b.** oven   **c.** classroom

Mixed Applications

18. This morning the temperature was 45°F. By this afternoon the temperature was 68°F. What was the change in the temperature?

19. Paula is wearing a ski jacket. It is 83°F outside. Is Paula wearing the right clothes? Why or why not?

Mixed Review

Compare. Write >, <, or =.

20. 332 ÷ 7 ● 523

21. 3 × 60 ● 4 × 40

22. $\frac{5}{8} + \frac{3}{8}$ ● $\frac{3}{4}$

23. 650 − 250 ● 350

24. $73.23 + $22.52 ● $100.00

25. 26 in. ● 2 ft

26. perimeter of a square with 4-in. sides ● 16 in.

EXTRA Practice, page 423

EXPLORING A CONCEPT

What Do You Measure?

The Jackson family plans to make a large bowl of lemonade to serve at the family reunion. What kinds of things do they need to know about the bowl?

What you need to know about an object determines what you should measure.

WORKING TOGETHER

1. Suppose you had a large bowl. What would you measure to find out how much lemonade it will hold?

2. **What if** you need to know if it will fit onto a small table? What would you measure?

3. **What if** you wanted to tell whether it was easier to carry the bowl or carry a stack of plates? What would you measure?

4. How could you check how cold the lemonade is?

5. Fill a large bowl with water. Estimate all of its measures. Find every measure you can.

SHARING IDEAS

6. Which of the measurements could the Jackson family estimate? Which should they actually measure? Why?

7. **Write a list** of situations where you need to measure the length, capacity, weight, or temperature of something. Tell if you can estimate or if you need to actually measure.

ON YOUR OWN

Write *capacity, length, weight,* or *temperature* to tell what should be measured in each of the following situations. Then tell what measuring tool should be used.

8. You need to know if a desk will fit through a door.

9. You want to know if the turkey is cooked enough.

10. You want to order some cheese from the deli counter.

11. You want to know how much water to add to a recipe.

Match the descriptions to the objects at the right.

12. Temperature: 32°F
Width: 6 in.
Length: 12 in.
Weight: 1 lb

13. Length: 12 in.
Width: 1 in.
Weight: 5 oz

14. Temperature: 145°F
Perimeter: 8 in.
Weight: $\frac{1}{4}$ lb

15. Temperature: 38°F
Weight: 3 lb
Capacity: 2 qt

16. Find some common object in your classroom or at home. Describe it with different measurements. Then ask another student to guess the object.

DECISION MAKING

Problem Solving: Planning a Hiking Trip

SITUATION

Greg and Sara and their mom and dad are planning a three-day hiking trip. They will camp out for two nights.

PROBLEM

What should each hiker carry?

DATA

Campmate Family Camping Checklist

	Pounds
sleeping bag	3
air mattress	1½
personal items (clothing, rain poncho, grooming kit, towel)	3½
or { jr. back pack	3
adult back pack	3½

Each carry own personal gear

cooking gear (utensils, dishes, pans, back pack stove, fuel) ..	5
trail meals for 4—	
2 dinner packs	5
2 lunch packs	4
2 breakfast packs	3
first-aid kit	1
survival kit (compass, map, flare, batteries, nylon cord) ...	2
or { small canteen	1
large canteen	1½
or { high intensity lantern	2
2 mini lights	1
tents—2-person (2)	15
4-person	11½

Who carries what? Can we take all?

Which should we take?

Other gear:

field glasses	1½
bird guide book	½
portable radio	1
folding camp seat	1½
camera	2

can we take?

Weight we can each carry:
Dad 32 lb   Mom 25 lb   Greg 14 lb   Sara 14 lb

USING THE DATA

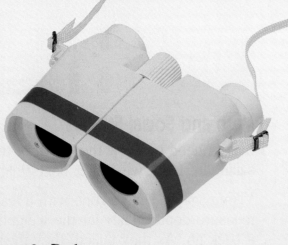

1. What is the weight of the personal gear that Greg and Sara will carry?

2. What is the weight of the personal gear that Mom and Dad will carry?

In addition to personal gear, how much more can each person carry?
3. Greg   4. Mom   5. Sara   6. Dad

7. How much do all the meals weigh?

MAKING DECISIONS

8. What are the things the family should think about when planning their camping trip?

9. Who should carry the food and cooking equipment? Why? How much more can each person carry now?

10. Besides personal gear, Greg and Sara can each carry a little more. What should they choose? Who should carry each item?

11. Should the family use two 2-person tents or one 4-person tent? Why?

12. The five gallons of water that the family needs for three days weigh 40 lb. Should they carry their water or camp near the stream? Why?

13. Does the family have room to take anything else? If so, what should they take?

14. **Write a list** of things you would take on a two-day camping trip. Give reasons for your answers.

Using Fractions   **417**

Math and Social Studies

Sailors need to know the depth of the water they are sailing in. Otherwise their boats might hit bottom.

Viking sailors used a rope with a lead weight on the end to measure depth. They let the weight hit bottom, then pulled it back up. As they pulled, they measured the length of the rope against their outstretched arms. The motion of opening and closing their arms reminded the Vikings of an embrace. So they called the length of their outstretched arms a **fathom,** the Viking word for embrace. For the Vikings, a fathom was about six feet. Today fathoms measure exactly six feet.

What if the distance to the bottom of a river is $\frac{1}{2}$ fathom? How many feet from the bottom is the boat?

Think: $\frac{1}{2}$ of 6 is 3.

The boat is 3 feet from the bottom.

ACTIVITY

1. Today we use sonar to measure depth. Find out what sonar is and how it works. How is sonar better than a rope? How is it worse?

Calculator: Finding Perimeter

A calculator can help you find the perimeter of many-sided figures.

Make a table like the one below. Examine each shape and fill in the table.

- Make an estimate of the perimeter of each shape.

- Use a calculator to find the exact perimeter of each shape.

- Check your estimated perimeter and your exact perimeter. Does your exact perimeter make sense?

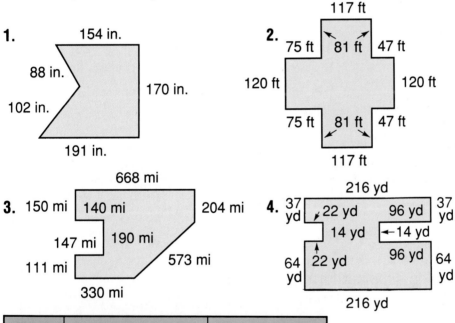

1. 154 in. 88 in. 102 in. 170 in. 191 in.

2. 117 ft 75 ft 81 ft 47 ft 120 ft 120 ft 75 ft 81 ft 47 ft 117 ft

3. 668 mi 150 mi 140 mi 204 mi 147 mi 190 mi 111 mi 573 mi 330 mi

4. 216 yd 37 yd 22 yd 96 yd 37 yd 14 yd 14 yd 64 yd 22 yd 96 yd 64 yd 216 yd

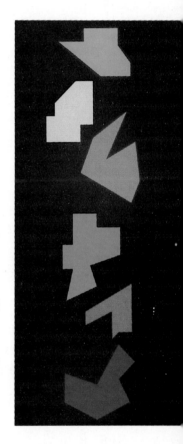

Figure	Estimated Perimeter	Exact Perimeter
1		
2		
3		
4		

USING THE CALCULATOR

5. Did your estimates confirm your exact answers?

6. Draw and label your own many-sided figure. Have others calculate the perimeter.

EXTRA PRACTICE

Finding Sums, page 389...

Add. Write the sum in simplest form.

1. $\frac{2}{5} + \frac{2}{5}$ **2.** $\frac{2}{8} + \frac{3}{8}$ **3.** $\frac{3}{6} + \frac{1}{6}$ **4.** $\frac{4}{7} + \frac{1}{7}$ **5.** $\frac{1}{4} + \frac{2}{4}$

6. $\frac{4}{10} + \frac{2}{10}$ **7.** $\frac{4}{9} + \frac{3}{9}$ **8.** $\frac{7}{12} + \frac{2}{12}$ **9.** $\frac{4}{8} + \frac{2}{8}$ **10.** $\frac{5}{16} + \frac{5}{16}$

11. $\begin{array}{r} \frac{1}{6} \\ + \frac{2}{6} \\ \hline \end{array}$ **12.** $\begin{array}{r} \frac{3}{10} \\ + \frac{5}{10} \\ \hline \end{array}$ **13.** $\begin{array}{r} \frac{3}{7} \\ + \frac{3}{7} \\ \hline \end{array}$ **14.** $\begin{array}{r} \frac{5}{12} \\ + \frac{3}{12} \\ \hline \end{array}$ **15.** $\begin{array}{r} \frac{1}{8} \\ + \frac{6}{8} \\ \hline \end{array}$

16. $\begin{array}{r} \frac{2}{10} \\ + \frac{3}{10} \\ \hline \end{array}$ **17.** $\begin{array}{r} \frac{5}{16} \\ + \frac{6}{16} \\ \hline \end{array}$ **18.** $\begin{array}{r} \frac{4}{8} \\ + \frac{3}{8} \\ \hline \end{array}$ **19.** $\begin{array}{r} \frac{2}{12} \\ + \frac{6}{12} \\ \hline \end{array}$ **20.** $\begin{array}{r} \frac{3}{9} \\ + \frac{3}{9} \\ \hline \end{array}$

Find the missing numerator.

21. $\frac{2}{9} + \frac{\blacksquare}{9} = \frac{5}{9}$ **22.** $\frac{3}{8} + \frac{1}{8} = \frac{\blacksquare}{8} + \frac{3}{8}$ **23.** $\frac{5}{10} + \frac{2}{10} + \frac{1}{10} = \frac{2}{10} + \frac{1}{10} + \frac{\blacksquare}{10}$

Finding Greater Sums, page 391...

Add. Write the sum in simplest form.

1. $\frac{1}{7} + \frac{6}{7}$ **2.** $\frac{2}{4} + \frac{3}{4}$ **3.** $\frac{3}{5} + \frac{4}{5}$ **4.** $\frac{3}{6} + \frac{4}{6}$ **5.** $\frac{4}{5} + \frac{4}{5}$

6. $\frac{5}{8} + \frac{7}{8}$ **7.** $\frac{3}{9} + \frac{8}{9}$ **8.** $\frac{11}{12} + \frac{7}{12}$ **9.** $\frac{2}{3} + \frac{2}{3}$ **10.** $\frac{8}{10} + \frac{8}{10}$

11. $\begin{array}{r} \frac{11}{12} \\ + \frac{1}{12} \\ \hline \end{array}$ **12.** $\begin{array}{r} \frac{3}{6} \\ + \frac{4}{6} \\ \hline \end{array}$ **13.** $\begin{array}{r} \frac{8}{9} \\ + \frac{3}{9} \\ \hline \end{array}$ **14.** $\begin{array}{r} \frac{6}{7} \\ + \frac{6}{7} \\ \hline \end{array}$ **15.** $\begin{array}{r} \frac{6}{8} \\ + \frac{7}{8} \\ \hline \end{array}$

16. $\begin{array}{r} \frac{4}{5} \\ + \frac{3}{5} \\ \hline \end{array}$ **17.** $\begin{array}{r} \frac{15}{16} \\ + \frac{15}{16} \\ \hline \end{array}$ **18.** $\begin{array}{r} \frac{6}{10} \\ + \frac{4}{10} \\ \hline \end{array}$ **19.** $\begin{array}{r} \frac{9}{12} \\ + \frac{9}{12} \\ \hline \end{array}$ **20.** $\begin{array}{r} \frac{7}{9} \\ + \frac{5}{9} \\ \hline \end{array}$

Problem Solving Strategy: Using Different Strategies, page 393..............

Solve the problem. Then use a different strategy to check your answer. Tell which strategies you used.

1. Joyce is buying supplies for an art class. She buys pastels for $26.50, an easel for $57.50, oil paints for $45.85, and some charcoals for $17. Will $150 be enough to pay for these items?

2. A number such as 22 has two digits that are the same. Find four different 2-digit numbers, in which the digits are the same, that have a sum of 165.

Subtracting Fractions, page 397 ....

Subtract. Write the difference in simplest form.

1. $\frac{3}{4} - \frac{1}{4}$ **2.** $\frac{5}{6} - \frac{1}{6}$ **3.** $\frac{6}{8} - \frac{5}{8}$ **4.** $\frac{7}{9} - \frac{3}{9}$ **5.** $\frac{4}{5} - \frac{3}{5}$

6. $\frac{8}{10} - \frac{6}{10}$ **7.** $\frac{6}{12} - \frac{3}{12}$ **8.** $\frac{4}{6} - \frac{1}{6}$ **9.** $\frac{15}{16} - \frac{5}{16}$ **10.** $\frac{3}{4} - \frac{3}{4}$

11. $\begin{array}{r} \frac{2}{8} \\ -\frac{1}{8} \\ \hline \end{array}$ **12.** $\begin{array}{r} \frac{5}{7} \\ -\frac{3}{7} \\ \hline \end{array}$ **13.** $\begin{array}{r} \frac{6}{10} \\ -\frac{4}{10} \\ \hline \end{array}$ **14.** $\begin{array}{r} \frac{8}{16} \\ -\frac{4}{16} \\ \hline \end{array}$ **15.** $\begin{array}{r} \frac{8}{9} \\ -\frac{6}{9} \\ \hline \end{array}$

16. $\begin{array}{r} \frac{6}{12} \\ -\frac{4}{12} \\ \hline \end{array}$ **17.** $\begin{array}{r} \frac{7}{8} \\ -\frac{3}{8} \\ \hline \end{array}$ **18.** $\begin{array}{r} \frac{9}{10} \\ -\frac{6}{10} \\ \hline \end{array}$ **19.** $\begin{array}{r} \frac{13}{16} \\ -\frac{4}{16} \\ \hline \end{array}$ **20.** $\begin{array}{r} \frac{6}{7} \\ -\frac{4}{7} \\ \hline \end{array}$

Find the missing numerator.

21. $\frac{4}{8} - \frac{\blacksquare}{8} = \frac{1}{8}$ **22.** $\frac{8}{10} - \frac{\blacksquare}{10} = \frac{3}{10}$ **23.** $\frac{\blacksquare}{7} - \frac{5}{7} = \frac{1}{7}$

24. $\frac{5}{9} - \frac{\blacksquare}{9} = \frac{2}{9}$ **25.** $\frac{7}{12} - \frac{\blacksquare}{12} = \frac{5}{12}$ **26.** $\frac{\blacksquare}{16} - \frac{5}{16} = \frac{3}{16}$

27. $\frac{6}{11} - \frac{\blacksquare}{11} = \frac{3}{11}$ **28.** $\frac{17}{19} - \frac{\blacksquare}{19} = \frac{6}{19}$ **29.** $\frac{3}{8} - \frac{\blacksquare}{8} = \frac{1}{8}$

EXTRA PRACTICE

Problem Solving Strategy: Solving a Simpler Problem, page 403 ....

Solve the problem. Solve a simpler problem if you need to.

1. The Day family drove 346 miles on Monday, 263 miles on Tuesday and 307 miles on Wednesday. How many miles did they drive in all?

2. The Day family used 4 tanks of gasoline on the trip. The gas tank of the car holds 22 gallons of gasoline. How many gallons of gasoline did they use?

Perimeter and Area, page 405 ....

Measure. Then find the perimeter and area.

1.

2.

Find the perimeter and area.

3.

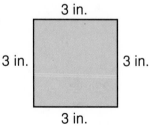

3 in. (top), 3 in. (left), 3 in. (right), 3 in. (bottom)

4.

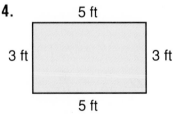

5 ft (top), 3 ft (left), 3 ft (right), 5 ft (bottom)

5.

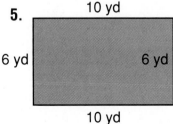

10 yd (top), 6 yd (left), 6 yd (right), 10 yd (bottom)

6. length = 30 ft
width = 6 ft

7. length = 18 in.
width = 7 in.

8. length = 29 yd
width = 7 yd

Capacity, page 407 ....

Choose the best estimate of capacity.

1. small bowl of soup   **a.** 1 c   **b.** 1 pt   **c.** 1 qt   **d.** 1 gal

2. can of gasoline   **a.** 1 c   **b.** 1 pt   **c.** 1 qt   **d.** 1 gal

3. large glass of milk   **a.** 1 c   **b.** 1 pt   **c.** 1 qt   **d.** 1 gal

Choose the best unit to use to measure the capacity. Write *cup, pint, quart,* or *gallon.*

4. swimming pool   **5.** mixing bowl   **6.** watering can

Weight, page 409 ....

Choose the best estimate of weight.

1. dinner plate   **a.** 1 oz   **b.** 10 oz   **c.** 3 lb   **d.** 10 lb

2. stick of butter   **a.** 1 oz   **b.** 4 oz   **c.** 1 lb   **d.** 4 lb

3. muffin   **a.** 2 oz   **b.** 5 oz   **c.** 1 lb   **d.** 4 lb

Choose the object closest to the given weight.

4. 1 oz   **a.** eraser   **b.** book   **c.** blanket

5. 4 lb   **a.** bicycle   **b.** yardstick   **c.** coat

Temperature, page 413 ....

Write the temperature in degrees Fahrenheit.

1. 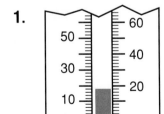   **2.** 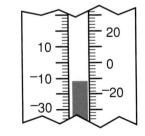   **3.**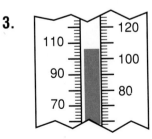

Choose the most reasonable temperature.

4. have a picnic   **a.** 25°F   **b.** 50°F   **c.** 75°F   **d.** ⁻68°F

5. build a snowman   **a.** 25°F   **b.** 50°F   **c.** 75°F   **d.** 110°F

PRACTICE PLUS

KEY SKILL: Finding Greater Sums (Use after page 391.)

Level A ....

Add. Write the sum in simplest form.

1. $\frac{3}{4} + \frac{1}{4}$ **2.** $\frac{4}{8} + \frac{5}{8}$ **3.** $\frac{6}{9} + \frac{3}{9}$ **4.** $\frac{4}{6} + \frac{3}{6}$

5. $\frac{2}{3} + \frac{1}{3}$ **6.** $\frac{1}{2} + \frac{1}{2}$ **7.** $\frac{11}{16} + \frac{8}{16}$ **8.** $\frac{6}{8} + \frac{4}{8}$

9. Gene floated $\frac{7}{8}$ mi to the dock. Then he floated $\frac{5}{8}$ mi to the island. How far did he float?

Level B ....

Add. Write the sum in simplest form.

10. $\frac{3}{8} + \frac{3}{8}$ **11.** $\frac{4}{6} + \frac{2}{6}$ **12.** $\frac{5}{7} + \frac{6}{7}$ **13.** $\frac{7}{12} + \frac{11}{12}$

14. $\begin{array}{r} \frac{1}{5} \\ + \frac{1}{5} \\ \hline \end{array}$ **15.** $\begin{array}{r} \frac{1}{4} \\ + \frac{3}{4} \\ \hline \end{array}$ **16.** $\begin{array}{r} \frac{1}{6} \\ + \frac{3}{6} \\ \hline \end{array}$ **17.** $\begin{array}{r} \frac{3}{4} \\ + \frac{3}{4} \\ \hline \end{array}$ **18.** $\begin{array}{r} \frac{5}{9} \\ + \frac{5}{9} \\ \hline \end{array}$

19. Pat ran for $\frac{3}{4}$ hour. Then she walked for $\frac{3}{4}$ hour. How much time did this take in all?

Level C ....

Add. Write the sum in simplest form.

20. $\begin{array}{r} \frac{4}{6} \\ + \frac{5}{6} \\ \hline \end{array}$ **21.** $\begin{array}{r} \frac{3}{7} \\ + \frac{6}{7} \\ \hline \end{array}$ **22.** $\begin{array}{r} \frac{7}{12} \\ + \frac{7}{12} \\ \hline \end{array}$ **23.** $\begin{array}{r} \frac{5}{9} \\ + \frac{4}{9} \\ \hline \end{array}$ **24.** $\begin{array}{r} \frac{4}{6} \\ + \frac{3}{6} \\ \hline \end{array}$

25. $\begin{array}{r} \frac{4}{8} \\ + \frac{4}{8} \\ \hline \end{array}$ **26.** $\begin{array}{r} \frac{8}{10} \\ + \frac{4}{10} \\ \hline \end{array}$ **27.** $\begin{array}{r} \frac{7}{16} \\ + \frac{10}{16} \\ \hline \end{array}$ **28.** $\begin{array}{r} \frac{9}{12} \\ + \frac{9}{12} \\ \hline \end{array}$ **29.** $\begin{array}{r} \frac{6}{8} \\ + \frac{7}{8} \\ \hline \end{array}$

30. The handle of the small shovel is $\frac{15}{16}$ in. long. Its blade is $\frac{7}{16}$ in. long. How long is the shovel altogether?

KEY SKILL: Finding Differences (Use after page 399.)

Level A ....

Subtract. Write the difference in simplest form.

1. $\frac{3}{4} - \frac{1}{4}$ 2. $\frac{9}{10} - \frac{3}{10}$ 3. $\frac{7}{8} - \frac{3}{8}$ 4. $\frac{5}{6} - \frac{1}{6}$

5. $\frac{4}{5} - \frac{3}{5}$ 6. $\frac{8}{12} - \frac{5}{12}$ 7. $\frac{9}{10} - \frac{4}{10}$ 8. $\frac{6}{16} - \frac{1}{16}$

9. Joan sold $\frac{3}{4}$ of a box of apples. Kate sold $\frac{1}{4}$ of a box. How much more did Joan sell?

Level B ....

Subtract. Write the difference in simplest form.

10. $\frac{3}{4} - \frac{2}{4}$ 11. $\frac{6}{9} - \frac{4}{9}$ 12. $\frac{14}{16} - \frac{8}{16}$ 13. $\frac{11}{12} - \frac{7}{12}$

14. $\frac{9}{12} - \frac{1}{12}$ 15. $\frac{4}{6} - \frac{2}{6}$ 16. $\frac{5}{8} - \frac{1}{8}$ 17. $\frac{7}{10} - \frac{3}{10}$

Find the missing numerator.

18. $\frac{4}{5} - \frac{\blacksquare}{5} = \frac{2}{5}$ 19. $\frac{7}{12} - \frac{\blacksquare}{12} = \frac{7}{12}$ 20. $\frac{9}{10} - \frac{\blacksquare}{10} = \frac{7}{10}$

21. Don mowed $\frac{4}{9}$ of the lawn. Tony mowed $\frac{5}{9}$ of the lawn. How much more did Tony mow?

Level C ....

Subtract. Write the difference in simplest form.

22. $\frac{9}{16}$ 23. $\frac{6}{7}$ 24. $\frac{5}{6}$ 25. $\frac{7}{8}$ 26. $\frac{8}{12}$
    $-\frac{1}{16}$ $-\frac{4}{7}$ $-\frac{2}{6}$ $-\frac{1}{8}$ $-\frac{4}{12}$

Find the missing numerator.

27. $\frac{6}{9} - \frac{\blacksquare}{9} = \frac{4}{9}$ 28. $\frac{9}{12} - \frac{\blacksquare}{12} = \frac{9}{12}$ 29. $\frac{15}{16} - \frac{\blacksquare}{16} = \frac{14}{16}$

30. Brenda spent $\frac{4}{10}$ of her time reading. She spent $\frac{2}{10}$ of her time writing. How much more of her time did she spend reading?

CHAPTER REVIEW

LANGUAGE AND MATHEMATICS

Complete the sentences. Use the words in the chart.

1. The ■ stays the same when you add fractions. *(page 388)*

2. To measure temperature in customary units, use a ■ thermometer. *(page 412)*

3. An ■ is a customary unit used to measure light objects. *(page 408)*

4. ■ is the distance around an object. *(page 404)*

5. *Write a definition* or give an example of the words you did not use from the chart.

> **VOCABULARY**
> **Fahrenheit**
> **numerator**
> **denominator**
> **area**
> **perimeter**
> **ounce**

CONCEPTS AND SKILLS

Add. Write the sum in simplest form. *(pages 388–391)*

6. $\frac{1}{2}$ $+\frac{1}{2}$

7. $\frac{2}{3}$ $+\frac{1}{3}$

8. $\frac{3}{5}$ $+\frac{3}{5}$

9. $\frac{3}{12}$ $+\frac{6}{12}$

10. $\frac{1}{10}$ $+\frac{3}{10}$

11. $\frac{5}{8}$ $+\frac{1}{8}$

12. $\frac{2}{6} + \frac{3}{6}$

13. $\frac{1}{9} + \frac{5}{9}$

14. $\frac{1}{4} + \frac{3}{4}$

15. $\frac{5}{6} + \frac{5}{6}$

Subtract. Write the difference in simplest form. *(page 398)*

16. $\frac{11}{12}$ $-\frac{5}{12}$

17. $\frac{3}{4}$ $-\frac{2}{4}$

18. $\frac{9}{3}$ $-\frac{6}{3}$

19. $\frac{5}{6}$ $-\frac{1}{6}$

20. $\frac{7}{8}$ $-\frac{1}{8}$

21. $\frac{9}{10}$ $-\frac{7}{10}$

22. $\frac{7}{10} - \frac{5}{10}$

23. $\frac{3}{4} - \frac{1}{4}$

24. $\frac{7}{8} - \frac{1}{8}$

25. $\frac{6}{7} - \frac{3}{7}$

Find the perimeter and area. *(page 404)*

26. 3 yd

27. 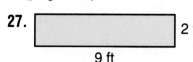 2 ft — 9 ft

28. 4 in. 6 in.

Choose the best estimate of capacity or weight. *(pages 406–409)*

29. bottle of maple syrup   **a.** 1 c   **b.** 10 pt   **c.** 1 qt   **d.** 1 gal

30. glass of milk   **a.** 2 c   **b.** 2 pt   **c.** 2 qt   **d.** 2 gal

31. magazine   **a.** 1 oz   **b.** 10 oz   **c.** 10 lb   **d.** 100 lb

32. big dictionary   **a.** 3 oz   **b.** 30 oz   **c.** 3 lb   **d.** 30 lb

Choose the object closest to the given temperature. *(page 412)*

33. 30°F   **a.** milkshake   **b.** apple pie   **c.** cheese

34. 95°F   **a.** shower   **b.** hot day   **c.** hot oven

Complete. *(page 410)*

35. 3 lb = ■ oz   **36.** 4 yd = ■ ft   **37.** 3 pt = ■ c

38. 10 qt = ■ gal   **39.** 5 ft = ■ in.   **40.** 80 oz = ■ lb

CRITICAL THINKING

41. Give two fractions that are both less than $\frac{1}{4}$ whose sum is greater than 1. *(page 390)*

42. Would Julie wash her car with 4 cups of water? Why or why not? *(page 406)*

MIXED APPLICATIONS

43. The ship has 3 decks. Each deck has 98 rooms. If each room can hold 4 passengers, how many passengers can the ship hold? *(page 402)*

44. Sandy has watched $\frac{1}{4}$ h of a film which is $2\frac{3}{4}$ h long. How much of the film is left for Sandy to watch? Write the answer in simplest form. *(page 400)*

45. The fruit is lined up in a row at the market. The apples are between the oranges and grapes. The bananas are to the right of the grapes. The grapes are to the left of the oranges. In what order is the fruit? *(page 392)*

CHAPTER TEST

Add or subtract. Write the answer in simplest form.

1. $\dfrac{2}{3}$ $+\dfrac{1}{3}$

2. $\dfrac{11}{12}$ $-\dfrac{9}{12}$

3. $\dfrac{3}{5}$ $+\dfrac{1}{5}$

4. $\dfrac{5}{8}$ $-\dfrac{1}{8}$

5. $\dfrac{2}{9} + \dfrac{4}{9}$

6. $\dfrac{8}{9} - \dfrac{7}{9}$

7. $\dfrac{7}{10} + \dfrac{9}{10}$

8. $\dfrac{5}{6} - \dfrac{3}{6}$

Find the perimeter and area.

9. square:
   side—9 in.

10. rectangle:
   length—25 ft
   width—8 ft

11. rectangle:
   length—12 yd
   width—6 yd

Choose the best estimate.

12. sink filled with water   **a.** 3 c   **b.** 3 pt   **c.** 3 qt   **d.** 3 gal

13. coffee in a pot   **a.** 1 c   **b.** 1 pt   **c.** 1 qt   **d.** 1 gal

14. box of tissues   **a.** 2 oz   **b.** 12 oz   **c.** 2 lb   **d.** 12 lb

15. typewriter   **a.** 5 oz   **b.** 15 oz   **c.** 5 lb   **d.** 15 lb

Choose the object closest to the given temperature.

16. 25°F   **a.** snowman   **b.** banana   **c.** yogurt

17. 80°F   **a.** milk   **b.** icicle   **c.** car heater

Solve.

18. Barb has 12 fewer plants than Judy. Judy has 8 more plants than Bob, who has twice as many as Chris. Chris has 14 plants. How many does Barb have?

19. For Arbor Day, the Outdoor Club is going to plant 22 trees in each of the city's 8 parks. How many trees will they plant?

20. Lori worked on her project for $\dfrac{5}{8}$ hour. Then she read for $\dfrac{3}{8}$ hour. For how long did she work and read?

TETROMINOES

Do you know what a tetromino is?
These are examples of tetrominoes.

A.

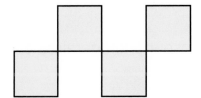

B.

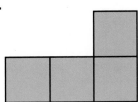

These are not tetrominoes.

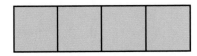

1. How could you describe a tetromino?

Turning a tetromino will not make a new one.
These two tetrominoes are the same.

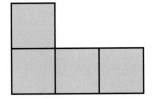

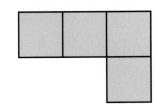

2. Is tetromino B the same as these two? Why or why not?

3. How many tetrominoes are there in all? Draw pictures to show your answer.

Suppose that each square has sides that measure 1 centimeter.

4. What is the perimeter of tetromino A? of tetromino B?

5. Find the perimeter of all the other tetrominoes. How are the squares arranged on the tetromino with the least perimeter?

CUMULATIVE REVIEW

Choose the letter of the correct answer.

1. Estimate the weight of a small frog.

   a. 8 oz   **c.** 8 lb
   b. 18 oz   **d.** 18 lb

2. Add $\frac{4}{5}$ plus $\frac{1}{5}$.

   a. 1   **c.** $1\frac{2}{5}$
   b. $1\frac{1}{5}$ **d.** not given

3. What is 52 divided by 6?

   a. 8 R3   **c.** 8 R5
   b. 8 R4   **d.** not given

4. Which polygon has 5 sides?

   a. pentagon   **c.** octagon
   b. hexagon   **d.** decagon

5. What is $\frac{7}{8} - \frac{5}{8}$ in simplest form?

   a. $\frac{2}{8}$ **c.** $\frac{1}{4}$
   b. $\frac{3}{8}$ **d.** not given

6. Name: $\overset{\bullet\rule[0.5ex]{3cm}{0.4pt}\bullet}{M \qquad\qquad N}$

   a. $\overrightarrow{MN}$ **c.** $\overleftrightarrow{MN}$
   b. $\overline{MN}$ **d.** not given

7. Compare: $4\frac{3}{8}$ ● $\frac{35}{8}$

   a. <   **c.** =
   b. >   **d.** not given

8. What is $\frac{5}{12} + \frac{5}{12}$ in simplest form?

   a. $\frac{10}{12}$ **c.** $1\frac{5}{6}$
   b. $\frac{5}{6}$ **d.** not given

9. What is $\frac{5}{6}$ of 30?

   a. 5   **c.** 15
   b. 25   **d.** not given

10. Choose the temperature of a hot day.

    a. 32°F   **c.** 92°F
    b. 62°F   **d.** 150°F

11. What is $\frac{7}{9} + \frac{5}{9}$?

    a. $\frac{2}{9}$ **c.** $1\frac{2}{9}$
    b. $\frac{1}{3}$ **d.** not given

12. Find the area of a square with 8-ft sides.

    a. 64 sq ft   **c.** 32 sq ft
    b. 32 ft   **d.** not given

13. Find the average: 7, 8, 8, 5.

    a. 4   **c.** 28
    b. 7   **d.** not given

14. 58 ÷ 7

    a. 8   **c.** 10
    b. 8 R2   **d.** not given

CHAPTER 11

Decimals and Probability

MATH CONNECTION: PROBLEM SOLVING

NEW RELEASES

CDs	Records	Cassettes
$13.99	$8.99	$8.99

1. What information do you see in this picture of a record store?

2. How can you use this information?

3. Write a problem about buying some music.

431

Decimals

Alicia sells cartons of solar panels in her construction supply shop. Each carton has 10 panels, and each panel has 10 solar cells. Alicia records the sales in this chart.

Day	Cartons	Panels	Single Cells
Monday	0	3	0
Tuesday	0	2	5
Wednesday	1	4	6
Thursday	1	0	8
Friday	1	1	1

WORKING TOGETHER

1. Think of a carton as the number 1. What fraction of a carton is 1 panel? What fraction of a carton is 1 cell?

2. Use place-value models. Think of a hundreds square as the model for 1 carton. Model the part of a carton sold on Monday. What fraction names this part?

You can also show three tenths as a **decimal**.

3. Model the part of a carton sold on Tuesday. What fraction names the part?

Ones		Tenths
0	.	3

You can also show twenty-five hundredths as a decimal.

Ones		Tenths	Hundredths
0	.	2	5

4. Model the number of cartons sold on Wednesday. What mixed number names the amount? Complete the table to show the decimal for the amount

Ones		Tenths	Hundredths
■	.	■	■

5. Model the number of cartons sold on Thursday. Write the mixed number for the amount. Write the decimal. Repeat for Friday.

SHARING IDEAS

6. How do decimal place values follow the same pattern as whole number place values?

7. What does a zero mean when it is written in a decimal?

ON YOUR OWN

8. You can use decimals with metric measures. Look at a meter stick. Think of the meter as the number 1. What part of a meter is 1 decimeter? What part of a meter is 1 centimeter?

9. Write the measure as a decimal. Show it as a number of meters.

a. 1 m 3 dm 5 cm **b.** 3 m 2 cm **c.** 6 m 8 dm
d. 4 dm **e.** 7 cm **f.** 15 dm

10. Money amounts are shown as decimals. What part of 1 dollar is 1 dime? What part of 1 dollar is 1 penny?

a. Put some dollar bills, dimes, and pennies in a bag. Take out some coins. Record the amount in a table like this. Repeat this 2 more times.

b. Take out some bills and coins. Record the amount. Repeat this 2 more times.

	Dollars	Dimes	Pennies
$	■	■	■
$	■	■	■
$	■	■	■
$	■	■	■
$	■	■	■
$	■	■	■

11. Look in almanacs for decimal numbers. Write about how the decimals are used.

DEVELOPING A CONCEPT

Tenths and Hundredths

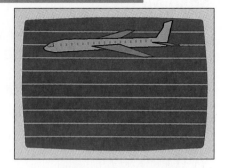

A. Devon used three tenths of his computer screen to draw a picture. How can you show and write the number three tenths?

WORKING TOGETHER

1. Use centimeter graph paper. Cut out a 10-cm by 10-cm square to model the whole screen. Draw lines to make 10 equal parts. What fraction of the whole is each part?

2. Color to show three tenths.

You can write three tenths as a fraction or as a decimal.

fraction → $\frac{3}{10}$ **0.3** ← decimal

Ones	Tenths
0 .	3

Read: three tenths

3. ***What if*** Devon used 35 hundredths of his screen? Make a model to show this.

You can write 35 hundredths as a fraction or as a decimal.

fraction → $\frac{35}{100}$ **0.35** ← decimal

Ones	Tenths	Hundredths
0 .	3	5

Read: thirty-five hundredths

B. You can write fractions greater than 1 as a mixed number or as a decimal.

Make a model of each of these numbers.

4. $1\frac{6}{10}$ 1.6   one and six tenths

5. $1\frac{15}{100}$ 1.15   one and fifteen hundredths

C. You can write equivalent decimals for the same number.

6. Make a model of each of these numbers: $1\frac{4}{10}$ and $1\frac{40}{100}$. Write the decimal for each number. What do you notice?

7. Model the number 2. Complete this sentence: 2 = $\frac{\blacksquare}{10}$ = $\frac{\blacksquare}{100}$ Show two decimals for the number 2.

SHARING IDEAS

8. How are fractions and decimals the same? How are they different?

9. Are the decimals 1.7 and 1.07 equivalent? Why or why not?

10. Are the decimals 0.4 and 0.40 equivalent? Why or why not?

PRACTICE

Write the decimal for the part that is shaded.

11. **12.** **13.** **14.**

Write the decimal.

15. $\frac{5}{10}$ **16.** $\frac{64}{100}$ **17.** $\frac{7}{100}$ **18.** $\frac{45}{100}$ **19.** $\frac{50}{100}$ **20.** $\frac{25}{100}$

21. $\frac{16}{10}$ **22.** $8\frac{2}{100}$ **23.** $15\frac{7}{10}$ **24.** $25\frac{25}{100}$ **25.** $100\frac{1}{10}$ **26.** $33\frac{3}{100}$

27. thirteen hundredths   **28.** two and four tenths

29. one tenth   **30.** sixteen and three hundredths

Write the word name.

31. 0.5   **32.** 0.09   **33.** 13.5   **34.** 0.60   **35.** 8.03

Tell whether the decimals are equivalent. Write *yes* or *no*.

36. 2.3 and 2.03   **37.** 2.30 and 2.3   **38.** 0.20 and 2.00

39. 15.40 and 15.04   **40.** 15.00 and 15.0   **41.** 15.4 and 15.40

Mixed Applications

42. Devon prints a report that uses 3 pages and 4 tenths of a page. What decimal shows how many pages he uses?

43. Joni wants to buy software that costs $79.95. She has saved $55.00. How much more money does she need?

44. Kay needs 100 copies of a school newsletter. She prints 45 copies. What decimal shows the part that she has printed?

UNDERSTANDING A CONCEPT

Comparing and Ordering Decimals

A. Carla uses a pedometer to measure how far she runs. Carla ran 1.8 km this morning and 1.6 km this afternoon. When did she run the greater distance?

You can use a number line to compare 1.8 and 1.6.

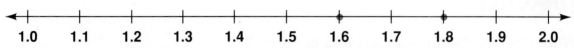

Think: 1.6 comes **before** 1.8.
1.6 is **less than** 1.8.

Write: $1.6 < 1.8$

Think: 1.8 comes **after** 1.6.
1.8 is **greater than** 1.6.

Write: $1.8 > 1.6$

Carla ran the greater distance this morning.

B. You can also compare decimals digit by digit.

Step 1	Step 2	Step 3
Line up the decimal points.	Begin at the left. Compare to find the first place where the digits are different.	Compare the digits.
1 . 8 1 . 6	1 . 8 1 . 6	**8 > 6 or 6 < 8** ***Think:*** $0.8 > 0.6; 0.6 < 0.8$

So $0.8 > 0.6$ or $0.6 < 0.8$

1. How is comparing decimals like comparing whole numbers?

C. You can compare 0.73, 2.78, and 2.76 to order them.

Step 1	Step 2	Step 3
Line up the decimal points.	Compare the other decimals.	Order the decimals.
0.73 2.78 2.76 The number without ones is the least.	2.78 2.76 ***Think:*** $0.06 < 0.08$ So $2.76 < 2.78$.	From least to greatest: 0.73, 2.76, 2.78 From greatest to least: 2.78, 2.76, 0.73

TRY OUT

Compare. Use >, <, or =.

2. 0.4 ● 0.40   **3.** 2.34 ● 2.23

Write in order from greatest to least.

4. 8.81, 18.18, 18.8   **5.** 1.4, 1.1, 1.7

PRACTICE

Compare. Write >, <, or =.

6. 0.6 ● 0.60   **7.** 2.4 ● 2.5   **8.** 0.51 ● 0.52   **9.** 2.6 ● 1.8

10. 1.7 ● 1.9   **11.** 0.75 ● 0.74   **12.** 4.26 ● 4.62   **13.** 3.52 ● 3.25

14. 6.78 ● 6.87   **15.** 9.27 ● 9.27   **16.** 10.10 ● 10.01   **17.** 9.09 ● 9.90

Write in order from greatest to least.

18. 0.25, 0.52, 0.50   **19.** 1.3, 1.4, 1.1   **20.** 0.88, 0.08, 0.80

Write in order from least to greatest.

21. 1.1, 0.7, 1.6   **22.** 3.61, 3.15, 3.09   **23.** 0.99, 1.09, 0.09

Mixed Applications

Solve. You may need to use the Databank on page 522.

24. What was the fastest time in the men's 100-meter freestyle? In what year was the race?

25. In which year was the women's 100-meter freestyle completed in a shorter time, 1980 or 1988?

CALCULATOR

You can compare fractions by changing them to decimals.

$\frac{3}{4}$ = ⬚3⬚ ⬚÷⬚ ⬚4⬚ ⬚=⬚ 0.75 $\frac{4}{5}$ = ⬚4⬚ ⬚÷⬚ ⬚5⬚ ⬚=⬚ 0.8

0.75 < 0.8, so $\frac{3}{4} < \frac{4}{5}$

Compare. Write >, <, or =. Use a calculator to change fractions to decimals.

1. $\frac{2}{5}$ ● $\frac{12}{25}$ **2.** $\frac{16}{20}$ ● $\frac{4}{5}$ **3.** $\frac{18}{50}$ ● $\frac{7}{20}$ **4.** $\frac{15}{20}$ ● $\frac{19}{25}$

PROBLEM SOLVING

Strategy: Making an Organized List

Eve is playing the electronic game Colorumber. She must choose one color and one number. In how many different ways can she choose a color and a number?

Eve plans to use a tree diagram to help her make an organized list of the color-number pairs.

If Eve chooses the color red, she can make two color-number pairs: red-1 and red-2.

1. If Eve chooses yellow, what color-number pairs can she make?

2. Make a tree diagram for the third color. What color-number pairs can she make?

3. In how many different ways can she choose a color and a number?

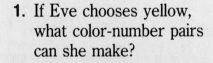

Tree Diagram	List
red $<$ 1 / 2	____ red-1 / ____ red-2
yellow $<$ 1 / 2	____ ■ / ____ ■
green $<$ ■ / ■	____ ■ / ____ ■

4. ***What if*** the game had a third number to choose from? In how many different ways could Eve choose a color and a number? Make a tree diagram and an organized list of all color-number combinations.

5. When is it useful to make a tree diagram and an organized list to solve a problem?

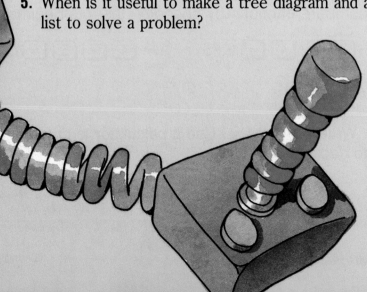

PRACTICE

Use a tree diagram to make an organized list. Then solve the problem.

6. When Lola works at Electro City, she wears a skirt and a blouse. She has 3 skirts and 4 blouses. How many different outfits can she wear?

7. Rob plans to watch TV at 8:00 and listen to music at 9:00. He will watch channel 2, 5, or 13 and then listen to a symphony or a concerto. How many different choices does Rob have?

8. At the Computer Corner each computer desk comes in a choice of oak or cherry wood. The desk can come with a shelf that pulls out or with one that does not move. How many kinds of desks does the store have?

9. Maria is ordering pizza for the people at the video store. She can order thin crust or thick crust and choose one topping from mushroom, pepper, zucchini, or tuna. How many different kinds of pizza can Maria order?

Strategies and Skills Review

Solve. Use mental math, a calculator, or paper and pencil.

10. Greg is setting the timer on his VCR to go on at 2:55 P.M. What other information do you need in order to find how long the tape will be running?

11. Beth is going to watch one comedy and then one news show. There are comedies on channels 2, 5, and 9. There are news shows on channels 2, 5, 7, and 9. In how many different ways can Beth choose which shows to watch?

12. Di's calculator has 4 rows of keys with 6 keys in each row. There are 10 digit keys. How many keys are not digit keys?

13. *Write a problem* that can be solved by making an organized list. Solve the problem. Ask others to solve your problem.

Adding and Subtracting Decimals

You can use place-value models to explore adding and subtracting decimals.

WORKING TOGETHER

Make two sets of cards with the digits 0 to 9. Shuffle the cards and place them facedown in a pile.

Step 1 Pick four cards. Use the digits to make an addition sentence, such as 0.7 + 2.3.

$$\boxed{0}.\boxed{7} + \boxed{2}.\boxed{3}$$

Step 2 Use place-value models to make a model of each decimal.

Step 3 Find the sum and record it. Repeat this activity three more times.

0.7	+	2.3	=	■

Step 4 Now pick six cards and make a sentence, such as 1.79 + 2.63. Use your place-value models to find the sum. Record it. Repeat this activity three more times.

1. Use your models to find the sum.

   a. 2.5 + 1.8   **b.** 1.8 + 2.5   **c.** 3.45 + 0

   d. 1.3 + 2.5 + 0.9   **e.** 2.5 + 0.9 + 1.3

2. Repeat Steps 1 to 4 for subtraction sentences, such as 3.5 − 1.7. Record your results.

$$\boxed{3}.\boxed{5} - \boxed{1}.\boxed{7}$$

3.5	−	1.7	=	■

3. Use your models to find the difference.

   a. 3.7 − 3.7   **b.** 1.25 − 0

SHARING IDEAS

4. How is regrouping in addition of decimals like regrouping in addition of whole numbers?

5. How is regrouping in subtraction of decimals like regrouping in subtraction of whole numbers?

6. Do you think the properties of addition and subtraction apply to decimals? Give examples.

7. Do you think the relationship between addition and subtraction applies to decimals? Give an example.

ON YOUR OWN

Solve. Use place-value models if needed.

8. Don's TV set has a mass of 42.81 kg. His VCR has a mass of 21.67 kg. What is their total mass?

9. Jan's computer screen measures 34.5 cm long. Her TV screen measures 54.3 cm long. How much longer is her TV screen?

10. Sal ordered videotapes priced at $15.95, $26.99, and $29.89. How much will the tapes cost in all?

11. *Write a problem* involving addition or subtraction of decimals. Solve your problem. Ask others to solve it.

UNDERSTANDING A CONCEPT

Estimating Decimal Sums and Differences

A. Dionne rode the bicycle her father invented 4.57 km from home to the library, 6.69 km from the library to the park, and 2.8 km from the park to home. About how far did she ride all together?

Here are two ways to estimate to solve the problem.

Estimate: 4.57 + 6.69 + 2.8

Rounding	Front-End Estimation	
Round to the nearest whole number.	Add the front digits.	Then adjust the estimate.

Rounding

4.57 ⟶	5
6.69 ⟶	7
+ 2.8 ⟶	+ 3
	15

Front-End Estimation

Add the front digits.

4.57
6.69
+ 2.8
12

Then adjust the estimate.

4.57 ⎫	*Think:*
6.69 ⎬	about 1
+ 2.8 ⎬	about 1
14	

Dionne rode about 14 or 15 km.

1. How will the exact answer compare with the estimate found by rounding?

2. How will the exact answer compare with the front-end estimate?

B. You can use two methods to estimate differences.

Rounding	Front-End Estimation	
Round to the nearest whole number.	Subtract the front digits.	Then adjust the estimate.

Rounding

7.64 ⟶	8
− 3.7 ⟶	− 4
	4

Front-End Estimation

Subtract the front digits.

7.64
− 3.7
4

Then adjust the estimate.

7.64	*Think:* 7 > 6
− 3.7	The exact answer
< 4	is less than 4.

3. Will the exact answer be greater than or less than the estimate found by rounding? How can you tell?

4. How would you estimate 5.4 − 3.28 using front-end estimation with adjusting? What is your estimate?

TRY OUT

Estimate by rounding to the nearest whole number.

5. 3.4 + 7.58 + 2.91   **6.** 4.39 − 2.2

Estimate. Use the front digits and adjust.

7. 5.67 + 1.35 + 4.43   **8.** 5.83 − 2.9

PRACTICE

Estimate by rounding to the nearest whole number.

9.	3.7	**10.**	6.35	**11.**	3.37	**12.**	6.19	**13.**	3.75
	4.3		− 1.92		6.49		− 2.9		4.13
	+ 2.5				+ 1.8				+ 6.7

14. 5.39 − 2.85   **15.** 8.4 − 3.97   **16.** 6.52 − 3.7

17. 3.7 + 2.14 + 7.3   **18.** 6.2 + 1.35 + 4.9   **19.** 3 − 2.86

Estimate. Use the front digits and adjust.

20.	3.45	**21.**	3.56	**22.**	8.16	**23.**	3.37	**24.**	6.59
	1.98		− 1.5		2.3		2.67		3.6
	+ 4.2				+ 1.27		+ 4		+ 2.98

25. 7.74 − 2.47   **26.** 6.4 − 2.85   **27.** 8.13 − 4.2

28. 3.17 + 3.9 + 4.1   **29.** 4.75 − 1.5   **30.** 2.9 + 3.01 + 5.16

Mixed Applications

31. Ali rides his bike 2.8 km each way to and from work. About how far does he ride in all?

32. Blank tapes cost $5.29 each. Sandy bought 4 of them. About how much did all of them cost?

Mixed Review

Find the answer. Which method did you use?

33.	$35.87	**34.**	390	**35.** 7)490	**36.**	354
	+ 76.95		− 240			× 8

MENTAL MATH
CALCULATOR
PAPER/PENCIL

Adding Decimals

A. Tara wants to buy a desk for her computer. The desk must be as wide as the combined width of the tables she uses now. Tara's computer table is 1.5 m wide. Her printer table is 0.65 m wide. What is the combined width of the two tables?

Estimate first. Then find the exact answer.

Tara used place-value models to find the sum.

Add: 1.5 + 0.65

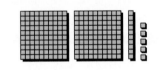

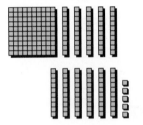

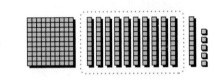

1. What is the combined width of the tables?

2. How did Tara combine the models to find the sum?

3. When did you have to regroup? Why?

B. You can add using paper and pencil.

Step 1	Step 2	Step 3
Line up the decimal points. Write an equivalent decimal if necessary.	Add the hundredths. Add the tenths. Regroup if necessary.	Add the ones. Regroup if necessary. Write the decimal point.
1.50 *Think:* 1.5 = 1.50 **+ 0.65**	¹ **1.50** **+ 0.65** ———— 1 5	¹ **1.50** **+ 0.65** ———— **2.15**

4. Is the answer reasonable? How do you know?

SHARING IDEAS

5. How is using models to find the sum the same as using paper and pencil? How is it different?

6. How is adding decimals similar to adding whole numbers? How is it different?

PRACTICE

Find the sum.

7. 6.4
    + 3.2

8. 3.45
    + 2.31

9. 5.17
    + 2.81

10. 5.4
    + 3.7

11. 6.23
    + 1.82

12. 3.4
    + 2.45

13. 7.16
    + 3.8

14. 1.43
    + 2.75

15. 3.9
    + 1.25

16. 6.27
    + 2.8

17. 3.1
     4.7
    + 2.4

18. 3.25
     1.6
    + 4.81

19. 1.36
     2.05
    + 0.92

20. 8.5
     1.36
    + 3.54

21. 2.9
     3.2
    + 5.88

22. 5.3 + 2.83 **23.** 7.50 + 2.25 **24.** 6.8 + 4.29 **25.** 3.14 + 2.99 + 0.7

26. 6.33 + 3.9 **27.** 8.9 + 4.07 **28.** 7.85 + 4.11 **29.** 2.73 + 4.3 + 5.4

Critical Thinking

30. Tara writes 20 + 4 + 0.7 + 0.08. What decimal is this? How do you know?

Mixed Applications Solve. Which method did you use?

31. Tara made a chart by taping together 2 sheet of paper. Each sheet was 21.6 cm wide. How wide was her chart?

32. Tara bought office supplies for $8.45. She gave the clerk $10.00. What was Tara's change?

33. Tara's study is 3.8 m long and 2.6 m wide. Is the perimeter of the room more than 10 m?

34. *Write a problem* that can be solved by adding decimals. Give it to another student to solve.

ESTIMATION
MENTAL MATH
CALCULATOR
PAPER/PENCIL

DEVELOPING A CONCEPT

Subtracting Decimals

A. The swim team coach uses a stopwatch to time Zach, Ty, and Angel in a race. Zach is the winner. He is 0.45 seconds faster than Angel and 1.6 seconds faster than Ty. How much faster than Ty is Angel?

Estimate first. Then find the exact answer.

Angel used place-value models to find the difference.

Subtract: 1.6 − 0.45

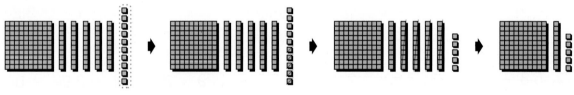

1. How much faster than Ty is Angel?

2. How does Angel use the models to find the difference? Does he regroup? Why?

B. You can use paper and pencil to subtract.

Step 1	Step 2	Step 3	Step 4
Line up the decimal points. Write an equivalent decimal if necessary.	Regroup if necessary. Subtract the hundredths.	Regroup if necessary. Subtract the tenths.	Subtract the ones. Write the decimal point.
$\begin{array}{r} 1.6\mathbf{0} \\ -\ 0.45 \\ \hline \end{array}$	$\begin{array}{r} {}^{5\,10} \\ 1\ .\ 6\ \mathbf{0} \\ -\ 0\ .\ 4\ 5 \\ \hline 5 \end{array}$	$\begin{array}{r} {}^{5\,10} \\ 1\ .\ 6\ 0 \\ -\ 0\ .\ 4\ 5 \\ \hline 1\ 5 \end{array}$	$\begin{array}{r} {}^{5\,10} \\ 1\ .\ 6\ 0 \\ -\ 0\ .\ 4\ 5 \\ \hline 1\ .\ 1\ 5 \end{array}$

Think: 1.6 = 1.60

3. Is the answer reasonable? How do you know?

4. How would you subtract 9.68 − 2.9? What is the difference?

SHARING IDEAS

5. How is using models to find the difference the same as using paper and pencil? How is it different?

6. How is subtracting decimals similar to subtracting whole numbers?

PRACTICE

Find the difference.

7. 0.9
   − 0.3

8. 6.7
   − 2.4

9. 8
   − 4.7

10. 7.1
   − 2.1

11. 5.3
   − 4.6

12. 6.75
   − 1.29

13. 7.13
   − 2.11

14. 8.3
   − 1.45

15. 9.11
   − 7.04

16. 3.06
   − 2.78

17. 7.8 − 5.4

18. 3.7 − 1.9

19. 5.75 − 1.63

20. 8.03 − 7.47

21. 7.04 − 6.3

22. 6.2 − 5.43

23. 7 − 2.51

24. 3 − 1.09

Mixed Applications

Solve. You may need to use the Databank on page 522.

25. In June, Roma could swim to the float in 45.3 seconds. By the end of August it took her 26.8 seconds. How much faster was her time in August?

26. How much faster was the women's winning time in the 100-meter freestyle race in 1984 than in 1988?

27. Write a problem comparing two winning times in the freestyle races. Have others solve your problem.

Mixed Review

Find the answer. Which method did you use?

28. 340
   + 510

29. $10.06
   − 9.68

30. 105
   × 3

31. 8)702

MENTAL MATH
CALCULATOR
PAPER/PENCIL

Logical Reasoning

A. Here is a puzzle for you to solve. First, get three slips of paper. Label them A, B, and C. Then make a stack of coins that has a quarter on the bottom, a nickel in the middle, and a penny on top. Place the stack of coins on paper slip A. Now try to move the stack to C. Follow these rules:

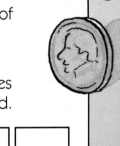

- You can move only one coin at a time.
- You can place coins on slip B.
- You cannot place a larger coin on a smaller coin.

1. Try the puzzle a few times. Write down the total number of moves you use each time. What was your least number of moves?

The shortest solution for this puzzle is 7 moves. These pictures show how it is done. Notice the way the moves are recorded.

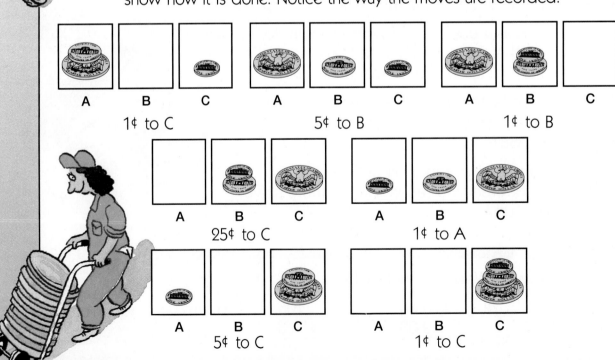

1¢ to C 　　 5¢ to B 　　 1¢ to B

25¢ to C 　　 1¢ to A

5¢ to C 　　 1¢ to C

2. Now try the same puzzle with 4 coins. Add a dime to the top of the beginning stack on slip A; the dime now becomes the smallest coin in the stack. How many moves will it take? Record your moves in the manner shown above.

B. When you play with 4 coins, the shortest solution takes 15 moves. If you needed more, try the puzzle again. This table shows the number of moves for the shortest solutions depending on how many coins are used.

Number of Coins	Number of Moves (Shortest Solution)	Pattern
3	7	$(2 \times 2 \times 2) - 1$
4	15	$(2 \times 2 \times 2 \times 2) - 1$
5		
6		

This table shows how the number of coins is related to the number of moves in the shortest solution. Can you find the pattern?

3. Look at the 3-coin puzzle in the table. How many times is 2 used as a factor in the last column? Now look at the 4-coin puzzle. How many times is 2 used as a factor in the last column?

4. What pattern do you see? What do you think the shortest solutions for 5 and 6 coins would be? Copy and complete the chart, following the pattern. Use a calculator to help you.

PROBLEM SOLVING

Strategy: Conducting an Experiment

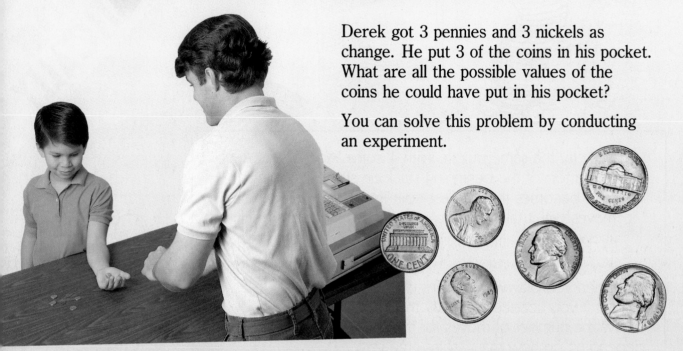

Derek got 3 pennies and 3 nickels as change. He put 3 of the coins in his pocket. What are all the possible values of the coins he could have put in his pocket?

You can solve this problem by conducting an experiment.

Plan your experiment.

Use red cubes to model pennies and blue cubes to model nickels. You can put 3 cubes of each color into a bag and draw 3 cubes. Record the total value of the cubes. Then put the cubes back into the bag. Repeat the experiment until you have drawn all possible combinations of coins.

Try your plan.

Put the cubes into the bag, draw 3 cubes, and record the results. For example, if you had drawn 2 red cubes and 1 blue cube, you would record the result like this:

$$1¢ + 1¢ + 5¢ = 7¢$$

1. What color are the cubes on your first draw? Record the results as pennies and nickels.

2. Repeat the experiment several times. What are all the possible values of the coins Derek could put in his pocket?

3. What other strategy could you have used to solve the problem?

PRACTICE

Do an experiment to solve the problem.

4. Louise has 1 nickel, 1 dime, and 1 quarter. She puts 2 coins in the jukebox. What are all the possible values of the coins she used?

5. Ollie has a record, a CD, and a cassette in a bag. He pulls 2 items out of the bag. How many different pairs of items could he have pulled out?

6. Rich just bought 4 CDs. One CD is dance music, 1 CD is rock, another is classical, and the last is folk. He shows Zelda 3 of the CDs. How many different sets of CDs could he show her?

7. Lee won the "guess the song" game. She can choose 1 CD from each of 2 groups. Each group has 3 different CDs. How many different pairs of CDs can she choose?

Strategies and Skills Review

Solve. Use mental math, a calculator, or paper and pencil.

8. Janet had $12.70. She spent $6.98 on a tape. Then her father gave her $9.00 for her birthday. How much does Janet have now?

9. Ned has 2 dimes and 3 pennies in his pocket. He pulls out 3 coins. What are all the possible values of these coins?

10. Zeke has 23 albums to put in boxes. Each box holds 6 albums. How many boxes will Zeke need?

11. **Write a problem** that can be solved by doing an experiment. Ask others to solve it.

DEVELOPING A CONCEPT

Probability

If you were to spin this spinner, what is the likelihood it would stop on blue? **Probability** is the likelihood that something will happen.

WORKING TOGETHER

You can use the spinner to do an experiment to explore probability.

1. How many colors are on the spinner? These are the **possible outcomes.**

2. How many parts of the spinner are blue? This is the **favorable outcome.**

3. Complete: The probability of the spinner stopping on blue is ■ chance out of ■.

4. Is each possible outcome equally likely? Why?

You can write the probability as a fraction.

$$\text{Probability} = \frac{\text{favorable outcomes}}{\text{possible outcomes}}$$

Write the probability as a fraction.

5. spinning blue   **6.** spinning red   **7.** spinning green   **8.** spinning yellow

9. Make a chart like the one to the right. Spin the spinner 20 times. Use tally marks to record how many times it stops on each color.

Red	
Green	
Blue	
Yellow	

10. How many colors are on the spinner to the right? What part of the spinner is each color? Are all of the outcomes equally likely? Why or why not?

11. What is the probability of spinning blue? spinning green? spinning red?

12. Spin the spinner 20 times. Use tally marks to record how many times it stops on each color.

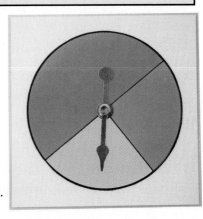

13. Compare the outcomes from both experiments. Are the results close to what you expected?

14. How do you know whether the outcomes of a probability experiment are equally likely?

15. If you spin the spinner you used in the first experiment, what is the probability of it stopping on one of the four colors? Why?

16. What is the probability of spinning orange? Why?

PRACTICE

Use a number cube numbered from 1 to 6. Find the probability of tossing:

17. 1.   **18.** 5.   **19.** 8.   **20.** 1 to 6.

21. Are the possible outcomes equally likely?

Find the probability of picking the color.

22. yellow   **23.** red   **24.** blue

25. Are the outcomes equally likely?

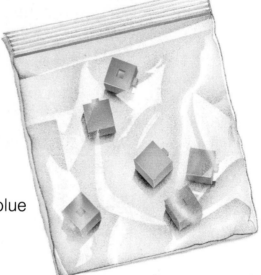

Critical Thinking

26. Ken spins the 4-color spinner. What is the probability of the spinner landing on blue or green? How do you know?

Mixed Applications

27. Lex has a penny, a nickel, and 3 dimes in his pocket. Which type of coin is he most likely to pull out of his pocket?

28. There are 2 white shirts and 1 red shirt in Ken's closet. What fraction of the shirts are white?

LOGICAL REASONING

Here are 2 views of the same number cube. What is the number on the bottom? How do you know?

DEVELOPING A CONCEPT
Prediction

Yaseff puts 3 red cubes, 2 blue cubes, and 1 yellow cube in a bag. He picks a cube from the bag, then returns it. If he does this 24 times, how many times will he pick a red cube? a blue cube? a yellow cube?

You can use probability to predict how many times something will happen.

1. What is the probability of his picking a red cube? a blue cube? a yellow cube?

2. How many times do you think he will pick each color?

WORKING TOGETHER

You can do an experiment to test your prediction.

Step 1 Put 3 red cubes, 2 blue cubes, and 1 yellow cube in a bag.

Step 2 Without looking, pick a cube from the bag. Record the outcome in a tally chart.

Red	
Blue	
Yellow	

Step 3 Put the cube back in the bag. Shake it, then pick again. Do this 24 times.

3. How many times did you pick a red cube? a blue cube? a yellow cube?

4. How do the outcomes compare to your predictions?

5. Suppose someone puts 6 cubes in a bag. You know that there are 3 of one color, 2 of another color, and 1 of a third color. This chart shows the outcomes of 24 picks. How can you predict what is in the bag without looking? Try your method.

Red	///
Blue	⅚ℋ ⅚ℋ //
Yellow	⅚ℋ ////

6. What is $\frac{1}{2}$ of 24? $\frac{1}{3}$ of 24? $\frac{1}{6}$ of 24? How do these numbers compare with the results of your first experiment?

7. Tell how you can use probability to predict outcomes.

PRACTICE

Look at the bag of pattern blocks to the right. Find the probability of picking the shape. Predict how many times you will pick it in 40 tries.

8. a hexagon   **9.** a square

10. a triangle   **11.** a rectangle

Find the probability of spinning the number. Predict how many times you will spin it in 30 tries.

12. 2   **13.** 8

14. one number 0 to 9

15. Marc puts 8 cubes in a bag. He puts in blue, red, and green cubes. There are 4 of one color, 3 of another color, and 1 of the third color. He picks cubes from the bag 40 times and graphs the outcomes. How many of each color are in the bag?

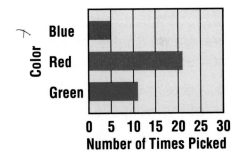

Mixed Review

Find the missing number.

16. $27 \div \blacksquare = 9$

17. A square's area is 16 sq. ft. Length of 1 side = $\blacksquare$

18. $\blacksquare + 237 = 638$

19. $\frac{3}{4} - \blacksquare = \frac{1}{2}$

20. $9\overline{)1{,}798}$

21. $2 \times \blacksquare = 860$

22. $3 \times \$1.27 = \blacksquare$

23. $312 + (192 - 44) = (192 + \blacksquare) - 44$

24. $3 + 1\frac{2}{3} = \blacksquare$

25. $\frac{4}{\blacksquare} = \frac{16}{36}$

Independent Events

Peggy has 2 bags. She puts 1 red cube and 1 blue cube in each bag. What is the probability of Peggy picking a blue cube from each bag?

The color of the cube Peggy picks from the second bag does not depend on the color of the cube she picks from the first bag. The two events are **independent** of each other.

WORKING TOGETHER

You can do an experiment to find the probability.

1st Bag	2nd Bag	Tally
Red	Blue	/

Step 1 Take 2 bags. Place 1 red cube and 1 blue cube in each bag.

Step 2 Pick 1 cube from each bag. Record each outcome in a table to make an organized list. Use tally marks to record how many times each outcome occurs.

Step 3 Place the cubes back in their bags. Shake the bags and pick again. Do this 40 times.

1. Look at your table. What are all of the possible outcomes?

2. What is the probability of picking a blue cube from both bags? How do you know?

3. How many times did you pick 2 blue cubes? Is this close to what you would have predicted based on the probability?

4. What are the probabilities of the other outcomes? How many times did each occur?

SHARING IDEAS

5. Look at your organized list of possible outcomes. Can you think of another way of finding what the possible outcomes are?

6. Are the possible outcomes of the experiment equally likely? How do you know?

ON YOUR OWN

7. What if you placed a yellow cube in each bag with the red and blue cubes? What are the possible outcomes if you still pick only 1 cube from each bag? What is the probability of each outcome?

8. Predict how many times 1 of the outcomes will occur if you pick 36 times. Then do the experiment to test your prediction.

9. *Make up* your own probability experiment with independent events. Find all of the possible outcomes and their probabilities. Predict how many times 1 of the outcomes will occur if you pick 40 times. Then do the experiment to test your prediction.

DECISION MAKING

Problem Solving: Choosing the Right Gift

SITUATION

Jamal has $20.00 to buy a birthday present for his sister, Miriam. Jamal goes to Murray's Musicland to buy the present.

PROBLEM

Should Jamal buy a record, a cassette, or a CD for Miriam?

DATA

CONSUMER NEWS

Records, Tapes, CD's–What's the Difference?

RECORDS
- Sound quality varies with sound system
- Easy to scratch or break

- Harder to clean
- Wear out more quickly than tapes
- Must turn over to hear whole record
- Cannot be played on a portable player
- Take up a lot of storage space

TAPES
- Sound quality varies with sound system
- Can be erased by contact with magnet
- Do not need cleaning
- Will wear out with use

- Must turn over to hear whole tape
- Can be played on a portable player
- Easy to store

CD'S
- Highest sound quality
- Very hard to damage
- Easy to clean Do not wear out

- No need to turn over to hear whole CD
- Can be played on a portable player
- Easiest to store

Miriam's favorite artists	CD (prices for new releases)	Record	Cassette
Jam Master Z	$13.99	$8.99	$8.99
Rap King Rapp	13.99	8.99	8.99
Winona	13.99	8.99	8.99
Johnny Morgan's X-rays	13.99	8.99	8.99
Record, cassette, or CD? Old or new release?			

USING THE DATA

1. How much would two CD's cost?

2. How much would two records or two cassettes cost?

3. How much would three records or cassettes cost?

4. How much would one CD and one record or cassette cost?

MAKING DECISIONS

5. Cassettes and records are less expensive than CD's. Will Jamal really get more for his money buying the cheaper item? What other factors should he consider in making his decision?

6. Miriam has Winona's new release, but not her first album. The first album is on sale for $5.99. How might that affect Jamal's decision?

7. **What if** Miriam has a cassette deck and turntable in her room, but the CD player is in the living room? How might it affect Jamal's decision?

8. There is a poster inside Rap King Rapp's new record, but not inside the cassette or CD. The CD has two songs that are not on the record or cassette. Which should Jamal buy?

9. **What if** Jamal knows that his parents are buying Miriam a portable CD player for her birthday? How would that help him to make up his mind?

10. Which format do you think Jamal should buy? Why?

11. If you were buying music for yourself, would you buy a record, a tape, or a CD? Why?

Math and Science

Electricity is the power that runs most household appliances. The amount of electricity that an appliance uses is measured in units called **kilowatt-hours.** A kilowatt-hour is the amount of electric energy needed to light ten 100-watt light bulbs for one hour.

Different appliances need different amounts of electricity to run properly. The table shows some common appliances and the amount of electricity they use in one hour.

Appliance	Kilowatt Usage
Stereo	0.1
Washing machine	0.25
Computer	0.08
Television	0.2
VCR	0.07
Dishwasher	1.0

What if you ran a stereo and computer at the same time? Would you use more electricity than if you used a television and VCR at the same time?

Think: Stereo and computer–0.1 + 0.08 = 0.18
Television and VCR–0.07 + 0.2 = 0.27
0.18 < 0.27

The stereo and computer would use less electricity than the television and VCR.

ACTIVITIES

1. Find out about one of the scientists who contributed to the development of electricity. Share your information with the class.

2. Many forms of energy use up limited natural resources. List some ways of saving energy, such as turning off lights when they are not in use.

Computer Simulations: Probability

If you toss a coin once, what is the probability of it landing heads up? tails up? What if you toss the coin 50 times? Predict how many times it will come up heads and tails. You can test your prediction by actually tossing a coin 50 times. You can also use a computer to **simulate,** or re-create, the coin tossing for you.

FREQUENCY TABLE

Number of Tosses	Heads	Tails
50	31	19

DATA

Toss a coin 50 times and record the results.

THINKING ABOUT COMPUTERS

1. How many times did your coin land heads up? tails up? How does this compare with the results from the computer program? Is this close to your prediction?

2. Your outcome is more likely to be close to your prediction if you toss the coin many times. What if the computer "tosses" the coin 500 times with these results: heads, 242 times; tails, 258 times? How do these numbers compare with what you would have predicted?

3. How many times would you toss the coin if you wanted an even closer result? Why would you use a computer?

EXTRA PRACTICE

Tenths and Hundredths, page 435...

Write the decimal for the part that is shaded.

1.   **2.**   **3.**   **4.**

5. 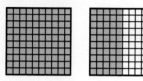   **6.**

Write the decimal.

7. $\frac{3}{10}$ **8.** $\frac{24}{100}$ **9.** $\frac{15}{100}$ **10.** $\frac{4}{100}$ **11.** $\frac{2}{100}$

12. $1\frac{5}{10}$ **13.** $12\frac{7}{10}$ **14.** $4\frac{6}{100}$ **15.** $11\frac{78}{100}$ **16.** $7\frac{3}{10}$

17. ten and four tenths   **18.** nine tenths

19. fifteen and six hundredths   **20.** two and eleven hundredths

Write the word name.

21. 0.5   **22.** 1.3   **23.** 0.32   **24.** 13.5   **25.** 0.09

Comparing and Ordering Decimals, page 437......................................

Compare. Use >, <, or =.

1. 4.6 ● 4.2   **2.** 3.9 ● 5.1   **3.** 6.5 ● 5.6

4. 2.30 ● 2.3   **5.** 0.43 ● 0.48   **6.** 7.26 ● 7.62

7. ten and four tenths ● thirty-three hundredths

8. five and seven tenths ● six and seven tenths

9. thirty-two hundredths ● three and nine tenths

10. thirty-nine and seven tenths ● ninety-four and two tenths

Problem Solving Strategy: Making an Organized List, page 439...............

Make an organized list. Then solve the problem.

1. Gail has 4 sweaters and 3 skirts. How many different outfits can she wear using these sweaters and skirts?

2. Kevin wants to order a plate of pasta with sauce for dinner. He can choose from 2 types of pasta and 3 types of sauce. How many different dinners can Kevin order?

3. Aisha makes ceramic bowls in 4 shapes. She can coat them with 2 different glazes. How many different bowls can Aisha make?

4. Lars buys a red balloon, a blue balloon, and a yellow balloon. He also buys 3 different pennants. He wants to give his sister a balloon and a pennant. How many different combinations can he make?

Estimating Decimal Sums and Differences, page 443...........................

Estimate by rounding to the nearest whole number.

1.	2.	3.	4.	5.
4.6 + 3.1	27.7 + 39.8	6.8 + 4.3	75.2 + 57.6	42.74 + 27.95

6.	7.	8.	9.	10.
50.74 − 39.85	4.07 − 2.8	32.40 − 23.95	70.40 − 36.28	1.98 − .90

Estimate. Use the front digits and adjust.

11.	12.	13.	14.	15.
14.26 + 1.96	26.13 + 33.87	56.76 + 28.69	47.98 + 7.14	39.07 + 15.25

16.	17.	18.	19.	20.
14.25 − 0.09	59.36 − 9.27	30.13 − 14.05	75.00 − 8.84	50.00 − 18.39

EXTRA PRACTICE

Adding Decimals, page 445 ....

Find the sum.

1. 1.4
   + 8.5

2. 42.5
   + 20.5

3. 10.4
   + 6.7

4. 19.6
   + 5.8

5. 6.3
   + 9.4

6. 6.81
   + 9.63

7. 8.52
   + 5.87

8. 1.49
   + 6.14

9. 0.37
   + 0.90

10. 8.64
   + 0.20

11. 12.5 + 13.8

12. 6.9 + 5.9

13. 9.8 + 40.9

14. 16.81 + 24.64

15. 8.73 + 9.86

16. 14.09 + 13.67

Subtracting Decimals, page 447 ....

Find the difference.

1. 57.3
   − 10.3

2. 26.6
   − 8.2

3. 35.0
   − 15.8

4. 4.1
   − 2.4

5. 9.2
   − 6.9

6. 0.94
   − 0.71

7. 3.52
   − 0.26

8. 7.3
   − 5.13

9. 9.61
   − 1.82

10. 6.0
   − 0.43

11. 3.5 − 2.9

12. 51.7 − 9.8

13. 84.8 − 83.9

14. 0.84 − 0.68

15. 9 − 3.76

16. 7.1 − 3.86

Problem Solving Strategy: Conducting an Experiment, page 451 ....

Use models to solve the problem.

1. Alan has 3 different shirts and 3 different pairs of slacks. How many different outfits can he make?

2. Sara makes sandwiches with tuna, egg salad, or cheese. She has whole wheat and white bread. How many different kinds of sandwiches can she make?

EXTRA PRACTICE

Probability, page 453..........

Find the probability of the spinner landing on

1. A.

2. B.

3. C.

4. D.

5. A, B, or C?

6. Which letter is most likely?

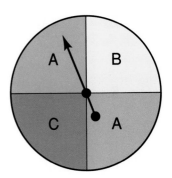

Write the probability.

7. A jar contains 4 balls. One is red, one is blue, one is white and one is green. What is the probability of picking a red ball?

8. What if a yellow ball is added to the jar. what is the probability of picking the yellow ball?

Prediction, page 455 ....

Use the following data for Questions 1 and 2.

Olivia put 7 marbles in a sack. Some were blue and some were green. There were 4 of one color and 3 of the other color. She picked marbles 21 times. These tally marks show the results.

Outcomes

blue	////	////	////
green	////	//	

1. Predict how many marbles of each color there are in the sack?

2. What is the probability of picking a blue marble? a green marble?

Practice PLUS

KEY SKILL: Subtracting Decimals (Use after page 447.)

Level A

Find the difference.

1. 7.28
    − 3.15

2. 24.6
    − 13.7

3. 44.86
    − 13.23

4. 5.81
    − 2.63

5. 64.86
    − 22.94

6. $35.64
    − 10.72

7. 72.3
    − 28.5

8. 36.4
    − 19.5

9. Chris jumped 4.27 m. Evan jumped 3.25 m.
   How much farther did Chris jump?

Level B

Find the difference.

10. 92.72
     − 31.35

11. 51.15
     − 20.08

12. 63.37
     − 29.29

13. 9.17
     − 3.03

14. $3.05
     − 1.12

15. 11.84
     − 10.5

16. 20.0
     − 2.76

17. 7.37
     − 4.91

18. Penny ran a race in 63.8 seconds. Audrey ran it in
    59.45 seconds. How much faster was Audrey's time.

Level C

Find the difference.

19. 7 − 4.19

20. 13.47 − 3.22

21. 6.28 − 4.84

22. 7.49 − 2.99

23. 10.15 − 8.97

24. 5 − 1.15

25. 6.9 − 1.78

26. 7.99 − 3.23

27. 5.4 − 3.91

28. Mike rode a total of 37.65 km in 2 hours. He rode
    20.9 km the first hour. How many kilometers did he
    ride the second hour?

KEY SKILL: Probability (Use after page 453.)

Level A

Find the probability of spinning:

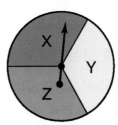

1. X.

2. Y.

3. Q.

4. A basket contains 2 blue and 3 green balls. What is the probability of picking a blue ball?

Level B

Find the probability of picking:

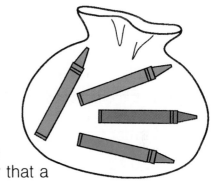

5. a red crayon.

6. a blue crayon.

7. a yellow crayon.

8. A spelling group has 4 boys and 5 girls. One student is chosen. How many possible outcomes are there? What is the probability that a girl will be chosen?

Level C

Find the probability of spinning:

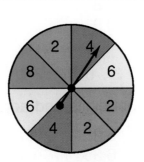

9. 2.

10. 4.

11. 8.

12. 10.

13. A box contains 13 marbles. Two of them are black. What is the probability of choosing a marble that is *not* black?

CHAPTER REVIEW

LANGUAGE AND MATHEMATICS

Complete the sentences. Use the words in the chart.

1. ■ decimals name the same number. *(page 434)*

2. Twenty-five hundredths written as a ■ is 0.25.

3. A penny is one ■ of a dollar. *(page 432)*

4. ■ is the chance that something will happen. *(page 452)*

5. **Write a definition** or give an example of the words you did not use from the chart.

> **VOCABULARY**
> hundredth
> probability
> fractions
> decimal
> equivalent
> outcomes

CONCEPTS AND SKILLS

Write the decimal. *(pages 434–437)*

6. $\frac{8}{100}$ **7.** $6\frac{23}{100}$ **8.** five and three-tenths

9. seven and sixteen hundredths   **10.** thirteen and nine-tenths

11.

Write in order from least to greatest. *(page 436)*

12. 2.70, 0.72, 1.27   **13.** .06, .66, .60   **14.** 1.31, 1.23, 1.13

Estimate by rounding to the nearest whole number. *(page 442)*

15.	**16.**	**17.**	**18.**	**19.**
5.8	7.32	9.2	6.19	3.5
3.1	2.8	− 4.76	− 3.8	2.83
+ 1.7	+ 3.91			+ 0.9

20. 7.3 + 1.9 + 3.2   **21.** 6.4 − 3.7   **22.** 2.6 + 3.4 + 5.8   **23.** 2 − 1.91

Estimate. Use the front digits and adjust. *(page 442)*

24.	**25.**	**26.**	**27.**	**28.**
2.81	5.2	3.1	8.67	6.32
3.57	6.35	− 1.72	− 5.9	− 3.93
+ 4.1	+ 1.7			

29. 4.9 + 5.1 + 2.8   **30.** 7.4 − 2.87   **31.** 2.9 + 1.5 + 3.5   **32.** 5.19 − 3.76

Find the sum or difference. *(pages 444–447)*

33.   4.9
  + 7.8

34. $7.96
  + $4.23

35.   4.98
    6.12
  + 0.6

36.   5.4
  − 3.6

37.   6.03
  − 4.78

38.   5.4
  + 2.3

39.   4.35
  + 3.12

40.   8.9
  − 4.7

41.   6.1
  − 5.9

42. $9.01
  − 2.98

43. 5.7 + 8.23 **44.** $4.12 + $8.29 + $.60 **45.** $9.02 − $8.56 **46.** 5 − 2.36

Use a bag of cubes with 3 red cubes, 2 yellow cubes, and
1 white cube. Pick a cube with your eyes closed. *(page 452)*

47. Are you more likely to pick a yellow cube or a red cube?

48. Find the probability of picking: **a.** yellow **b.** white **c.** red or yellow

49. You pick a cube from the bag 20 times. Predict how
many times you will pick a red cube. *(page 454)*

CRITICAL THINKING

Write *true* or *false*. Tell how you made your choice.

50. If 2 decimals are both less than
0.5, then their sum is greater
than 1. *(page 444)*

51. If 2 decimals are both greater
than 1, the difference between
them can be greater than 1.
(page 446)

52. If a coin turns up heads 5 times in a row, then it is
more likely that it will turn up tails the next time it is
tossed. *(page 452)*

MIXED APPLICATIONS

53. The lumber yard has oak,
cedar, and pine. Ned wants to
make a table. He has 4 models
to choose from. How many
choices does he have for the
table? *(page 438)*

54. Mike has 4 dimes and 3
nickels. He gives 3 coins to his
sister. What are all possible
values of the 3 coins?
(page 450)

CHAPTER TEST

Write the decimal.

1. $\frac{19}{100}$

2. $4\frac{3}{10}$

3. six hundredths

Compare. Write >, <, or =.

Order from greatest to least.

4. 8.7 ● 8.07

5. 9.32 ● 9.23

6. 4.41; 4.05; 4.5

Estimate by rounding to the nearest whole number.

7. 7.7
3.9
+ 1.2

8. 7.92
− 2.4

9. 6.8 + 3.2 + 4.9

10. 5.6 − 3.18

Estimate. Use the front digits and adjust.

11. 1.02
3.34
+ 4.95

12. 8.42
− 3.01

13. 5.9 + 3.01 + 3.1

14. 5.83 − 2.9

Find the sum or difference.

15. 8.9
+ 4.76

16. 9.03
+ 6.58

17. 6.4
− 2.9

18. 8.3
− 2.75

19. 3.76 + 9.4 + 0.81

20. 7 − 2.93

Use a number cube labeled 1–6. Find the probability of tossing:

21. 4

22. 9

23. Predict how many times you will toss a 5 if you toss the cube 18 times.

Solve.

24. Jan has 3 types of cheese and 2 types of bread. She wants only one type of cheese and one type of bread in each sandwich. How many different sandwiches can she make?

25. Gina has 2 red beads, 1 blue bead, and 3 white beads. She strings 3 of them together. What are all the possible ways she could string the 3 beads?

ODDS

The odds of picking a card marked with a B are 2 to 9, or $\frac{2}{9}$.

$$\text{ODDS} = \frac{\text{NUMBER OF FAVORABLE OUTCOMES}}{\text{NUMBER OF UNFAVORABLE OUTCOMES}}$$

1. Why are there two favorable outcomes?

2. Why are there nine unfavorable outcomes?

3. How many possible outcomes are there?

4. What do you notice about the sum of the numbers in the ratio?

What if you wanted to find the odds of picking a card not marked with a B?

5. How many favorable outcomes are there?

6. How many unfavorable outcomes are there?

7. What are the odds of picking a card not marked with a B?

Find the odds of spinning the color.

8. green

9. not red

10. red, yellow, or green

11. not blue and not green

12. Make your own experiment. You find the probability of each outcome. Have another student find the odds.

Cumulative Review

Choose the letter of the correct answer.

1. Which figure is a polygon?

    a. curve   **c.** triangle

    b. angle   **d.** circle

2. Which is eight hundredths?

    a. 0.8   **c.** 0.80

    b. 0.08   **d.** 800

3. Which is 0.20?

    a. two tenths   **c.** twenty tenths

    b. twenty   **d.** two hundredths

4. What is 8.26 plus 4.08?

    a. 13.24   **c.** 12.36

    b. 12.34   **d.** not given

5. Compare: 3.9 ● 3.90

    a. <   **c.** =

    b. >   **d.** not given

6.

What figure is shown?

    a. square   **c.** cube

    b. rectangle   **d.** not given

7. Estimate by rounding:
6.2 + 3.81

    a. 10   **c.** 14

    b. 12   **d.** 15

8. What is 3.76 less than 5?

    a. 1.24   **c.** 2.24

    b. 2.76   **d.** not given

9. Compare: $5\frac{1}{2}$ ● $5\frac{1}{5}$

    a. <   **c.** =

    b. >   **d.** not given

10. What is $\frac{3}{8}$ of 40?

    a. 5   **c.** 18

    b. 12   **d.** not given

11. 9.2 − 5.63

    a. 3.57   **c.** 4.63

    b. 3.67   **d.** not given

12. Add $\frac{3}{4} + \frac{3}{4}$.

    a. $\frac{6}{8}$ **c.** $1\frac{1}{2}$

    b. $1\frac{3}{4}$ **d.** not given

13. Find the perimeter of a square with 8-in. sides.

    a. 16 in.   **c.** 64 in.

    b. 32 in.   **d.** not given

14. Rename: $\frac{18}{4}$

    a. $3\frac{1}{2}$ **c.** $4\frac{1}{2}$

    b. 4   **d.** not given

Multiplying and Dividing by 2-Digit Numbers

MATH CONNECTION: PROBLEM SOLVING

Fund Raising Car Wash Today

Outside wash $4.00
Inside cleaning $2.00

SAVE!!!
Outside and inside
$5.00

-GOAL-
$500.00

1. What is happening in this picture?

2. What information do you see?

3. How can you use this information?

4. Write a problem about the picture.

Mental Math: Multiply 10s; 100s; 1,000s

Naima and Miles organize a drive to collect newspapers for a recycling project. This chart shows their goal for collecting papers. How many newspapers will they collect each month?

Month	Number of Newspapers Collected
June	30 newspapers from 80 homes
July	30 newspapers from 800 homes
August	30 newspapers from 8,000 homes

You can find out how many by multiplying.

June: 30 × 80 = 2,400
July: 30 × 800 = 24,000
August: 30 × 8,000 = 240,000

So they will collect 2,400 newspapers in June, 24,000 newspapers in July, and 240,000 newspapers in August.

1. Compare the number of zeros in each product with the number of zeros in the factors. What pattern do you see?

2. Does the pattern work for these products? Why or why not?

   50 × 60 = 3,000
   500 × 60 = 30,000

TRY OUT Write the letter of the correct answer.
Multiply mentally.

3. 80 × 90   **a.** 720   **b.** 7,200   **c.** 72,000   **d.** 720,000

4. 30 × 60   **a.** 180   **b.** 1,800   **c.** 18,000   **d.** 180,000

5. 80 × 500   **a.** 400   **b.** 4,000   **c.** 40,000   **d.** 400,000

6. 50 × 2,000   **a.** 100   **b.** 1,000   **c.** 10,000   **d.** 100,000

PRACTICE

Multiply mentally.

7.	10 × 10	8.	100 × 10	9.	1,000 × 10	10.	50 × 10	11.	60 × 10

12.	70 × 10	13.	40 × 20	14.	60 × 40	15.	200 × 70	16.	3,000 × 50

17. 80 × 10   18. 100 × 70   19. 50 × 1,000   20. 30 × 90

21. 50 × 20   22. 70 × 70   23. 80 × 100   24. 800 × 50

25. 400 × 80   26. 60 × 2,000   27. 5,000 × 60   28. 2,000 × 70

29. Multiply by 20.

30	40	50	60	70	80	90
■	■	■	■	■	■	■

30. Multiply by 50.

30	40	50	60	70	80	90
■	■	■	■	■	■	■

Compare. Write >, <, or =.

31. 90 × 600 ● 900 × 600

32. 40 × 50 ● 10 × 100

33. 30 × 700 ● 300 × 700

34. 40 × 400 ● 4 × 4,000

Tell which pair of products are the same in the row.

35. **a.** 50 × 4,000   **b.** 90 × 400   **c.** 80 × 40   **d.** 60 × 600

36. **a.** 80 × 30   **b.** 40 × 600   **c.** 60 × 44,000   **d.** 300 × 8

Mixed Applications

37. In May Naima and Miles collect 40 newspapers from 90 homes. How many newspapers do they collect in May?

38. Glenn collects from 10 homes. Each home gives her 400 sheets of newspaper. How many sheets does Glenn get?

39. The students collect 300 lb of newspapers in March, 500 lb in April, and 250 lb in May. How many lb of newspapers do they collect in three months?

40. Harvey collects 30 papers from Morgan St. and 400 papers from Third St. How many more papers does he collect from Third St.?

UNDERSTANDING A CONCEPT

Estimating Products

The booklets for the summer sports program will be passed out by 47 students. If each student passes out 88 booklets, about how many booklets will be passed out in all?

You can estimate to find the answer.

Rounding	**Front-End Estimation**
Round each factor to its greatest place.	Use the front digits.

47 × 88	47 × 88
↓ ↓	↓ ↓
Think: 50 × 90 = 4,500	*Think:* 40 × 80 = 3,200
The estimate is 4,500.	The estimate is 3,200.

Between 3,200 and 4,500 booklets will be passed out.

1. Is the estimate found by rounding greater than or less than the exact answer? How do you know?

2. Is the estimate found by using the front digits greater than or less than the exact answer? How do you know?

3. Estimate 62 × 739 by rounding and front-end estimation. How do the estimates compare? Why?

TRY OUT Write the letter of the correct answer.

Estimate. Use rounding.

4. 78 × 25   **a.** 160   **b.** 240   **c.** 1,400   **d.** 2,400

5. 53 × $663   **a.** $3,000   **b.** $3,500   **c.** $35,000   **d.** $350,000

Estimate. Use the front digits.

6. 38 × $94   **a.** $2,700   **b.** $270   **c.** $3,600   **d.** $4,000

7. 77 × 852   **a.** 7,200   **b.** 56,000   **c.** 72,000   **d.** 560,000

PRACTICE

Estimate by rounding.

8. 38
×41

9. 93
× 37

10. 56
× 66

11. $82
× 74

12. 58
× 52

13. 705
× 48

14. 593
× 23

15. 457
× 81

16. 915
× 77

17. $629
× 43

18. 69 × 84

19. 93 × $32

20. 88 × $405

21. 94 × 516

Estimate by using the front digits.

22. 83
× 52

23. 71
× 67

24. $45
× 57

25. 91
× 65

26. 67
× 63

27. 625
× 48

28. $904
× 87

29. 290
× 31

30. 565
× 96

31. 718
× 61

32. 56 × $48

33. 83 × 62

34. 74 × $525

35. 26 × 546

Compare the products. Write >, <, or =.

36. 45 × 362 ● 61 × 346

37. 77 × 566 ● 68 × 487

38. 55 × 306 ● 62 × 317

39. 88 × 11 ● 66 × 33

40. 93 × 216 ● 142 × 85

41. 649 × 58 ● 584 × 69

Mixed Applications

42. Each school can have 175 students in its summer sports program. There are 12 schools. About how many students can be in the program?

43. The sports program runs from 8:15 A.M. until 12 noon each day. How long does the program last each day?

Mixed Review

Find the missing number.

44. ■ × 9 = 72

45. $\frac{4}{8} = \frac{■}{2}$

46. $\frac{3}{4} + \frac{■}{4} = \frac{3}{4}$

47. 48 ÷ ■ = 8

UNDERSTANDING A CONCEPT

Multiplying by Multiples of Ten

A. The Valley Baseball Boosters are selling season tickets to their team's games. If each of the 20 members sells 35 tickets, how many tickets will the Boosters sell?

Multiply: 20×35

Step 1	Step 2
Multiply by the ones.	**Multiply by the tens.**

Step 1
$$\begin{array}{r} 35 \\ \times\ 20 \\ \hline 0 \end{array}$$
Think: $0 \times 35 = 0$

Step 2
$$\begin{array}{r} \overset{1}{3}5 \\ \times\ 20 \\ \hline 700 \end{array}$$
Think: 2 tens $\times$ 35 = 70 tens

The Boosters will sell 700 season tickets.

1. What if each member sells 86 tickets? How many season tickets would be sold?

B. You can multiply larger numbers by multiples of ten.

Multiply: 30×246

Step 1	Step 2
Multiply by the ones.	**Multiply by the tens.**

Step 1
$$\begin{array}{r} 246 \\ \times\ 30 \\ \hline 0 \end{array}$$
Think: $0 \times 246 = 0$

Step 2
$$\begin{array}{r} \overset{1}{2}\overset{1}{4}6 \\ \times\ 30 \\ \hline 7,380 \end{array}$$
Think: 3 tens $\times$ 246 = 738 tens

TRY OUT Write the letter of the correct answer.

2. 20×56 **a.** 112   **b.** 1,020   **c.** 1,120   **d.** 11,200

3. 80×41 **a.** 328   **b.** 3,280   **c.** 3,281   **d.** 32,800

4. 30×124 **a.** 372   **b.** 462   **c.** 3,720   **d.** 3,724

5. 50×245 **a.** 25   **b.** 125   **c.** 1,225   **d.** 12,250

PRACTICE

Multiply.

6. $\begin{array}{r} 41 \\ \times\ 40 \end{array}$	**7.** $\begin{array}{r} 83 \\ \times\ 30 \end{array}$	**8.** $\begin{array}{r} 92 \\ \times\ 40 \end{array}$	**9.** $\begin{array}{r} 81 \\ \times\ 90 \end{array}$	**10.** $\begin{array}{r} 64 \\ \times\ 20 \end{array}$
11. $\begin{array}{r} 406 \\ \times\ \ 70 \end{array}$	**12.** $\begin{array}{r} 314 \\ \times\ \ 20 \end{array}$	**13.** $\begin{array}{r} 911 \\ \times\ \ 90 \end{array}$	**14.** $\begin{array}{r} 703 \\ \times\ \ 30 \end{array}$	**15.** $\begin{array}{r} 625 \\ \times\ \ 40 \end{array}$

16. 60×45 **17.** 50×18 **18.** 60×37 **19.** 30×115

20. 20×347 **21.** 60×419 **22.** 20×73 **23.** 30×42

24. 70×68 **25.** 250×60 **26.** 340×40 **27.** 50×440

Mixed Applications   Solve. Which method did you use?

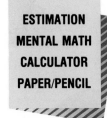

ESTIMATION
MENTAL MATH
CALCULATOR
PAPER/PENCIL

28. The Boosters want to raise money to buy 40 baseball bats. Each bat costs $15. How much money will they need?

29. Derek sells $80 worth of tickets. Carol sells $212 worth of tickets. Who sells more? How much more?

30. Sam sells 50 season tickets. Each ticket costs $25. Does Sam sell over $1,000 worth of tickets?

MENTAL MATH

You can use the properties of multiplication to help you find products mentally.

$$5 \times (18 \times 2) = 5 \times (2 \times 18)$$
$$= (5 \times 2) \times 18$$
$$= 10 \times 18$$
$$= 180$$

Multiply.

1. $4 \times 12 \times 5$ **2.** $20 \times 17 \times 5$ **3.** $25 \times 4 \times 22$

PROBLEM SOLVING

UNDERSTAND
PLAN
TRY
CHECK
EXTEND

Strategies Review

Use these problem-solving strategies to solve the problems. Remember that some problems can be solved using more than one strategy.

- Using Number Sense
- Choosing the Operation
- Drawing a Diagram
- Solving a Two-Step Problem
- Using Estimation
- Making an Organized List
- Conducting an Experiment

- Finding a Pattern
- Solving a Multistep Problem
- Working Backward
- Guess, Test, and Revise
- Solving a Simpler Problem
- Making a Table

Solve. Tell which strategy you used.
Use mental math, a calculator, or paper and pencil.

1. Every student in Mrs. Rose's class is on one fund-raising team. The book-fair team has 12 fewer students than the bake-sale team. The bake-sale team has twice as many students as the movie team. The movie team has 8 students. How many students are in Mrs. Rose's class?

2. The books for the fair are stacked in piles. There are 16 books in the first pile, 14 in the second pile, and 12 in the third pile. If this pattern continues, how many books will be in the next two piles?

3. The movie team sold 320 tickets. The tickets come in books of 12 tickets. How many complete books did the team sell?

4. At the book fair packages of 5 paperbacks and 2 hardcovers are sold at a discount. One woman bought enough packages to have 35 paperbacks. How many books did she buy?

5. The book-fair team hopes that 2,000 books will be donated by the end of May. The team received 450 books in March and 970 books in April. Will it meet its goal if 700 books are donated in May?

6. For the bake sale the team plans to bake chocolate and vanilla cupcakes with a topping of nuts, sprinkles, or chocolate chips. How many different kinds of cupcakes will they bake?

7. Louise, Lester, Larry, and Lulu each bought a different kind of book. They bought books on sports, fairy tales, poetry, and short stories. A boy bought a book of poetry. Lulu did not buy a book of fairy tales. Lester bought a book of short stories. Who bought the book on sports?

8. The bake-sale team meets every fourth day, and the book-fair team meets every third day. This week the bake-sale team meets on Monday, and the book-fair team meets on Tuesday. What is the next day on which both teams will meet?

9. The Rosas family is going to the movies. They pay for 2 adults' tickets and 2 children's tickets. Adults' tickets cost $1.00. Children's tickets cost $.50. How much does it cost them to go to the movies?

10. There are 10 clowns at the school fair. Some are on bicycles, and the rest are on tricycles. If there are 24 wheels in all, how many bicycles and tricycles are there?

11. One parent bought 6 muffins for $.45 each and 6 brownies for $.35 each. How much did she pay?

12. *Write a problem* that can be solved by using one of the strategies. Solve the problem. Ask others to solve it.

Multiplying and Dividing by 2-Digit Numbers **481**

Multiplying by 2-Digit Factors

Mrs. Mack is donating a large quilt to a charity auction. The quilt is made of colored squares sewn together. It is shaped like a rectangle that is 36 squares long and 24 squares wide. How many squares are in the quilt?

Estimate. Then draw a diagram to solve the problem.

WORKING TOGETHER

Step 1 On graph paper draw a rectangle that is 36 squares long and 24 squares wide.

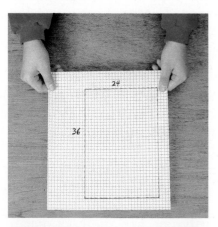

Step 2 Separate the rectangle into 4 sections like this:

Section A: 6 squares long by 4 squares wide

Section B: 6 squares long by 20 squares wide

Section C: 30 squares long by 4 squares wide

Section D: 30 squares long by 20 squares wide

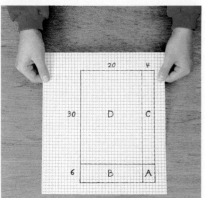

How many squares are in each section?

1. A   **2.** B   **3.** C   **4.** D

5. How many squares are there in the diagram?

6. How many squares are there in the quilt?

SHARING IDEAS

7. How did you find the number of squares in each section?

8. Write a multiplication sentence to show how to find the number of squares in each section.

9. How did separating the entire rectangle into smaller rectangular sections make it easier to find the total number of squares?

ON YOUR OWN

Draw a diagram to solve. Write multiplication sentences to show the number of squares in each section.

10. David and his mother are working on a quilt that is 21 squares long and 23 squares wide. How many squares are in their quilt?

11. Mrs. Henderson's quilt won an award at the quilting bee. Her quilt is 42 squares long and 56 squares wide. How many squares are in her quilt?

12. *Write a problem* that involves multiplying two 2-digit numbers. Solve your problem. Ask others to solve it.

Multiplying 2-Digit Numbers

A. Janet is selling tickets to the talent show. The theater has 24 rows of seats. There are 32 seats in each row. How many tickets can Janet sell for the show?

You can draw a diagram. Then write multiplication sentences to find the area of each section. Add to find the total.

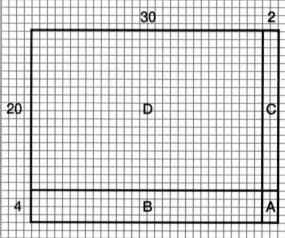

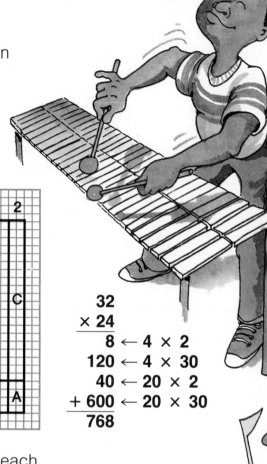

$$
\begin{array}{r}
32 \\
\times\, 24 \\
\hline
8 \leftarrow 4 \times 2 \\
120 \leftarrow 4 \times 30 \\
40 \leftarrow 20 \times 2 \\
+\, 600 \leftarrow 20 \times 30 \\
\hline
768
\end{array}
$$

1. What part of the diagram does each multiplication sentence show?

Janet can sell 768 tickets.

B. Here is another way to multiply to find the answer.

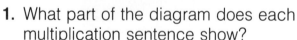

Step 1	Step 2	Step 3
Multiply the ones.	**Multiply the tens.**	**Add the products.**
$\begin{array}{r} 32 \\ \times\, 24 \\ \hline 128 \end{array}$ ← 4 × 32	$\begin{array}{r} 32 \\ \times\, 24 \\ \hline 128 \\ 640 \end{array}$ ← 20 × 32	$\begin{array}{r} 32 \\ \times\, 24 \\ \hline 128 \\ +\, 640 \\ \hline 768 \end{array}$

SHARING IDEAS

2. In the first method, why do you add 4 numbers to find the total?

3. In the second method, why do you add 2 numbers to find the total?

PRACTICE

Multiply.

4. 19 $\times$ 11	**5.** 42 $\times$ 21	**6.** 33 $\times$ 13	**7.** 21 $\times$ 24	**8.** 32 $\times$ 22
9. 12 $\times$ 12	**10.** 31 $\times$ 15	**11.** 58 $\times$ 11	**12.** 81 $\times$ 32	**13.** 71 $\times$ 62

14. 22 $\times$ 51 **15.** 13 $\times$ 62 **16.** 44 $\times$ 22 **17.** 13 $\times$ 13

18. 72 $\times$ 11 **19.** 12 $\times$ 24 **20.** 32 $\times$ 23 **21.** 53 $\times$ 21

22. 13 $\times$ 43 **23.** 42 $\times$ 22 **24.** 33 $\times$ 33 **25.** 41 $\times$ 12

26. What is the product of thirty-four times twelve?

Mixed Applications

27. The Lark Theater has 12 rows of seats. There are 21 seats in each row. How many seats are in the theater?

28. The Computer Club bought tickets worth $75.25. The Crafts Club spent $123.75 on tickets. How much did the two clubs spend on tickets?

29. During the first week of sales 12 of the students sold 20 tickets each. How many tickets did they sell?

30. *Write a problem* that can be solved by multiplying 2-digit numbers. Solve your problem. Ask others to solve it.

Mixed Review

Find the answer. Which method did you use?

31. 4,001 − 1,999 **32.** 7,896 + 6,947 **33.** 578 ÷ 4

MENTAL MATH
CALCULATOR
PAPER/PENCIL

EXTRA Practice, page 507

More Multiplying 2-Digit Numbers

A. To raise money for a track meet, Sarah's track team sold 24 cartons of light bulbs. Each carton contained 58 light bulbs. How many light bulbs did her team sell?

Multiply: 58 × 24
Estimate first. Then multiply.

Step 1	Step 2	Step 3
Multiply the ones.	**Multiply the tens.**	**Add the products.**

Step 1
Multiply the ones.

$$\begin{array}{r} \overset{3}{5}\,8 \\ \times\,2\,4 \\ \hline 2\,3\,2 \end{array} \leftarrow 4 \times 58$$

Step 2
Multiply the tens.

$$\begin{array}{r} \overset{1}{\overset{3}{5}}\,8 \\ \times\,2\,4 \\ \hline 2\,3\,2 \\ 1\,1\,6\,0 \end{array} \leftarrow 20 \times 58$$

Step 3
Add the products.

$$\begin{array}{r} \overset{1}{\overset{3}{5}}\,8 \\ \times\,2\,4 \\ \hline 2\,3\,2 \\ +\,1\,1\,6\,0 \\ \hline 1,3\,9\,2 \end{array}$$

Sarah's team sold 1,392 light bulbs.

1. Is this answer reasonable? How do you know?

2. *What if* you multiply 24 × 48? How would Steps 1 and 2 be different? Would the product in Step 3 be the same?

B. You can multiply money amounts the same way you multiply whole numbers.

Multiply: 24 × $.89

Step 1
Multiply as you would whole numbers.

$$\begin{array}{r} \overset{1}{\overset{3}{\$}}.8\,9 \\ \times\quad 2\,4 \\ \hline 3\,5\,6 \\ 1\,7\,8\,0 \\ \hline 2\,1\,3\,6 \end{array}$$

Step 2
Write a dollar sign and decimal point in the answer.

$$\begin{array}{r} \overset{1}{\overset{3}{\$}}.8\,9 \\ \times\quad 2\,4 \\ \hline 3\,5\,6 \\ 1\,7\,8\,0 \\ \hline \$2\,1.3\,6 \end{array}$$

3. 36 × 24   **a.** 764   **b.** 816   **c.** 864   **d.** 1,344

4. 75 × $.42   **a.** $4.50   **b.** $4.50   **c.** $29.40   **d.** $31.50

5. 41 × 18   **a.** 369   **b.** 738   **c.** 7,380   **d.** $31.50

6. 23 × $.31   **a.** $.92   **b.** $6.13   **c.** $7.13   **d.** $69.23

PRACTICE

Multiply.

7. 52
× 41

8. 31
× 26

9. 45
× 91

10. $.76
× 18

11. 98
× 52

12. 58
× 12

13. $.85
× 22

14. 25
× 33

15. 64
× 53

16. 48
× 45

17. 57 × 35

18. 48 × $.48

19. 83 × 99

20. 87 × 37

21. 72 × 66

22. 75 × 75

23. 85 × 89

24. 38 × $.19

Mixed Applications

Solve. Which method did you use?

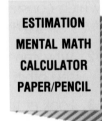

ESTIMATION
MENTAL MATH
CALCULATOR
PAPER/PENCIL

25. The members of the archery team bring their own arrows to the National Archery Finals. They can fit 18 cases of arrows into the team van. Each case holds 32 arrows. How many arrows can they bring to the contest?

26. At the track meet, Dar runs in a 100-meter race, a 500-meter race, and a 1,000-meter race. How many meters does he run?

27. The swim team had a goal of selling 3,000 buttons. They sold 27 cases of buttons that had 80 buttons in each case. Did they meet their goal?

28. There were 1,289 fans at the first meet, 2,378 fans at the second meet, and 3,897 fans at the third. What was the total attendance at the meets?

OUT OF THIS WORLD

Logical Reasoning

A. These problems are truly out in orbit! But with careful reasoning you will be able to solve them all.

1. Ada, Byron, and Clare wrote reports about constellations. Each wrote about one of the following: the Big Dipper, Canis Major, and Orion. Can you look at the clues and decide which constellation each student wrote about?

	Big Dipper	Canis Major	Orion
Ada			
Byron			
Clare			

- Byron's constellation has only one word in its name.
- Ada did not write about the Big Dipper.

Use the chart to organize what you know. Look at the first clue. Which constellation must Byron have written about? Put a ✔ for "yes" in the box where Byron and that constellation meet.

2. Why can you put Xs for "no" in the boxes where Byron and the other constellations meet? Where else can you put Xs? Why?

3. Look at the second clue. Which constellation must Ada have written about? How do you know? What did Clare write about? Complete the chart.

B. Solve the puzzles.

4. Lars, Bruto, and Weptune are creatures from other planets. Which creature comes from which planet?

	Mars	Pluto	Neptune
Lars			
Bruto			
Weptune			

- No creature has a name that rhymes with its home planet.
- Lars and the creature from Neptune are both purple.

5. Quatron, Baxiol-6, and Plim are newly discovered planets. One of the planets has 1 moon. One planet has 2 moons. One planet has 3 moons. How many moons does each planet have?

	1 Moon	2 Moons	3 Moons
Quatron			
Baxiol-6			
Plim			

- Plim has more moons than Quatron.
- Baxiol-6 has an even number of moons.

6. The Acme Spaceship company sells four different models. The brands and prices are shown in the chart. All prices are in interplanetary money—zix. What is the cost of each spaceship?

- The Astro Rocket and the Planet Rover are the most expensive ships.
- The Stellar Flyer is more expensive than the Star Car.
- The Astro Rocket costs more than three times as much as the Star Car.

	6,000 zix	10,000 zix	15,000 zix	20,000 zix
Astro Rocket				
Star Car				
Stellar Flyer				
Planet Rover				

UNDERSTANDING A CONCEPT

Multiplying 3-Digit Numbers

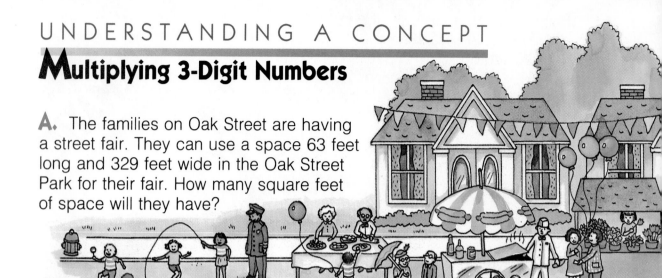

A. The families on Oak Street are having a street fair. They can use a space 63 feet long and 329 feet wide in the Oak Street Park for their fair. How many square feet of space will they have?

Multiply: 63 × 329

Step 1	Step 2	Step 3
Multiply the ones.	**Multiply the tens.**	**Add the products.**
$\begin{array}{r} 2 \\ 329 \\ \times\ \ 63 \\ \hline 987 \end{array}$ ← 3 × 329	$\begin{array}{r} 1\ 5 \\ 2 \\ 329 \\ \times\ \ 63 \\ \hline 987 \\ 19740 \end{array}$ ← 60 × 329	$\begin{array}{r} 1\ 5 \\ 2 \\ 329 \\ \times\ \ 63 \\ \hline 987 \\ +19740 \\ \hline 20{,}727 \end{array}$

The Oak Street families will have 20,727 square feet of space.

1. If 63 × 329 = 20,727, what does 63 × $3.29 equal?

B. You must be careful when there is a zero in a factor

Multiply: 47 × 507

$$\begin{array}{r} 2 \\ 4 \\ 507 \\ \times\ \ 47 \\ \hline 3549 \\ +20280 \\ \hline 23{,}829 \end{array}$$

2. Tell how the multiplication was done.

3. *What if* you need to find 507 × 47? How could you more easily find the product?

TRY OUT   Write the letter of the correct answer.

4. 47 × $6.72   **a.** $31.38   **b.** $73.92   **c.** $315.84   **d.** $3,138.40

5. 52 × 508   **a.** 2,606   **b.** 3,556   **c.** 26,416   **d.** 264,160

6. 36 × $8.03   **a.** $289.08   **b.** $73.08   **c.** $298.18   **d.** $730.80

PRACTICE

Multiply.

7. 132 × 13

8. $2.05 × .42

9. 312 × 32

10. 407 × 41

11. $3.22 × 14

12. 306 × 78

13. $4.92 × 47

14. 860 × 73

15. 804 × $.49

16. 384 × 80

17. $5.35 × 44

18. 647 × 65

19. 507 × $.36

20. 770 × 16

21. 908 × 75

22. 72 × 345   **23.** 16 × $8.02   **24.** 41 × 527   **25.** 65 × 526

26. 329 × 29   **27.** 486 × 53   **28.** 704 × $.49   **29.** 84 × 708

Critical Thinking

30. Miguel made $123.67 by selling sandwiches at the fair. He sold 83 sandwiches for $1.49 each. How many sandwiches would Miguel need to sell at $.83 a sandwich in order to make the same amount of money? How can you find the answer without multiplying?

Mixed Applications

31. Meg sold 262 glasses of apple juice, 92 glasses of grape juice, and 354 glasses of orange juice. How many glasses of juice did she sell?

32. If 875 people come to the street fair and each person spends $16, how much money will be spent?

33. At the fair 108 people paid $.75 each to ride on Kevin's old "bicycle built for two." How much did Kevin earn?

EXPLORING A CONCEPT
Multiplying Larger Numbers

The Highland Helpers want to buy 12 computers for the volunteer fire department. The computers will cost $1,123 each. How much must they raise to buy the computers?

WORKING TOGETHER

You can use mental math, paper and pencil, or a calculator to find the answer. Work with a group. Have each person in your group use a different method to solve the problem.

1. Can you find the product mentally? Why or why not?

2. Tell how you can use paper and pencil to find the product. What is the product?

3. How can you be sure your product is reasonable?

4. Which method was fastest in finding the product?

5. **What if** the computers cost $1,000 each? Which method would be fastest?

Multiply. Which method did you use?

6. 40 × 4,000   **7.** 58 × 6,978   **8.** 33 × 6,211

SHARING IDEAS

9. How is multiplying 4-digit numbers like multiplying 2-digit or 3-digit numbers?

10. How did you decide which method to use to find the answers for Exercises 6–8?

11. When do you think it is easier to use mental math? paper and pencil? a calculator? Why?

12. How many places can there be when you multiply a 4-digit number by a 5-digit number? How do you know?

ON YOUR OWN

Solve. Which method did you use?

MENTAL MATH
CALCULATOR
PAPER/PENCIL

13. The Highland Helpers use fire trucks and hoses that spout 2,375 liters of water each minute. They spent 35 minutes putting out a fire. How much water did they use?

14. A box of computer diskettes costs $22. Is $500 enough to buy 20 boxes of diskettes?

15. *Write a problem* using large numbers that can be solved by multiplication. Choose which method you think should be used to solve it. Trade problems with other students and compare methods.

Mental Math: Divide 10s; 100s; 1,000s

The members of the Safe Home Society want to visit 24,000 homes in Trouble City. If there are 60 members, how many homes will each member have to visit?

You can use a pattern to find the answer.

$$24 \div 6 = 4$$
$$240 \div 60 = 4$$
$$2,400 \div 60 = 40$$
$$24,000 \div 60 = 400$$

Each member will have to visit 400 homes.

1. Count the number of zeros in the dividend, the divisor, and the quotient. What pattern do you see?

2. Does the pattern work for $2,000 \div 50$? Why or why not?

TRY OUT Write the letter of the correct answer.

3. $80 \div 10$	**a.** 8	**b.** 10	**c.** 80	**d.** 800
4. $4,000 \div 50$	**a.** 8	**b.** 80	**c.** 20,000	**d.** 200,000
5. $36,000 \div 40$	**a.** 9	**b.** 90	**c.** 900	**d.** 9,000
6. $40,000 \div 80$	**a.** 5	**b.** 20	**c.** 200	**d.** 500

PRACTICE

Divide. Use mental math.

7. $20\overline{)20}$ **8.** $70\overline{)140}$ **9.** $50\overline{)3,500}$ **10.** $90\overline{)9,000}$ **11.** $20\overline{)60,000}$

12. $80 \div 40$ **13.** $60 \div 20$ **14.** $60 \div 30$ **15.** $90 \div 30$

16. $250 \div 50$ **17.** $320 \div 40$ **18.** $420 \div 60$ **19.** $640 \div 80$

20. $3,600 \div 90$ **21.** $8,100 \div 90$ **22.** $4,500 \div 50$ **23.** $2,100 \div 30$

24. $12,000 \div 20$ **25.** $18,000 \div 60$ **26.** $64,000 \div 80$ **27.** $40,000 \div 40$

28. Rule: Divide by 20.

100	120	800	1,800	80,000
■	■	■	■	■

29. Find the rule.

150	180	1,200	27,000	30,000
5	6	40	900	1,000

Find the missing number.

30. $50 \times \blacksquare = 35,000$ **31.** $36,000 \div \blacksquare = 90$ **32.** $\blacksquare \div 70 = 0$

33. $\blacksquare \times 300 = 6,000$ **34.** $15,000 \div \blacksquare = 500$ **35.** $\blacksquare \div 20 = 800$

Mixed Applications

36. Claude is mailing 60 boxes of letters on home safety. If each box holds 30 letters, how many letters does Claude mail?

37. Sally ships about 40,000 safety posters to 20 cities. If she ships the same number of posters to each city, how many does each city get?

38. It took 10 hours to print all of the letters to be mailed. Brooke worked for 5 hours. What fraction of the total time did she work?

39. *Write a problem* that can be solved by dividing mentally. Solve your problem. Ask others to solve it.

Mixed Review

Find the answer. Which method did you use?

MENTAL MATH
CALCULATOR
PAPER/PENCIL

40. $\frac{3}{8} + \frac{3}{8}$ **41.** 50×50 **42.** $20,304 - 19,458$

DEVELOPING A CONCEPT
Dividing by Multiples of Ten

A. There are 126 students at a meeting about the foreign exchange program. Each classroom holds 40 students. How many classrooms are filled with students? How many students are in an unfilled classroom?

WORKING TOGETHER
Estimate first. Then use place-value models to solve the problem.

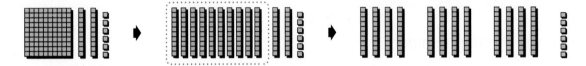

1. How many times can you take 40 away from 126? How many groups of 40 can you make?

2. How many models cannot be put in a group?

3. How many classrooms are filled?

4. How many students are in the unfilled classroom?

B. You can divide without using models.

Step 1	Step 2
Decide where to place the first digit in the quotient.	**Divide the ones. Write the remainder.**

$40\overline{)126}$ **Think:** 40 > 1 Not enough hundreds
40 > 12 Not enough tens
40 < 126 Divide the ones.

$\begin{array}{r} 3\text{ R6} \\ 40\overline{)126} \\ -120 \\ \hline 6 \end{array}$ **Think:** $4\overline{)12}^{\,3}$
Multiply: $3 \times 40 = 120$
Compare: $120 < 126$
Subtract: $126 - 120 = 6$

5. **What if** there were 88 students at a meeting? How would you divide to find how many classrooms would be filled and how many students would be in the unfilled classroom?

SHARING IDEAS

6. How does the method using place-value models compare with the method using numbers?

7. How can you use division facts to help you divide by multiples of ten?

PRACTICE

Divide.

8. $40\overline{)48}$ **9.** $20\overline{)82}$ **10.** $10\overline{)76}$ **11.** $90\overline{)95}$ **12.** $30\overline{)99}$

13. $20\overline{)32}$ **14.** $50\overline{)72}$ **15.** $40\overline{)97}$ **16.** $30\overline{)83}$ **17.** $20\overline{)94}$

18. $70\overline{)143}$ **19.** $50\overline{)158}$ **20.** $60\overline{)184}$ **21.** $90\overline{)725}$ **22.** $80\overline{)647}$

23. $20\overline{)198}$ **24.** $40\overline{)340}$ **25.** $70\overline{)334}$ **26.** $50\overline{)385}$ **27.** $60\overline{)419}$

28. $89 \div 40$ **29.** $62 \div 20$ **30.** $43 \div 30$ **31.** $70 \div 30$

32. $256 \div 50$ **33.** $330 \div 40$ **34.** $527 \div 60$ **35.** $539 \div 90$

Critical Thinking

36. What is the greatest remainder you can have when you are dividing by a 2-digit number? How do you know?

Mixed Applications

Solve. You may need to use the Databank on page 522.

37. Alice is in Spain and exchanges $35 for pesetas. She sees a pair of shoes she likes for 3,780 pesetas. Does she have enough money to buy the shoes?

38. In France, Jessica has 360 francs to buy souvenirs for 20 classmates. If she buys the same thing for each friend, how much can each present cost?

39. If Sean has 2,880 Belgian francs, how many U.S. dollars can he get in exchange?

40. *Write a problem* using the information in the Databank. Ask others to solve it.

PROBLEM SOLVING

Strategy: Choosing the Operation

In order to plan which operation to choose to solve a problem you have to understand what is happening in the problem.

Addition: Combine unequal groups.
Subtraction: Separate into unequal groups.
Multiplication: Combine equal groups.
Division: Separate into equal groups.

The students in Mrs. Cruz's class held a rummage sale to raise money for some local charities. They made $103.25 on clothes, $55.35 on books, and $85.40 on household items. How much money did they make in all?

1. Which operation can you use to solve the problem? How much money did they make in all?

They plan to give $45 to each of 4 charities. How much money will they give to charities?

2. Which operation can you use to solve the problem? How much money will they give to charities?

They want to use the rest of the money to buy books for the class library. How much will they spend on books for the library?

3. Which operation can you use to solve the problem? How much will they spend on books?

What if they had given all the money to the 4 charities, in equal amounts? How much would each charity get?

4. Which operation can you use to solve the problem? How much would each charity get?

PRACTICE

Decide which operation to use. Then solve the problem.
Use mental math, a calculator, or paper and pencil.

5. Joan bought a plant for $1.35, a poster for $2.50, and a dress for $4.80. How much did she spend in all?

6. At the rummage sale Pat bought a box of 12 tapes. If each tape cost $.35, how much did she pay?

7. The school is using the money earned at the rummage sale to buy a computer for $430. It will be paid for in 10 installments. How much is each installment?

8. When the sale started David had 24 quarts of punch to sell. When it was over he had 5 quarts left. How many quarts of punch did he sell?

Strategies and Skills Review

9. Bob's class is putting on a play to raise money for playground equipment. If Bob sells 6 tickets a day and Seth sells 4 tickets a day, how many tickets will Seth have sold when Bob has sold 36?

10. They had 5 performances of the class play. There were 90 tickets for each performance. All tickets were sold for the first 4 performances. All but 27 tickets were sold for the last performance. How many tickets were sold in all?

11. The students are making posters for the rummage sale. They will use 10 feet of streamers on each poster. They have 160 feet of streamers. How many posters can they make?

12. The parents of the students in Mrs. Cruz's class held a dinner as a fund-raiser. The dinner cost $12 a person. There were 125 people at the dinner. How much money did they raise?

13. At the rummage sale David, Edie, and 5 of their friends bought 63 records. If each bought the same number of records, how many records did each one buy?

14. *Write a problem* that can be solved by addition, subtraction, multiplication, or division. Solve the problem. Ask others to solve your problem.

Dividing by 2-Digit Divisors

There are 167 tickets left for the summer sports festival. Hector wants to send 52 tickets to each of the local gyms. How many gyms can Hector send tickets to? How many tickets will be left?

You can divide to find the number of gyms.

WORKING TOGETHER

Estimate. Then solve the problem using place-value models.

Step 1 Make a model of the number of tickets.

Step 2 Separate your model into groups of 52.

1. How did you model the number of tickets?

2. Did you have to regroup before you could make the groups? Why or why not?

3. How many groups of 52 did you make? How many models were left?

4. How many gyms can Hector send tickets to? How many tickets will be left?

Use models to find each quotient and remainder.

5. $53 \div 12$

6. $94 \div 14$

7. $96 \div 26$

8. $159 \div 42$

9. $226 \div 22$

10. $564 \div 46$

SHARING IDEAS

11. How is dividing by a 2-digit divisor like dividing by a 1-digit divisor? How is it different?

12. Is the quotient always a 1-digit number when you divide a larger 2-digit number by a 2-digit number? Why or why not?

13. Is the quotient always a 1-digit number when you divide a 3-digit number by a 2-digit number? Why or why not?

ON YOUR OWN

Use place-value models to solve the problem.

14. The food stand at the festival took in $384 in one hour. If the average amount spent was $16, how many people bought food at the stand?

15. There are 140 tickets set aside for the members of the teams at the festival. Each team has 25 members. How many tickets should each member get so that they each have the same number of tickets? How many tickets will be left?

16. *Write a problem* that can be solved by dividing by a 2-digit divisor. Solve your problem. Ask others to solve it.

Problem Solving: Planning a Car Wash

SITUATION

The students of Greenvale Elementary School are planning a car-wash fund-raiser. They want to plant sapling maple trees in the new East City Park. Each tree is $39. Their goal is to plant at least five trees.

PROBLEM

How should the class organize the car wash?

DATA

Car Wash
Day / Time: Saturday 8:00-5:00
Steps: ① Wash
② Rinse with hose
③ Dry
Location: School parking lot
(2 outdoor faucets)
School provides water
and hoses
Teacher supervisors provide
cash box and change

BIG AL'S AUTO SUPPLY

Softee car soap
5 lb for
$5.75
Enough for
60 washes!

Window squeegees
$2.75 each

Bucket
$2.50 each

Wheel scrubber
$2.50 each

Softee car sponge
$1.25 each

Drying/finishing cloth
$4.50 each

USING THE DATA

How much will these supplies cost?

1. 4 buckets
2. 4 sponges
3. 4 drying towels
4. 2 boxes of soap
5. 2 squeegees
6. 2 wheel scrubbers

7. If they buy all the items listed above, how much will the students spend on supplies?

8. How much must they make to buy 5 saplings and cover the cost of the supplies?

9. If they expect 50 customers, how much should they charge each customer to reach their goal? If they expect 100 customers, how much should they charge?

10. If a team of students takes 10 minutes to wash a car, how many cars can the team wash in an hour? How many can the team wash in one 9-hour day?

MAKING DECISIONS

11. What kinds of jobs are there in each washing crew? Which of the supplies are needed to do each job?

12. How many students should be in each washing crew? How many washing crews should there be? Why?

13. What other jobs are there besides the washing crews?

14. *What if* the students work 3-hour shifts? How many students are needed to work a 9-hour day? Why?

15. *What if* one worker uses the hose and works with two crews? How would this change the planning for the car wash?

16. *Write a plan* for a fund-raising car wash. List how you would organize it, how many students you would have, and how much you would charge for each car.

CURRICULUM CONNECTION

Math and Social Studies

Since 1851 the cost of sending a letter has been based on weight. Back in 1860 it could cost you $10 to send a 1-oz letter—if you sent it by Pony Express from Missouri to California. In 1863 it cost 6¢ to mail a 1-oz letter by first-class mail. Rates fell to 2¢ in 1885, but they have been rising ever since.

The chart shows the cost from 1863 to 1988 of mailing a 1-oz first-class letter anywhere in the United States. In about 100 years the cost of mailing a 1-oz first-class letter rose only 23¢. Rates rose the most in the 1970s and 1980s.

COST OF MAILING A FIRST-CLASS LETTER

Year	Amount	Year	Amount
1863	6¢	1971	8¢
1885	2¢	1974	10¢
1932	3¢	1981	20¢
1963	5¢	1988	25¢

How much more did it cost to mail 25 letters in 1988 than it did in 1963?

Think: In 1963 each letter cost 5¢ to mail.
25×5 cents $= 125$¢ or $1.25

In 1988 each letter cost 25¢ to mail.
25×25 cents $= 625$¢ or $6.25

$6.25 - $1.25 = $5.00

It cost $5.00 more to mail 25 letters in 1988 than it did in 1963.

ACTIVITY

1. Find out about one of the following: the postal service in Colonial times, the Pony Express, or rare stamps. Share this information with your class.

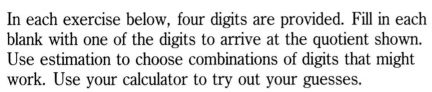

Calculator: Missing Digits

You can use a calculator to help improve your number sense.

In each exercise below, four digits are provided. Fill in each blank with one of the digits to arrive at the quotient shown. Use estimation to choose combinations of digits that might work. Use your calculator to try out your guesses.

1. $\boxed{2,\ 3,\ 0,\ 7}$ ___)‾2‾‾9‾ ___ ___ ___

(*Hint:* There are no tens in the dividend.)

2. $\boxed{8,\ 2,\ 2,\ 6}$ ___)‾4‾‾7‾ ___ ___ ___

(*Hint:* The dividend has the same number of ones as hundreds.)

3. $\boxed{8,\ 8,\ 2,\ 8}$ ___)‾3‾‾6‾ ___ ___ ___

4. $\boxed{7,\ 1,\ 4,\ 4}$ ___)‾6‾‾3‾ ___ ___ ___

5. $\boxed{5,\ 2,\ 2,\ 4}$ ___)‾6‾‾3‾ ___ ___ ___

6. $\boxed{6,\ 4,\ 8,\ 9}$ ___)‾9‾‾4‾ ___ ___ ___

7. $\boxed{5,\ 3,\ 8,\ 7}$ ___)‾5‾‾5‾ ___ ___ ___

8. $\boxed{2,\ 8,\ 9,\ 8}$ ___)‾9‾‾2‾ ___ ___ ___

USING THE CALCULATOR

9. Tell how you chose the dividends and divisors. How did using a calculator help?

EXTRA PRACTICE

Mental Math: Multiply 10s; 100s; 1,000s, page 475.............................

Multiply mentally.

1. 70 × 10   **2.** 30 × 100   **3.** 40 × 20   **4.** 30 × 400

5. 30 × 800   **6.** 300 × 50   **7.** 800 × 40   **8.** 200 × 50

Compare. Write >, <, or =.

9. 20 × 400 ● 60 × 300   **10.** 90 × 300 ● 50 × 500

Estimating Products, page 477 ....

Estimate. Use rounding.

1. 543	**2.** 38	**3.** 496	**4.** 312	**5.** 449
× 64	× 24	× 56	× 72	× 26

6. 705 × 42   **7.** 550 × 35   **8.** 29 × 68   **9.** 31 × 77

Estimate. Use the front digits.

10. 625	**11.** 12	**12.** 68	**13.** 45	**14.** 47
× 52	× 23	× 19	× 88	× 24

15. 71 × 67   **16.** 20 × 24   **17.** 47 × 56   **18.** 32 × 422

Multiplying by Multiples of Ten, page 479 ....

Multiply.

1. 81	**2.** 65	**3.** 46	**4.** 73	**5.** 54
× 90	× 20	× 20	× 30	× 30

6. 911	**7.** 314	**8.** 406	**9.** 604	**10.** 917
× 90	× 20	× 70	× 40	× 20

11. 30 × 115   **12.** 60 × 27   **13.** 40 × 16   **14.** 50 × 45

15. 342 × 20   **16.** 348 × 30   **17.** 400 × 70   **18.** 90 × 312

Problem Solving: Strategies Review, page 481 ....

Solve. Tell which strategy you used.

1. Book labels are sold in packages of 12 or 16. Nancy bought 5 packages and got 76 labels. How many labels were in each package?

2. Mark made some cookies. He gave 24 to Tina and 12 to Ariel. He had 24 left. How many cookies did he make?

3. Ray bought 3 books for $2.95 each. He bought a poster for $5.95. He paid with a $20 bill. What was his change?

4. Gloria buys 6 flowers at the flower booth. They are either red or white. How many different combinations could she have?

Multiplying 2-Digit Numbers, page 485 ....

Multiply.

1. $\begin{array}{r} 52 \\ \times 13 \end{array}$	**2.** $\begin{array}{r} 64 \\ \times 12 \end{array}$	**3.** $\begin{array}{r} 81 \\ \times 18 \end{array}$	**4.** $\begin{array}{r} 32 \\ \times 14 \end{array}$	**5.** $\begin{array}{r} 43 \\ \times 13 \end{array}$
6. $\begin{array}{r} 74 \\ \times 21 \end{array}$	**7.** $\begin{array}{r} 31 \\ \times 31 \end{array}$	**8.** $\begin{array}{r} 61 \\ \times 58 \end{array}$	**9.** $\begin{array}{r} 52 \\ \times 34 \end{array}$	**10.** $\begin{array}{r} 91 \\ \times 12 \end{array}$
11. $\begin{array}{r} 64 \\ \times 22 \end{array}$	**12.** $\begin{array}{r} 15 \\ \times 11 \end{array}$	**13.** $\begin{array}{r} 22 \\ \times 24 \end{array}$	**14.** $\begin{array}{r} 31 \\ \times 13 \end{array}$	**15.** $\begin{array}{r} 72 \\ \times 34 \end{array}$

16. 81×74 **17.** 91×34 **18.** 31×25 **19.** 22×32

20. 62×44 **21.** 81×37 **22.** 53×32 **23.** 74×22

24. 33×33 **25.** 42×92 **26.** 71×43 **27.** 23×52

EXTRA PRACTICE

More Multiplying 2-Digit Numbers, page 487 ....

Multiply.

1. 87 × 20	**2.** 73 × 59	**3.** 41 × 70	**4.** 48 × 73	**5.** 79 × 45
6. 56 × 47	**7.** 72 × 25	**8.** 81 × 29	**9.** 63 × 53	**10.** 52 × 72

11. 27 × 52   **12.** 26 × 38   **13.** 39 × 23   **14.** 37 × 85

15. 39 × 61   **16.** 38 × 17   **17.** 26 × 36   **18.** 65 × 92

Multiplying 3-Digit Numbers, page 491 ....

Multiply.

1. 352 × 14	**2.** 423 × 15	**3.** 179 × 46	**4.** 346 × 16	**5.** 664 × 73
6. 243 × 24	**7.** 327 × 19	**8.** 256 × 85	**9.** 727 × 17	**10.** 743 × 56
11. 558 × 63	**12.** 652 × 95	**13.** 567 × 15	**14.** 347 × 36	**15.** 465 × 44

16. 35 × 146   **17.** 24 × 308   **18.** 72 × 716   **19.** 45 × 320

20. 18 × 651   **21.** 83 × 429   **22.** 57 × 601   **23.** 67 × 562

Mental Math: Divide 10s; 100s; 1,000s, page 495 ....

Divide. Use mental math.

1. 80 ÷ 20   **2.** 48,000 ÷ 80   **3.** 450 ÷ 50   **4.** 360 ÷ 90

5. 630 ÷ 70   **6.** 56,000 ÷ 80   **7.** 80 ÷ 40   **8.** 18,000 ÷ 30

9. 2,400 ÷ 30   **10.** 4,500 ÷ 90   **11.** 40 ÷ 20   **12.** 12,000 ÷ 60

Dividing by Multiples of Ten, page 497 ....

Divide.

1. $20\overline{)480}$ 2. $60\overline{)668}$ 3. $40\overline{)860}$ 4. $50\overline{)750}$ 5. $30\overline{)659}$

6. $20\overline{)23}$ 7. $30\overline{)62}$ 8. $20\overline{)86}$ 9. $40\overline{)85}$ 10. $30\overline{)95}$

11. $50\overline{)79}$ 12. $30\overline{)243}$ 13. $20\overline{)165}$ 14. $40\overline{)280}$ 15. $80\overline{)666}$

16. $376 \div 40$ 17. $630 \div 20$ 18. $538 \div 50$ 19. $886 \div 60$

20. $254 \div 30$ 21. $429 \div 70$ 22. $716 \div 80$ 23. $679 \div 30$

24. $927 \div 40$ 25. $608 \div 30$ 26. $751 \div 50$ 27. $840 \div 60$

Problem Solving Strategy: Choosing the Operation, page 499 ....

Decide which operation to use. Then solve the problem.

1. The fourth grade class has 125 students. All but 7 of them helped work on the class project. How many worked on the project?

2. The cafeteria can serve 5 people in one minute. How many minutes does it take to serve 45 people?

3. Eric eats 3 meals each day. How many meals does he eat in one week?

4. Amy spent 37¢ on supplies for the project. Joe spent $1.97. How much did they spend all together?

5. The School Store made a profit of $528 this year. The school wants to use this money to buy some new bushes for their garden. Each bush costs $16. How many bushes can they buy?

6. The school is 12 km from Brian's home. The library is 9 km away. How much further is the school than the library?

Practice PLUS

KEY SKILL: Multiplying 3-Digit Numbers (Use after page 491.)

Level A ....

Multiply.

1. 537 × 42	**2.** 328 × 43	**3.** 234 × 54	**4.** 314 × 81	**5.** 706 × 36
6. 109 × 69	**7.** 444 × 23	**8.** 482 × 34	**9.** 824 × 72	**10.** 680 × 59

11. The Book Club sold 12 books for $2.25 each. How much did they make?

Level B ....

Multiply.

12. 426 × 43	**13.** 952 × 47	**14.** 667 × 23	**15.** 235 × 47	**16.** 926 × 42
17. 564 × 33	**18.** $182 × 74	**19.** 343 × 34	**20.** 486 × 42	**21.** 251 × 96

22. 84 × 602   **23.** 39 × 648   **24.** 365 × 74   **25.** 520 × 58

26. Jake made 14 long-distance calls. Each call cost $1.90. How much did he spend?

Level C ....

Multiply.

27. 38 × 725   **28.** 64 × 977   **29.** 42 × 814   **30.** 18 × 473

31. 647 × 60   **32.** 753 × 45   **33.** 469 × 52   **34.** 308 × 75

35. Alice sold 83 sandwiches for $2.95 each. How much did she make?

PRACTICE PLUS

KEY SKILL: Dividing by Multiples of Ten (Use after page 497.)

Level A
Divide.

1. $30\overline{)65}$ **2.** $70\overline{)97}$ **3.** $10\overline{)88}$ **4.** $40\overline{)91}$ **5.** $20\overline{)79}$

6. $50\overline{)85}$ **7.** $60\overline{)96}$ **8.** $80\overline{)84}$ **9.** $20\overline{)55}$ **10.** $30\overline{)48}$

11. $129 \div 40$ **12.** $890 \div 90$ **13.** $685 \div 50$ **14.** $555 \div 20$ **15.** $798 \div 70$

16. Eric had 86 books to pack. He put 20 in each box. How many boxes did he fill? How many books were left?

Level B
Divide.

17. $40\overline{)86}$ **18.** $30\overline{)55}$ **19.** $20\overline{)48}$ **20.** $50\overline{)71}$ **21.** $10\overline{)29}$

22. $419 \div 40$ **23.** $760 \div 50$ **24.** $377 \div 40$ **25.** $351 \div 10$ **26.** $569 \div 20$

27. $761 \div 30$ **28.** $805 \div 60$ **29.** $901 \div 70$ **30.** $387 \div 30$ **31.** $686 \div 40$

32. Margie had 342 marbles. She put them in groups of 30 each. How many groups did she have? How many marbles were left?

Level C
Divide.

33. $858 \div 60$ **34.** $679 \div 40$ **35.** $458 \div 30$ **36.** $915 \div 70$ **37.** $333 \div 10$

38. $654 \div 90$ **39.** $469 \div 50$ **40.** $578 \div 80$ **41.** $190 \div 20$ **42.** $451 \div 50$

43. $572 \div 60$ **44.** $308 \div 70$ **45.** $250 \div 50$ **46.** $278 \div 40$ **47.** $555 \div 30$

48. Oliver saved 427 stamps. He kept them in envelopes of 20 stamps each. How many envelopes did he have? How many stamps were left?

CHAPTER REVIEW

LANGUAGE AND MATHEMATICS

Complete the sentences. Use the words in the chart.

1. When multiplying 2-digit numbers, multiply by the ones and then the ■. *(page 486)*

2. When you multiply by 1,000, the number of zeros in the ■ equals the number of zeros in both factors. *(page 474)*

3. You can use ■ to estimate a product. *(page 476)*

4. When dividing 72,000 by 90, the number of zeros in the ■ is less than the number of zeros in the dividend. *(page 494)*

5. ***Write a definition*** or give an example of the words you did not use from the chart.

> **VOCABULARY**
> add
> quotient
> rounding
> tens
> hundreds
> product
> subtract

CONCEPTS AND SKILLS

Estimate. Use rounding. *(page 476)*

6.	7.	8.	9.	10.
82	91	39	814	$762
× 19	× 47	× 78	× 55	× 28

11. $208 × 76   12. 233 × 75   13. 321 × 58   14. 469 × 82

Estimate. Use the front digits *(page 476)*

15.	16.	17.	18.	19.
93	78	$450	781	$409
× 74	× 39	× 51	× 62	× 87

20. 55 × $84   21. 77 × 22   22. 47 × $255   23. 46 × 622

Multiply. *(pages 474, 478, 484–487, 490)*

24.	25.	26.	27.	28.
80	40	200	3,000	5,000
× 70	× 90	× 30	× 80	× 50

29.	30.	31.	32.	33.
31	93	82	71	54
× 30	× 40	× 50	× 90	× 20

Multiply. *(pages 474, 478, 484–487, 490)*

34. 30 × 50

35. 400 × 60

36. 20 × 9,000

37. 30 × 4,000

38. 42 × 61

39. $53 × 29

40. 87 × 36

41. 15 × $95

42. 782 × 19

43. 28 × $6.37

44. 59 × $4.32

45. 16 × 92

Divide. *(pages 494–497)*

46. 560 ÷ 70

47. 3,600 ÷ 60

48. 18,000 ÷ 20

49. 630 ÷ 90

50. 16,000 ÷ 40

51. 450 ÷ 50

52. 40$\overline{)327}$

53. 70$\overline{)633}$

54. 30$\overline{)241}$

55. 40$\overline{)288}$

56. 90$\overline{)288}$

57. 80$\overline{)609}$

58. 50$\overline{)354}$

59. 60$\overline{)565}$

60. 20$\overline{)99}$

61. 50$\overline{)350}$

62. 70$\overline{)4,900}$

63. 90$\overline{)800}$

CRITICAL THINKING

64. Ian estimates that the product of two 2-digit factors is 4,000. What are the greatest factors possible? *(page 478)*

65. How would you model 240 ÷ 20? What is the quotient? *(page 492)*

MIXED APPLICATIONS

66. Kim bought 15 daisies. If each daisy cost 65¢, how much did he pay? Which operation did you choose? *(page 498)*

67. Jon's uncle owns a sock store and gives Jon 15 pairs of new socks every month. How many pairs of socks does Jon receive from his uncle in a year? *(page 486)*

68. Steve has to put 130 shingles on his roof. If the shingles are sold in boxes of 8, how many boxes does he need to buy? *(page 480)*

CHAPTER TEST

Estimate. Use rounding.

1.   53
      × 47

2. $572
      ×   33

3. 811 × 39

4. 49 × 669

Estimate. Use the front digits.

5.   31
      × 47

6.   461
      × 57

7. 460 × 83

8. 22 × $927

Multiply or divide mentally.

9.   60
      × 70

10. 90)810

11. 500 × 30

12. 2,700 ÷ 90

Multiply.

13.   26
       × 23

14. $3.86
       ×   52

15.   7,000
       ×   40

16. 19 × $32

17. 58 × 47

Divide.

18. 50)917

19. 30)882

20. 729 ÷ 20

21. 204 ÷ 60

Solve.

22. Nina collects rocks and shells. She has 78 more rocks than shells. If she has 43 shells, how many rocks does she have?

23. Chris has 57 shells and 106 rocks in his collection. How many more rocks than shells does he have?

24. Carla keeps her shells in a special case. It has 12 rows. There are 12 shells in each row. How many shells are in the case?

25. Aaron has 70 sand dollars. He needs 10 sand dollars for each picture he is making. How many pictures can he make?

CASTING OUT NINES

Tim uses a method called **casting out nines** to check multiplication. He follows these steps.

Step 1

Add the digits of the factors. If the sum is greater than 9, add the digits of the sum. Continue until you get a sum of 9 or less. Then, multiply the results. If necessary, add the digits of the product to get a sum of 9 or less.

$$637 \quad 6 + 3 + 7 = 16 \rightarrow 1 + 6 = 7$$
$$\times \ 69 \quad 6 + 9 = 15 \rightarrow 1 + 5 = 6$$
$$7 \times 6 = 42 \rightarrow 4 + 2 = 6$$

Step 2

Add the digits of the product. If the sum is greater than 9, add the digits of the sum. Continue until you get a sum of 9 or less.

$$4 + 3 + 9 + 5 + 3 = 24$$
$$2 + 4 = 6$$

$$2 + 4 = 6$$

Compare. The numbers should be the same.
Tim's product is correct. $637 \times 69 = 43,953$.

Check each product by casting out nines. Tell if the product checks or not. If it doesn't check, write the correct answer.

1. 82
      × 19
      ‾‾‾‾
      1,558

2. 38
      × 72
      ‾‾‾‾
      2,616

3. 677
      × 29
      ‾‾‾‾‾
      17,503

4. 548
      × 56
      ‾‾‾‾‾
      30,688

5. 62
      × 46
      ‾‾‾‾
      2,852

CUMULATIVE REVIEW

Choose the letter of the correct answer.

1. What is 60 times 90?

   a. 540   **c.** 54,000
   b. 5,400   **d.** not given

2. Compare: $\frac{32}{6}$ ● $5\frac{1}{3}$

   a. <   **c.** =
   b. >   **d.** not given

3. What is 28,000 divided by 70?

   a. 40   **c.** 4,000
   b. 400   **d.** not given

4. Subtract 9.06 from 12.

   a. 21.06   **c.** 3.06
   b. 2.94   **d.** not given

5. Find 36 multiplied by 417.

   a. 14,012   **c.** 15,012
   b. 14,912   **d.** not given

6. 63 × $5.49

   a. $345.87 **c.** $49.41
   b. $355.87 **d.** not given

7. Estimate the width of a car.

   a. 6 in.   **c.** 6 yd
   b. 6 ft   **d.** 6 mi

8. Estimate by rounding: 689 × 82.

   a. 56,000   **c.** 48,000
   b. 54,000   **d.** 63,000

9. Find the area of a square with 9-yd sides.

   a. 18 sq yd **c.** 36 sq yd
   b. 36 yd   **d.** not given

10. 92 × $47

    a. $3,724 **c.** $4,314
    b. $4,224 **d.** not given

11. Divide 86 by 20.

    a. 5   **c.** 4 R6
    b. 4 R16   **d.** not given

12. Estimate the weight of a rake.

    a. 1 oz   **c.** 1 lb
    b. 60 oz   **d.** 10 lb

13. Compare: 1.10 ● 1.01

    a. <   **c.** =
    b. >   **d.** not given

14. Find the product of 80 and 692.

    a. 55,360 **c.** 48,360
    b. 55,260 **d.** not given

PROFITS OF FOUR COMPUTER COMPANIES

Statement of Profits		
Company	Year	
	1990	1991
Fina Computers	$486,424,920	$794,806,050
Kane Computers Inc.	$163,725,908	$207,918,420
Mavis Software	$ 23,078,090	$ 39,649,750
Beekman Computersoft	$110,319,998	$119,139,099

YEARLY RAINFALL
(in inches)

San Francisco, CA	35 in.
Moscow, USSR	24 in.
Berlin, Germany	23 in.
Paris, France	22 in.
Fairbanks, AK	13 in.
Phoenix, AZ	10 in.
Los Angeles, CA	8 in.
Cairo, Egypt	1 in.

TALL BUILDINGS IN THE UNITED STATES

City	Building	Height		
		Stories	Feet	Meters
Chicago	Sears Tower	110	1,454	436
New York	World Trade Center	110	1,350	405
New York	Empire State Building	102	1,250	375
Chicago	Amoco	80	1,136	341
Chicago	John Hancock Center	100	1,127	338
New York	Chrysler Building	77	1,046	314
Los Angeles	Library Tower	73	1,017	305
Houston	Texas Commerce Tower	75	1,002	301
Houston	Allied Bank	71	992	298
Seattle	Columbia Seafirst Center	76	954	286

DATABANK

PLUM TREE CAFÉ MENU

Sandwiches
Tuna Salad. $2.35
Tuna Melt. $3.10
Shrimp Salad. $3.55
Chicken Salad. $2.75

Entrées
Roast Chicken. $8.50
Sirloin Steak $9.25
Broiled Fish $9.50
Broiled Pork Chops
  with applesauce. $8.75

Above served with potato, vegetable, and salad

Side Orders
French Fries $1.10
Cole Slaw . . . . $ .95
Cheese Sticks $2.25

Salads
Lettuce and Tomato $1.75
Chef's Salad $4.50
Health Salad $3.70

Beverages
Orange Juice $1.00
Fruit Punch $1.50
Milk . . . . $ .80
Apple Juice. $1.00
Lemonade. . . . . $ .75
Chocolate Milk. . . . . $ .90

Apple — 80 calories, 120 Vit. A units
Ice cream (1 cup) — 270 calories, 540 Vit. A units
Milk (1 cup) — 150 calories, 310 Vit. A units
Lemonade (1 cup) — 105 calories, 10 Vit. A units
Apple pie (1 piece of pie) — 345 calories, 40 Vit. A units
Green beans (1 cup) — 35 calories, 780 Vit. A units
Hamburger — 185 calories, 20 Vit. A units
Fried egg — 85 calories, 290 Vit. A units
Bacon (2 slices) — 85 calories, 0 Vit. A units
Corn (1 ear) — 70 calories, 310 Vit. A units
Muffin (corn) — 125 calories, 120 Vit. A units

CALORIES AND VITAMIN A* IN SOME FOODS * International Units

DATES OF SOME INVENTIONS

	1990
	1985
	1980
	1975
HOME COMPUTER ▶	1971
	1965
LASER ▶	1958
SOLAR BATTERY ▶	1954
ELECTRONIC COMPUTER ▶	1946
RADAR ▶	1940
	1935
ROCKET ▶	1926
TELEVISION ▶	1923
	1920
	1915
CELLOPHANE ▶	1908
AIRPLANE ▶	1903
	1900
RADIO ▶	1895
ZIPPER ▶	1891
MOVIE CAMERA ▶	1889
GASOLINE ▶	1887
AUTOMOBILE	
LIGHT BULB ▶	1879
TELEPHONE ▶	1876
	1870
TYPEWRITER ▶	1867
	1865

GREATEST KNOWN SPEEDS OF VARIOUS ANIMALS
(in miles per hour)

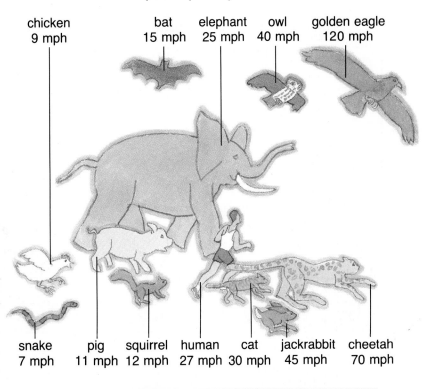

chicken 9 mph
bat 15 mph
elephant 25 mph
owl 40 mph
golden eagle 120 mph

snake 7 mph
pig 11 mph
squirrel 12 mph
human 27 mph
cat 30 mph
jackrabbit 45 mph
cheetah 70 mph

SPACE MUSEUM GIFT SHOP PRICES	
Item	Price
Astronaut food packet	$2.15
Astronaut ice cream	$1.78
Frisbees	$4.79
Kites	$8.75
Robot models	$6.35
Rocket models	$8.59
Spaceship models	$9.98

DATABANK

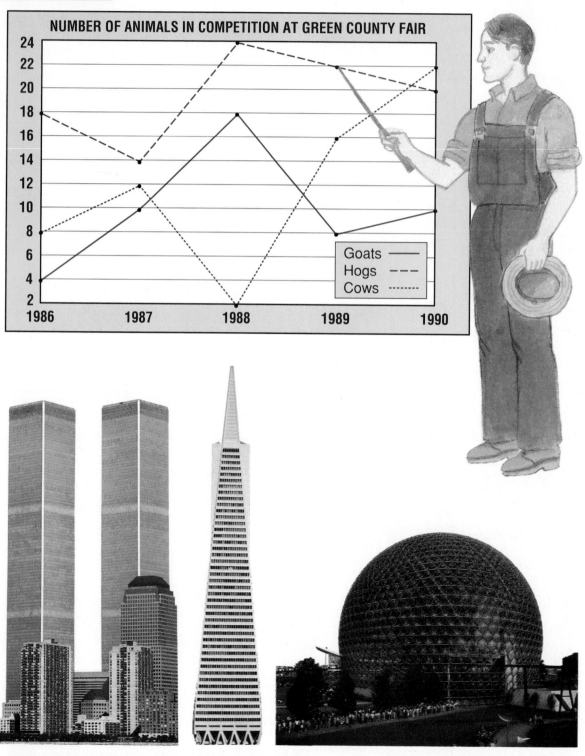

NUMBER OF ANIMALS IN COMPETITION AT GREEN COUNTY FAIR

Legend:
- Goats ——
- Hogs – –
- Cows ·····

Y-axis: 2, 4, 6, 8, 10, 12, 14, 16, 18, 20, 22, 24

X-axis: 1986, 1987, 1988, 1989, 1990

World Trade Center
(New York City)

TransAmerica Building
(San Francisco)

Expo '67 Dome
(Montreal)

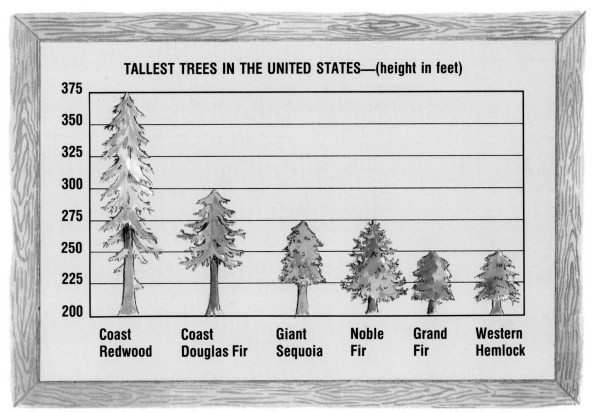

TALLEST TREES IN THE UNITED STATES—(height in feet)

	Coast Redwood	Coast Douglas Fir	Giant Sequoia	Noble Fir	Grand Fir	Western Hemlock

(Height scale: 200, 225, 250, 275, 300, 325, 350, 375)

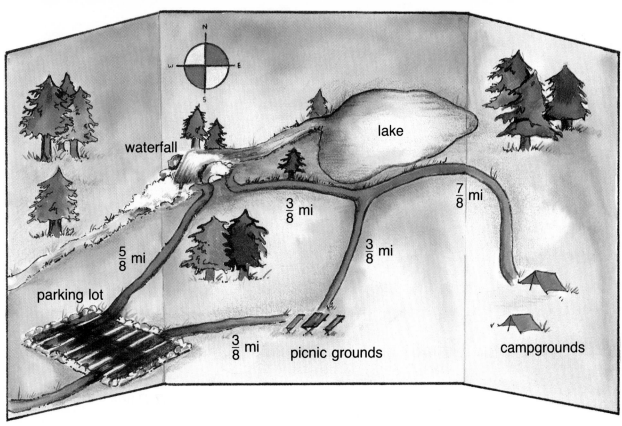

waterfall

lake

$\frac{7}{8}$ mi

$\frac{3}{8}$ mi

$\frac{3}{8}$ mi

$\frac{5}{8}$ mi

parking lot

$\frac{3}{8}$ mi   picnic grounds

campgrounds

WINNING TIMES IN OLYMPIC SWIMMING
100-METER FREESTYLE (in seconds)

Year	Men	Women
1976	49.99	55.65
1980	50.40	54.79
1984	49.50	55.92
1988	48.63	54.93

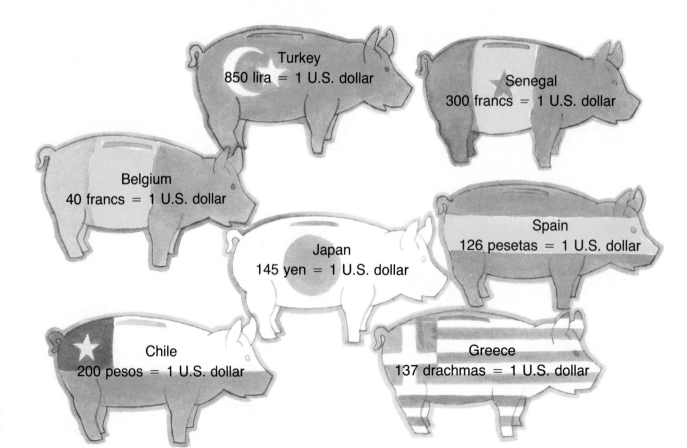

Turkey
850 lira = 1 U.S. dollar

Senegal
300 francs = 1 U.S. dollar

Belgium
40 francs = 1 U.S. dollar

Japan
145 yen = 1 U.S. dollar

Spain
126 pesetas = 1 U.S. dollar

Chile
200 pesos = 1 U.S. dollar

Greece
137 drachmas = 1 U.S. dollar

GLOSSARY

A

addend A number that is added. In the addition sentence $5 + 4 = 9$, 5 and 4 are addends.

addition An operation on two or more numbers that tells *how many in all* or *how much in all*.

$$\begin{array}{r} 9 \leftarrow \text{addend} \\ + 3 \leftarrow \text{addend} \\ \hline 12 \leftarrow \text{sum} \end{array}$$

A.M. A label showing that a time occurs between 12:00 midnight and 12:00 noon.

angle A figure formed by two *rays* with the same endpoint.

area The number of square units it takes to cover the surface of a *plane figure*.

 The area of the square is 9 square units.

average A statistic about two or more numbers found by adding the numbers and dividing their sum by the total number of addends.

B

bar graph A *graph* that displays data using bars of different lengths.

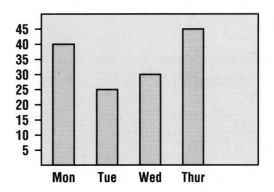

C

capacity The *volume* of a *space figure* expressed as the amount of liquid the figure can hold.

centimeter (cm) A metric unit of length. There are 100 centimeters in a *meter*.

⊢——⊣ 1 centimeter

circle A closed plane figure.

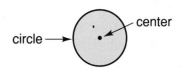

common factor A number that is a factor of two or more numbers. The common factors of 4 and 6 are 1 and 2.

common multiple a number that is a multiple of two or more numbers. A common multiple of 2 and 3 is 6.

composite number A whole number that has factors other than itself and 1.

congruent Figures that have the same size and shape, and lines and angles that have the same measure are congruent.

customary system A method of measurement that uses the following units:
Length/Distance: inch, foot, yard, mile
Capacity: cup, pint, quart, gallon
Weight: ounce, pound, ton
Temperature: degree Fahrenheit

D

decimal A number that uses place value and a decimal point to express mixed numbers or fractions with denominators of multiples of ten.

tens	ones	.	tenths	hundredths
2	4	.	7	5

$$24.75 = 24\frac{75}{100}$$
↑
decimal point

decimeter (dm) A metric unit of length. A decimeter equals 10 *centimeters*. There are 10 decimeters in a *meter*.

degree Celsius The unit used to measure temperature in the metric system of measurement.

degree Fahrenheit The unit used to measure temperature in the customary system of measurement.

denominator The number beneath the line in a fraction.

$$\frac{3}{4} \begin{array}{l} \leftarrow \text{numerator} \\ \leftarrow \text{denominator} \end{array}$$

difference The answer to a subtraction problem. In the subtraction sentence $10 - 6 = 4$, 4 is the difference.

dividend The number that is divided into equal parts in a division problem.

$$5\overline{)40}^{\,8} \leftarrow \text{dividend}$$

division An operation on two numbers that tells *how many groups* or *how many in each group*. Division can also tell *how many are left over*.

$46 \div 5 = 9$ R1 There are nine 5's in 46 with a *remainder* of 1.

divisor The number that the *dividend* is divided by to find the *quotient* in a division problem.

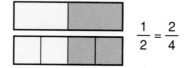

E

equivalent decimals Two or more decimals that name the same number. 3.1 and 3.10 are equivalent decimals.

equivalent fractions Two or more fractions that name the same number. $\frac{1}{3}$, $\frac{2}{6}$, and $\frac{3}{9}$ are equivalent fractions.

$$\frac{1}{2} = \frac{2}{4}$$

estimate To find *about how many* or *about how much* by using rounded numbers or only the front digits to compute. Estimating gives an answer that is close to the exact answer.

even number Any whole number that can be divided by 2 evenly. All even numbers have a 0, 2, 4, 6, or 8 in the ones place.

expanded form A method of writing a number as the sum of the values of its digits. Expanded form for 389 is 300 + 80 + 9.

F

fact family A group of related number sentences.

$$2 \times 4 = 8 \quad 4 \times 2 = 8$$
$$8 \div 2 = 4 \quad 8 \div 4 = 2$$

$$3 + 2 = 5 \quad 2 + 3 = 5$$
$$5 - 2 = 3 \quad 5 - 3 = 2$$

factor The numbers that are multiplied in a multiplication problem.

$$7 \times 8 = 56$$
— Factors

fraction A number that names part of a whole or part of a set.

$\frac{1}{2}$ of the circle is shaded.

front-end estimation A method of finding an answer that is close to the exact answer by computing with only the front digits of numbers.

G

graph A picture that organizes data to make it easy to understand and use. Types of graphs are *bar graph, line graph,* and *pictograph.*

I

intersecting lines Lines or line segments that meet or cross at a common point.

K

kilometer A metric unit of length. There are 1,000 *meters* in a kilometer.

L

line A straight figure that extends in both directions.

line segment A straight figure with two endpoints.

line of symmetry A line on which a figure can be folded so that its two halves match exactly.

M

mass The mass of an object tells how much of it there is.

median The middle number in a series of numbers ordered from the least to the greatest. The median of 4, 5, and 7 is 5.

meter A metric unit of length. There are 100 *centimeters* in a meter.

metric system A method of measurement that uses the following units:
Length/Distance: centimeter, decimeter, meter, kilometer
Capacity: milliliter, liter
Mass: gram, kilogram
Temperature: degree Celsius

mixed number A number made up of a whole number and a fraction. All mixed numbers are greater than 1. $3\frac{1}{2}$ is a mixed number.

multiple The product of a number and any other whole number. Some multiples of 4 are 4, 8, 12, 16, 24, 32, and 48.

multiplication An operation that tells *how many in all* when equal groups are combined.

$$8 \leftarrow \text{factor}$$
$$\underline{\times 5} \leftarrow \text{factor}$$
$$40 \leftarrow \text{product}$$

N

number sentence A mathematical fact written horizontally with numbers and symbols.

$$4 \times 5 < 23 \qquad 42 \div 7 = 6$$

numerator The number above the line in a fraction.

$$\frac{3}{5} \begin{array}{l} \leftarrow \text{numerator} \\ \leftarrow \text{denominator} \end{array}$$

O

odd number Any whole number that has a 1, 3, 5, 7, or a 9 in the ones place.

ordered pair A pair of numbers that describes the location of a point on a graph, map, or grid. The ordered pair $(5, 2)$ describes where the star is located on the grid.

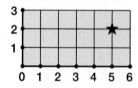

order property A characteristic of addition and multiplication that gives the same result no matter how the addends or factors are positioned.

$$3 + 4 = 7 \qquad 4 \times 5 = 20$$
$$4 + 3 = 7 \qquad 5 \times 4 = 20$$

P

parallel lines Lines in the same plane that never meet.

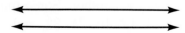

perimeter The distance around an object. The perimeter of the rectangle is 16.

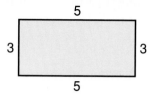

place value The value of a digit based on its position in a number. In 9,275, the 7 stands for 7 tens, or 70. In 784, the 7 stands for 7 hundreds or 700.

plane A flat surface extending in all directions.

plane figure Any figure that exists on a flat surface. Plane figures have no depth.

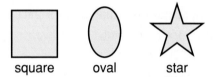

square   oval   star

P.M. A label showing that a time occurs between 12 noon and 12 midnight.

polygon A closed plane figure formed by line segments.

prime number A whole number greater than 1 with only itself and 1 as factors. 7 is a prime number. 2 is the only even number that is prime.

probability A number from 0 to 1 that measures the likelihood of an event occurring.

product The answer to a multiplication problem.

$$5$$
$$\underline{\times 4}$$
$$20 \leftarrow \text{product}$$

Q

quotient The answer to a division problem.

$$\begin{array}{r} 5 \\ 7\overline{)35} \end{array} \leftarrow \text{quotient}$$

R

range The difference between the least number and the greatest number in a set of data.

ray A straight line that has one endpoint and extends in one direction.

regroup To exchange equal amounts, such as 1 ten for 10 ones or 10 tens for 1 hundred, when computing.

remainder The amount left over when a number cannot be divided into equal parts.

$$\begin{array}{r} 6 \text{ R1} \\ 7\overline{)43} \\ \underline{42} \\ 1 \end{array} \quad \text{remainder}$$

Roman numerals Symbols used by the ancient Romans to name whole numbers.

I = 1	V = 5	X = 10	L = 50
C = 100	D = 500		M = 1,000

rounding Replacing the exact number with the nearest multiple of 10, 100, 1,000, and so on. You compute with rounded numbers to find out about how many.

2,648 to the nearest 10 is 2,650
2,648 to the nearest 100 is 2,600
2,648 to the nearest 1,000 is 3,000

S

space figure A figure that exists in more than one plane.

cube   sphere   pyramid

sphere A space figure in the shape of a ball.

subtraction An operation on two numbers that tells *how many are left* when some are taken away. Subtraction is also used to compare two numbers.

$$\begin{array}{r} 14 \\ -\ 8 \\ \hline 6 \end{array} \leftarrow \text{difference}$$

sum The answer to an addition problem. In the addition sentence $9 + 6 = 15$, 15 is the sum.

V

vertex The common point of the two rays of an angle or the two sides of a polygon.

volume The number of cubic units a space figure contains. The rectangular prism has 3 cubic units.

W

whole numbers The numbers that tell how many: 0, 1, 2, 3, . . .

TABLE OF MEASURES

TIME

60 minutes (min)	=	1 hour (hr)
24 hours	=	1 day (d)
7 days	=	1 week (wk)
12 months (mo)	=	1 year (y)
about 52 weeks	=	1 year
365 days	=	1 year
366 days	=	1 leap year

METRIC UNITS

LENGTH

1 centimeter (cm)	=	10 millimeters (mm)
10 centimeters	=	1 decimeter (dm)
10 decimeters	=	1 meter (m)
1,000 meters	=	1 kilometer (km)

MASS   1 kilogram (kg) = 1,000 grams (g)

CAPACITY   1 liter (L) = 1,000 milliliters (mL)

TEMPERATURE   0° Celsius (°C) . . . Water freezes
100° Celsius . . . Water boils

CUSTOMARY UNITS

LENGTH

1 foot (ft)	=	12 inches (in.)
1 yard (yd)	=	36 inches
1 yard	=	3 feet
1 mile (mi)	=	5,280 feet
1 mile	=	1,760 yards

WEIGHT   1 pound (lb) = 16 ounces (oz)

CAPACITY

1 cup (c)	=	8 fluid ounces
1 pint (pt)	=	2 cups
1 quart (qt)	=	2 pints
1 gallon (gal)	=	4 quarts

TEMPERATURE   32° Fahrenheit (°F) . . . Water freezes
212° Fahrenheit . . . Water boils

SYMBOLS

$<$	is less than	$\overrightarrow{AB}$	ray AB
$>$	is greater than	$B\!\!<^{\!A}_{\!C}$	angle ABC or angle CBA or angle B
$=$	is equal to	$\overleftrightarrow{LM} \parallel \overleftrightarrow{OP}$	line LM parallel to line OP
°	degree	$\overleftrightarrow{UZ} \perp \overleftrightarrow{WY}$	line UZ perpendicular to line WY
$\overleftrightarrow{AB}$	line AB	$(5,3)$	ordered pair 5,3
$\overline{AB}$	line segment AB		

INDEX

529